Organ Regeneration Based on Developmental Biology

Takashi Tsuji
Editor

Organ Regeneration Based on Developmental Biology

Springer

Editor
Takashi Tsuji
Laboratory for Organ Regeneration
RIKEN Center for Developmental Biology
Kobe, Hyogo, Japan

ISBN 978-981-10-3766-5 ISBN 978-981-10-3768-9 (eBook)
DOI 10.1007/978-981-10-3768-9

Library of Congress Control Number: 2017935574

Printed on acid-free paper

This Springer imprint is published by Springer Nature
The registered company is Springer Nature Singapore Pte Ltd.
The registered company address is: 152 Beach Road, #21-01/04 Gateway East, Singapore 189721, Singapore

This book is dedicated to the memory of Yoshiki Sasai, a scientist who made a great contribution to the advancement of developmental biology.

Preface

Organogenesis, the generation of functional tissues and organs following an established body plan during development, is a complex process that involves tissue self-organization, cell–cell interactions, and the regulation of cell-signaling molecules and cell movements. Various lineages of stem cells are produced and play central roles in organ development. Stem cell research thus not only yields new fundamental insights into biology but also can contribute to the development of regenerative medicine.

In the first generation of regenerative medicine, stem cell transplantation approaches have been developed or are under development for diseases such as hematopoietic malignancies, Parkinson's disease, and myocardial infarction. The second generation of regenerative medicine is tissue engineering using cell sheets and cell aggregates of homogeneous cell populations such as skin, cartilage, and bone. The next generation of regenerative therapy will be the development of fully functioning bioengineered organs that can replace lost or damaged organs following disease, injury, or aging. Efforts are now under way to develop bioengineering technologies to reconstruct fully functional organs in vitro through the precise arrangement of several different cell types. Although attempts to make functional organs by tissue engineering using scaffolds, cytokines, and various lineages of cells have been made over the past several decades, clinical applications have remained elusive.

In recent years, significant advances in techniques for organ regeneration, including neuroectodermal, ectodermal, and endodermal organs, have been made using three-dimensional stem cell culture in vitro by using various stem cell lineages and pluripotent stem cells, such as embryonic stem cells and induced pluripotent stem cells. Several groups have reported the generation of neuroectodermal and endodermal organoids by harnessing the regulation of complex patterning signals during embryogenesis and self-organization of pluripotent stem cells in three-dimensional stem cell culture. Other groups have sought to generate functional organs that develop by reciprocal epithelial and mesenchymal interactions using embryonic organ inductive stem cells. Several groups have also reported the generation of three-dimensional organoids/tissues by the reproduction of stem cells and their niches. These studies provide a better understanding of organogenesis in developmental biology and represent a breakthrough in bioengineered organ regeneration which shows real promise for uses in next-generation organ regenerative therapy.

Here, we have collected a series of reviews on recent studies of organ regeneration from stem cells using strategies inspired by developmental biology. These reviews have contributed to a better understanding of the recent advances in the study of organogenesis in both developmental biology and in the development of clinical applications for next-generation regenerative therapies.

I sincerely thank all of the authors for their contributions. I am also grateful to Momoko Asawa and Yasutaka Okazaki, editors of the Springer Review Books.

Kobe, Japan Takashi Tsuji

Contents

List of Contributors

Mototsugu Eiraku RIKEN Center for Developmental Biology, Kobe, Hyogo, Japan

Alhasan N. Elghouche Department of Otolaryngology-Head & Neck Surgery, Indiana University School of Medicine, Indianapolis, IN, USA

Thomas A. Forbes Murdoch Childrens Research Institute, Parkville, VIC, Australia

Department of Paediatrics, University of Melbourne, Parkville, VIC, Australia

Royal Children's Hospital, Parkville, VIC, Australia

Eri Hashino Department of Otolaryngology-Head & Neck Surgery, Indiana University School of Medicine, Indianapolis, IN, USA

Masatoshi Hirayama Department of Ophthalmology, Keio University School of Medicine, Shinjuku-ku, Tokyo, Japan

Laertis Ikonomou Center for Regenerative Medicine (CReM) of Boston University and Boston Medical Center, Boston, MA, USA

Taisuke Kadoshima RIKEN Center for Developmental Biology, Kobe, Hyogo, Japan

Asubio Pharma Co., Ltd., Kobe, Hyogo, Japan

Melissa H. Little Murdoch Childrens Research Institute, Parkville, VIC, Australia

Department of Paediatrics, University of Melbourne, Parkville, VIC, Australia

Keiko Muguruma Laboratory for Cell Asymmetry, RIKEN Center for Developmental Biology, Kobe, Hyogo, Japan

Jorge O. Múnera Division of Developmental Biology Cincinnati Children's Hospital Research Foundation, Cincinnati, OH, USA

Rick F. Nelson Department of Otolaryngology-Head & Neck Surgery, Indiana University School of Medicine, Indianapolis, IN, USA

Ryuichi Nishinakamura Department of Kidney Development, Institute of Molecular Embryology and Genetics, Kumamoto University, Kumamoto, Japan

Miho Ogawa Organ Technologies Inc., Tokyo, Japan

RIKEN Center for Developmental Biology, Kobe, Hyogo, Japan

Masamitsu Oshima Department of Oral Rehabilitation and Regenerative Medicine, Graduate School of Medicine, Dentistry and Pharmaceutical Sciences, Okayama University, Okayama, Japan

RIKEN Center for Developmental Biology, Kobe, Hyogo, Japan

Chikafumi Ozone Department of Endocrinology and Diabetes, Graduate School of Medicine, Nagoya University, Nagoya, Aich, Japan

RIKEN Center for Developmental Biology, Kobe, Hyogo, Japan

Hideya Sakaguchi RIKEN Center for Developmental Biology, Kyoto, Hyogo, Japan

Center for iPS Cell Research and Application (CiRA), Kyoto University, Kyoto, Japan

Keisuke Sekine Department of Regenerative Medicine, Yokohama City University Graduate School of Medicine, Yokohama, Kanagawa, Japan

Joanne Y.-C. Soo Murdoch Childrens Research Institute, Parkville, VIC, Australia

Department of Paediatrics, University of Melbourne, Parkville, VIC, Australia

Hidetaka Suga Department of Endocrinology and Diabetes, Graduate School of Medicine, Nagoya University, Nagoya, Aich, Japan

Atsuhiro Taguchi Department of Kidney Development, Institute of Molecular Embryology and Genetics, Kumamoto University, Kumamoto, Japan

Minoru Takasato Murdoch Childrens Research Institute, Parkville, VIC, Australia

RIKEN Center for Developmental Biology, Kobe, Hyogo, Japan

Takanori Takebe Department of Regenerative Medicine, Yokohama City University Graduate School of Medicine, Yokohama, Kanagawa, Japan

Advanced Medical Research Center, Yokohama City University, Yokohama, Kanagawa, Japan

PRESTO, Japan Science and Technology Agency, Saitama, Japan

Department of Pediatrics, Cincinnati Children's Hospital Medical Center, University of Cincinnati, Cincinnati, OH, USA

Hideki Taniguchi Department of Regenerative Medicine, Yokohama City University Graduate School of Medicine, Yokohama, Kanagawa, Japan

Advanced Medical Research Center, Yokohama City University, Yokohama, Kanagawa, Japan

Koh-ei Toyoshima Department Regenerative Medicine, Plastic and Reconstructive Surgery, Kitasato University School of Medicine, Sagamihara, Kanagawa, Japan

RIKEN Center for Developmental Biology, Kobe, Hyogo, Japan

Organ Technologies Inc., Tokyo, Japan

Kazuo Tsubota Department of Ophthalmology, Keio University School of Medicine, Shinjuku-ku, Tokyo, Japan

Takashi Tsuji Team Leader, Laboratory for Organ Regeneration, RIKEN Center for Developmental Biology, Kobe, Hyogo, Japan

James M. Wells Division of Developmental Biology Cincinnati Children's Hospital Research Foundation, Cincinnati, OH, USA

Andrew Wilson Center for Regenerative Medicine (CReM) of Boston University and Boston Medical Center, Boston, MA, USA

Chapter 1
Telencephalic Tissue Formation in 3D Stem Cell Culture

Taisuke Kadoshima, Hideya Sakaguchi, and Mototsugu Eiraku

Abstract The telencephalon is one of the most complex regions of the brain, and it plays important roles in higher brain functions in humans. However, due to the poor accessibility of human brain tissue for research, it has been difficult to explore the mechanisms of human brain development and function. Recent advances in stem cell technology now enable us to generate telencephalic tissues from human pluripotent stem cells. This may become a powerful tool for examining developmental processes or disease mechanisms related to the telencephalon. In this review, we provide an overview of recent advances in the recapitulation of telencephalic tissues in vitro, mainly focusing on three-dimensional (3D) culture of human embryonic stem cells (ESCs). Ventral and dorsal telencephalic tissues can now be sufficiently induced using specific media that support the differentiation of each tissue type. The generation of telencephalic tissues from human pluripotent stem cells (PSCs) provides a new experimental platform for biomedical research, such as disease modeling, drug screening, and cell-based transplantation therapy.

Keywords SFEBq culture • Human ESCs • Telencephalon • Cerebral cortex • Ganglionic eminence • Medial pallium • Hippocampus

---

Authors "Taisuke Kadoshima" "Hideya Sakaguchi" are contributed equally to this work.

T. Kadoshima
RIKEN Center for Developmental Biology, Kobe, Hyogo 650-0047, Japan

Asubio Pharma Co., Ltd., Kobe, Hyogo 650-0047, Japan

H. Sakaguchi
RIKEN Center for Developmental Biology, Kobe, Hyogo 650-0047, Japan

Center for iPS Cell Research and Application (CiRA), Kyoto University, Kyoto 606-8507, Japan

M. Eiraku (✉)
RIKEN Center for Developmental Biology, Kobe, Hyogo 650-0047, Japan
e-mail: eiraku@cdb.riken.jp

T. Tsuji (ed.), *Organ Regeneration Based on Developmental Biology*,
DOI 10.1007/978-981-10-3768-9_1

1.1 Introduction

The brain is the most complex organ in the human body. It has increased in size and complexity throughout human evolution, supporting capabilities such as language, planning, and social behavior. During early neural development, the anlage of the central nervous system first forms as a neural plate in the dorsal ectoderm. Then the neural plate folds and rolls up into a neural tube (Rubenstein et al. 1998; Lumsden and Krumlauf 1996). Subsequently, the neural tube is regionalized into distinct progenitor domains along the anterior–posterior (AP) and dorsal–ventral (DV) axes by the effects of secreted factors and their downstream transcription factors (Lee and Jessell 1999). The distinct progenitor domains give rise to a variety of neurons and glia, which have different morphological and functional properties (Shen et al. 2006). The regional pattern along the AP axis is determined at an early embryonic stage, and subsequently the regional pattern along the DV axis is fixed (Levine and Brivanlou 2007). In vivo, the telencephalon develops from the forebrain, which is derived from the rostral portion of the neural tube (Wilson and Houart 2004) (Fig. 1.1a). The telencephalon further subdivides into dorsal and ventral telencephalon along the DV axis. The dorsal telencephalon consists of the choroid plexus, cortical hem, medial pallium (giving rise to the hippocampus) and dorsolateral pallium (giving rise to the cerebral cortex), whereas the ventral telencephalon consists of the lateral, medial, and caudal ganglionic eminence (LGE, MGE, and CGE; giving rise to the striatum and globus pallidus as well as telencephalic GABAergic interneurons) (Sousa and Fishell 2010; Suzuki and Vanderhaeghen 2015) (Fig. 1.1b).

Classical embryological studies in amphibians proposed the "neural default model" in which ectodermal cells undergo differentiation into neural progenitors in the absence of exogenous inhibitory factors (Levine and Brivanlou 2007; Godsave and Slack 1989; Grunz and Tacke 1989; Sasai et al. 1995; Piccolo et al. 1996; Sasai and De Robertis 1997; Sato and Sargent 1989). This model has been thought to be applicable to in vitro differentiation; neural differentiation spontaneously occurs in embryonic stem (ES) cells cultured in medium containing minimal extrinsic signals (Tropepe et al. 2001; Ying et al. 2003; Watanabe et al. 2005; Smukler et al. 2006). Based on this idea, we developed a versatile 3D differentiation system, called SFEB/SFEBq (serum-free floating culture of embryoid body-like aggregates with quick reaggregation) (Watanabe et al. 2005, 2007; Wataya et al. 2008; Eiraku et al. 2008, Sasai 2013a, b) (Fig. 1.1c). In the SFEBq culture, dissociated mouse and human ESCs are reaggregated using low-cell-adhesion 96-well culture plate. The floating aggregates cultured in serum-free medium that contains no or minimal growth factors can efficiently differentiate into neural progenitors.

The telencephalon is one of the most interesting regions of the brain because its dysfunction is linked to various neurological and neuropsychiatric diseases such as dementia, autism, schizophrenia, and mood disorders (depression and bipolar disease). Using a variety of animal models, we have increased our knowledge about the development, function, and structures of the telencephalon. However, the human brain has not been sufficiently examined, partly because of limited access to human tissues. Recent advances in 3D differentiation technologies of PSCs (ES cells and

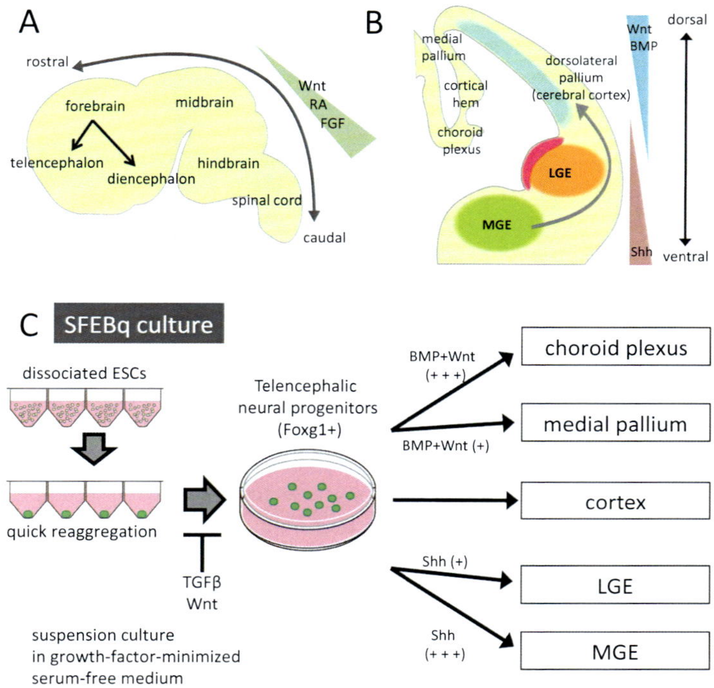

Fig. 1.1 In vivo neural development and differentiation strategy of SFEBq culture. (**a**) Regional patterning of neural plate. During early neural development, the neural tube is subdivided into the forebrain, midbrain, hindbrain, and spinal cord along the AP axis. The forebrain comprises the telencephalon and diencephalon. The early neural progenitors exhibit the anterior identity initially and are subsequently patterned by caudalizing signals such as Wnt, RA, and FGF. (**b**) Schematic of the developing fetal telencephalon. Along the DV axis, ventralization of telencephalic progenitors is promoted by the Shh signal, and dorsalization is caused by BMP and Wnt signals. (**c**) In SFEBq culture of human ESCs, 9000 cells are quickly reaggregated using a low-cell-adhesion 96-well culture plate. The floating aggregates are cultured in growth factor-minimized serum-free medium. In the absence of Wnt signals, the tissue differentiates into Foxg1^{+} telencephalic precursors. For the dorsalization of telencephalic tissues, aggregates are treated with BMP4 ligand plus CHIR 99021 (GSK3 inhibitor, also known to accelerate Wnt signaling). For the ventralization of telencephalic tissues, aggregates are treated with the hedgehog agonist SAG

induced pluripotent stem (iPS) cells), which can recapitulate complex organ structures, offer us a new opportunity to understand human brain development and apply our findings to regenerative medicine and drug discovery (Eiraku et al. 2008; Kadoshima et al. 2013; Lancaster et al. 2013; Kamiya et al. 2011).

Here we review recent progress in the 3D organ generation of various telencephalic regions from PSCs based on developmental biology. First, we review the basic features of in vivo development, along with recent progress in in vitro cortical tissue generation from PSCs. Then, we focus on the generation of ventral telencephalic regions, followed by the generation of dorsal telencephalic regions including the hippocampus.

1.2 The Cerebral Cortex

1.2.1 Development of the Cerebral Cortex In Vivo

Among the various telencephalic regions, the cerebral cortex is the center of human-specific integral neural activity. The cerebral cortex exhibits a highly complex structure, which is characterized by a six-layered laminar structure (Northcutt and Kaas 1995). This complex structure is established through precisely designed processes, including cell proliferation, differentiation, and migration (Kriegstein et al. 2006; Kriegstein and Noctor 2004). During the early stages of cortical development, cortical neuroepithelial (NE) cells are located in the ventricular zone (VZ) facing the ventricle and form a highly polarized, pseudostratified layer structure (Huttner and Brand 1997) (Fig. 1.2). They first self-renew by symmetric divisions, and then they convert into neurogenic apical progenitors (called radial glia) (Noctor et al. 2001; Miyata et al. 2001). Apical progenitors, which possess an apical process to the pial surface, generate cortical pyramidal neurons by asymmetric divisions either directly or through transit progenitors (called intermediate progenitors; $Tbr2^{+}$) in the subventricular zone (SVZ) (Noctor et al. 2004). Different types of cortical pyramidal neurons are born in a sequential order from these progenitors and migrate along the apical process of the progenitors to the pial surface (Kriegstein and Noctor 2004). Then, the newborn neurons settle in the appropriate laminae of the cortical plate in an inside-out manner (in which late-born neurons migrate to more superficial layers than early-born neurons) (Angevine and Sidman 1961). Finally, the cortical plate forms six neuronal layers that differ markedly in gene expression, physiological aspects, and projection patterns (Molyneaux et al. 2007). The pyramidal neurons are glutamatergic/excitatory neurons, and they comprise the majority (approximately 80%) of cortical neurons. GABAergic/inhibitory neurons, comprising approximately 20% of the cortical neurons, originate from the ventral telencephalon and then migrate tangentially to the cortex and integrate into the network (DeFelipe et al. 2002).

The mammalian cortex can be divided into several areas, such as visual or motor areas, which have distinct anatomy and function (Wree et al. 1983). Two models of the mechanism that controls cortical arealization have been proposed from mouse studies: the protomap model and the protocortex model. The protomap model postulates that cortical arealization is patterned by the graded expression of certain transcriptional factors in apical progenitors at early stages (Rakic 1988; Sansom et al. 2005). The protocortex model postulates that cortical arealization is patterned by the thalamocortical axon (TCA) inputs at later stages (O'Leary 1989). These mechanisms are not mutually exclusive; they may play complementary roles for rigorous regulation (Sur and Rubenstein 2005).

The basic mechanisms of corticogenesis seem to be conserved in the mammalian cortex, but the primate cortex exhibits distinct features compared with rodent cortex (Smart et al. 2002). The most striking histological difference is the thickness of the

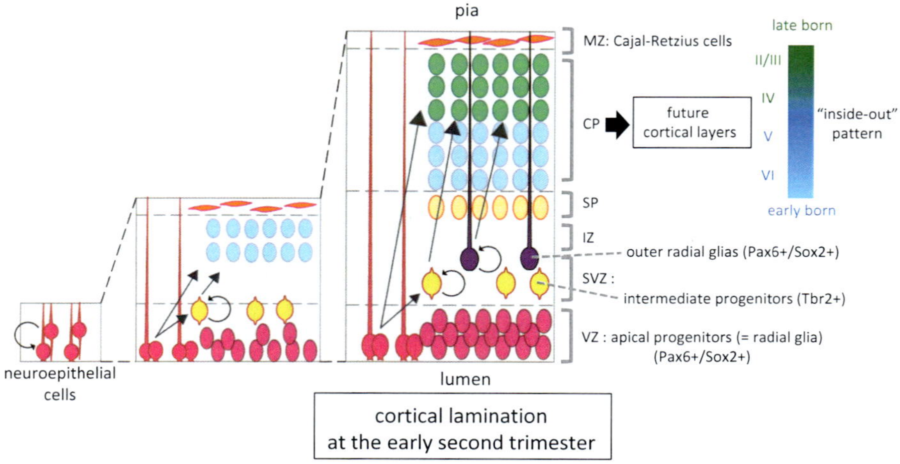

Fig. 1.2 Development of fetal cortical neuroepithelium. During corticogenesis, the neuroepithelial cells convert into neurogenic apical progenitors. The progenitors generate cortical neurons either directly or through transit progenitors (intermediate progenitors). Outer radial glia is a newly discovered type of neurogenic progenitor that exists abundantly in human fetal cortex. The newborn neurons migrate and settle in the appropriate laminae in an inside-out pattern. The cortical plate finally forms six neuronal layers. *LGE* lateral ganglionic eminence; *MGE* medial ganglionic eminence; *MZ* marginal zone; *CP* cortical plate; *SP* subplate; *IZ* intermediate zone; *SVZ* subventricular zone, *VZ* ventricular zone

SVZ: the primate (especially human) cortex is characterized by a highly expanded SVZ, which is subdivided into inner and outer SVZ (ISVZ and OSVZ) (Smart et al. 2002). Recent studies have shown that a novel type of neurogenic progenitor exists in the human OSVZ at midgestation, which is distinct from apical progenitors and intermediate progenitors (Fietz et al. 2010; Hansen et al. 2010). These progenitors, termed outer radial glia (oRG) (or OSVZ progenitors), show similar marker expression to apical progenitors ($Pax6^+$, $Sox2^+$, $Tbr2^-$) but a distinct morphology: oRG cells have a basal process extending to the pial surface and lack an apical process unlike apical progenitors (Fietz et al. 2010; Hansen et al. 2010) (Fig. 1.2). oRG cells undergo multiple rounds of self-renewal and generate neurons indirectly through their daughter cells, which also have an increased capacity for transit amplification. oRG cells are present in the mouse fetal brain, but the population is very small compared to the human fetal brain (Shitamukai et al. 2011; Wang et al. 2011). Therefore, oRG cells are thought to contribute to the species-specific increased number and diversity of human cortical neurons (Lui et al. 2011).

1.2.2 3D Cortical Tissue Generation In Vitro

Using SFEBq methods, mouse ESC-derived aggregates can be differentiated into cortical progenitors and self-organize a stratified structure while recapitulating early steps of corticogenesis (Eiraku et al. 2008; Nasu et al. 2012). We have also induced human ESC-derived cortical tissue that resembles the very early cortical anlage during the first trimester of human development (Eiraku et al. 2008). Moreover, recent advances have allowed us to evaluate human PSC-derived cortical tissues in more detail (Kadoshima et al. 2013). It is well recognized that the default state of neural progenitors for regional specification along the neural AP axis is that of the rostral forebrain, including the telencephalon and hypothalamus, and subsequently patterned by caudalizing signals such as Wnt, RA, and FGF (Nordström et al. 2002; Durston et al. 1998; Muhr et al. 1999; Cox and Hemmati-Brivanlou 1995) (Fig. 1.1a). As seen in vivo, mouse and human ESC-derived neural progenitors can differentiate into telencephalic precursors (Foxg1^{+}) efficiently in the absence of caudalizing signals (Gaspard et al. 2008; Espuny-Camacho et al. 2013). Specifically, the active inhibition of Wnt signaling by the Wnt antagonist Dkk1 or a chemical inhibitor is effective for inducing the differentiation of telencephalic precursors (Watanabe et al. 2005; Eiraku et al. 2008; Kadoshima et al. 2013). In the absence of any patterning signals along the DV axis, the telencephalic precursors derived from human ESCs tend to have dorsal identities (Pax6^{+} cerebral cortex), whereas in the case of mouse cells, they tend to have more ventral identities (Gaspard et al. 2008). During in vivo corticogenesis, cortical apical progenitors produce different types of neurons (corresponding to different cortical layers) in a sequential manner. This sequential generation of layer-specific pyramidal neurons is controlled by intrinsic mechanisms in cortical progenitors along a time axis (Hevner et al. 2003). This sequential generation of layer-specific neurons can be seen even in primary mouse cell culture of isolated cortical progenitors (Shen et al. 2006). Consistent with this, mouse ESC-derived cortical progenitors in both 2D and 3D cultures also generate layer-specific neurons in a sequential manner (Gaspard et al. 2008; Eiraku et al. 2008). In the case of human PSCs, it takes longer to induce neural tissues than it does in mouse cells. Human ESC-derived telencephalic progenitors exhibit a NE structure and start to express Foxg1 around day 18–20 (~75% of total cells). Most of the Foxg1^{+} NE cells express Pax6, suggesting that they acquire cortical identities. In the human SFEBq aggregates, Reelin^{+}/Tbr1^{+} Cajal–Retzius cells (layer I neurons) are born around day 35, followed by the generation of Reelin^{-}/Tbr1^{+}/Ctip2^{+} deep-layer neurons (layer V and VI neurons) around day 46, and then Satb2^{+}/Brn2^{+} upper-layer neurons (layer II–IV neurons) differentiate around day 91 (Eiraku et al. 2008; Kadoshima et al. 2013), recapitulating the sequential generation observed in vivo. Other groups have reported similar sequential generation of cortical neurons in both 2D and 3D human PSC cultures (Espuny-Camacho et al. 2013; Lancaster et al. 2013).

One of the striking aspects of in vitro human corticogenesis is the protracted differentiation process compared with mice, which reflects the species-specific difference in developmental speed in vivo. Indeed, the emergence of each subtype of cortical pyramidal neuron in vitro substantially coincides with the predicted timing of that observed in vivo (Clancy et al. 2007). Interestingly, such protracted development of human PSC-derived neurons is also observed after transplantation into the mouse brain (Espuny-Camacho et al. 2013). This suggests that the progenitors and differentiating neurons sense and respond precisely to the differentiation processes via an intrinsic timer.

Although the sequential generation of cortical neurons can be observed in both 2D and 3D cultures (Gaspard et al. 2008; Eiraku et al. 2008), 3D aggregate culture has an advantage in the formation of complex tissue structures (Eiraku et al. 2008; Kadoshima et al. 2013). In mouse SFEBq culture, cortical tissues generate a self-organized laminar structure including four distinct zones along the apical–basal axis: $Pax6^{+}$ VZ, $Tbr2^{+}$ SVZ, $Tbr1^{+}$ early cortical plate, and $Reelin^{+}$ marginal zone (Eiraku et al. 2008) (Fig. 1.3a). In the case of human SFEBq culture, the apical side ($aPKC^{+}$) of the $Foxg1^{+}$ NE is located on the surface of the aggregate at an early phase (Kadoshima et al. 2013). After day 20, the structure starts to break into several large domains and subsequently becomes apically concave. After dynamic rolling morphogenesis, $Foxg1^{+}$ NE generates a $Pax6^{+}/Sox2^{+}$ semispherical cortical structure with a lumen at around day 30 (Fig. 1.3b). The cortical NE derived from human ESCs shows a large, continuous structure surrounding a ventricle-like cavity inside. The apical side markers are localized on the inner side facing the cavity, while basal side markers are located outside. $Pax6^{+}$ and $Sox2^{+}$ cortical progenitors form cell dense VZ on the luminal side, whereas pH 3^{+} progenitors undergoing mitosis are found exclusively in its innermost part. $TuJ1^{+}$ neurons occupy the zone outside of the VZ and express markers of early cortical neurons such as Reelin, Tbr1, and Ctip2 on day 42 (Eiraku et al. 2008; Kadoshima et al. 2013) (Fig. 1.3b). In addition, live imaging of cortical NE derived from human ESCs shows that the $Pax6^{+}$ cortical progenitors are highly proliferative and undergo characteristic interkinetic nuclear migration, typical for pseudostratified epithelium, similar to the early embryonic cortex (Miyata et al. 2001). Taken together, ESC-derived cortical NE spontaneously forms a typical apicobasal polarity as seen in early embryonic cortex with interkinetic nuclear migration.

Generally, the prolonged culture of human ESC-derived cortical tissues leads to oxygen and nutrient deprivation in the deep portion of the aggregates, and typically the breakdown of cortical structure and cell death occur after about 50 days. But when the 3D cortical tissues are cultured with 40% O_2 and nutrient supplementation to the medium, they can grow for several months, which allows the induction of cortical tissues that are equivalent to the second trimester of human development (Kadoshima et al. 2013). The human ESC-derived cortical NE increases in thickness and starts to show morphologically stratified structure on day 70 (Fig. 1.3b). Layer marker expression patterns reveal complex zone separations in the cortical tissues. On the luminal side, in addition to the VZ ($Pax6^{+}/Sox2^{+}$), $Tbr2^{+}$ intermediate progenitors form a SVZ basally adjacent to the VZ. The superficial-most portion is the marginal zone (MZ), which contains $Reelin^{+}$ Cajal–Retzius (CR) cells and has accumulated laminin in its surface. Beneath the MZ, cortical pyramidal neurons

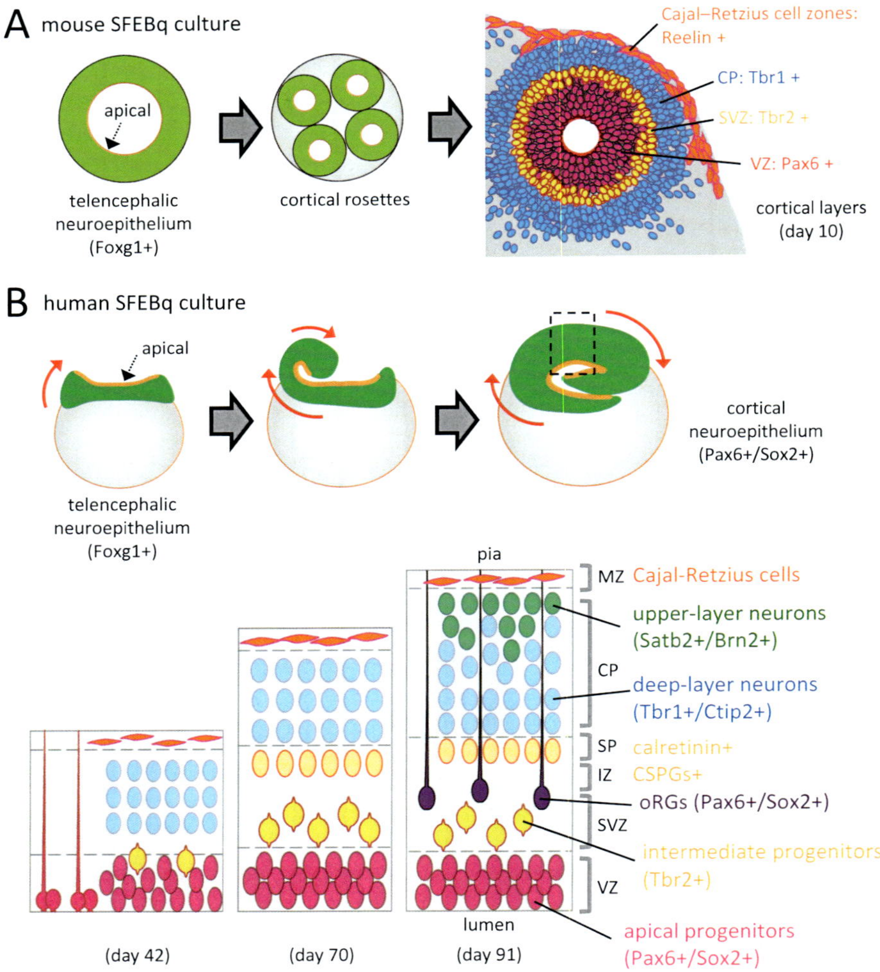

Fig. 1.3 Self-organizing 3D formation of layered cortical tissues. (**a**) In mouse ESC culture, the telencephalic neuroepithelium forms an apical surface inside the SFEBq aggregate. The neuroepithelium transforms into small cortical rosettes, which form multiple layers of neurons and progenitors. (**b**) In human ESC culture, the apical side of the telencephalic neuroepithelium is located on the surface of the SFEBq aggregate at an early phase. The neuroepithelium becomes apically concave in curvature and finally generates a semispherical cortical structure with a lumen inside. The cortical neuroepithelium self-forms a multilayered structure including *MZ* (marginal zone), *CP* (cortical plate), *SP* (subplate), *IZ* (intermediate zone), *SVZ* (subventricular zone), and *VZ* (ventricular zone) in the same apical-to-basal order as seen in fetal cortex. Within the CP, the upper-layer neurons preferentially localize more superficially to deep-layer neurons. $Pax6^{+}/Sox2^{+}$ oRG (outer radial glia)-like progenitors that accumulate in the outer portion of the SVZ have a long basal process extending toward the pia, similar to in vivo oRG

form a cortical plate (CP). Between the CP and SVZ, a calretinin$^+$ and chondroitin sulfate proteoglycan (CSPG)-rich subplate and intermediate zone are formed. At day 70, the CP is mostly composed of Tbr1$^+$ and Ctip2$^+$ early-born/deep-layer neurons, but by day 91, it contains a number of Satb2$^+$ and Brn2$^+$ late-born/superficial-layer neurons in addition to Tbr1$^+$ and Ctip2$^+$ neurons (Fig. 1.3b). Remarkably, within the cortical plate derived from human ESCs, late-born neurons (Satb2$^+$/Brn2$^+$) tend to preferentially localize more superficially than early-born neurons (Tbr1$^+$/Ctip2$^+$) on day 91 (Fig. 1.3b), which is reminiscent of the inside-out pattern during fetal corticogenesis. Furthermore, after prolonged culture, one of the maturation markers of neurons, CaMKII, is preferentially expressed in Tbr1$^+$ neurons located in the deep portion of the cortical plate on day 112, indicating the advanced maturation of early-born neurons. Using a nonselective differentiation method, a similar observation of the advanced stratified cortical tissue generated from human PSCs has been reported (Lancaster et al. 2013). These data suggest that human ESC-derived cortical NE spontaneously forms multilayered structures as seen in human fetal cortex at the beginning of the second trimester. However, the ESC-derived cortical tissues don't have distinct layer with inside-out pattern formation as in vivo cortex. This may suggest the necessity of not only more prolonged culture but also extrinsic factors from the periphery, such as the meninges, cerebrospinal fluid, and vascular networks, which is lacking in the in vitro cortical tissues (Tiberi et al. 2012).

In early corticogenesis, COUP-TF1 is expressed in a caudal-high to rostral-low gradient across the cortex (O'Leary et al. 2007; Zhou et al. 2001), whereas Sp8 is expressed in an opposing gradient (Sahara et al. 2007) (Fig. 1.4a). In vivo studies using mutant mice showed that Coup-TF1 and Sp8 have a crucial role for the caudal and rostral specification of the cortex, respectively (Armentano et al. 2007; Zembrzycki et al. 2007). FGF8, which is secreted from anterior neural ridge and the commissural plate (Crossley and Martin 1995), is widely recognized as a key player of cortical arealization through the regulation of expression of these transcriptional factors (Fig. 1.4a). Indeed, overexpression of FGF8 in the mouse cortex induces a caudal shift of graded expression of the transcriptional factors followed by the expansion of rostral areas, while reduction of FGF8 induces a rostral shift of graded expression of them followed by the reduction of rostral areas (Fukuchi-Shimogori and Grove 2001, 2003; Garel et al. 2003). With the advance of 3D culture, cortical arealization is also recapitulated. In the case of mouse SFEBq culture, extrinsic FGF8 treatment of ESC-derived cortical NE strongly suppresses COUP-TF1 expression. These cells express Tbx21, Reelin, and Tbr1, which is characteristic of the olfactory bulb projection neurons in the rostral-most cortex (Yoshihara et al. 2005). In contrast, COUP-TF1$^+$ cells substantially increase in number after treating the NE with the Fgf inhibitor FGFR3-Fc. In a mouse study, ESC-derived cortical NE gradually broke into several epithelial vesicles called neural rosettes in the aggregate (Fig. 1.3a), which makes it difficult to study continuous tissue formation. On the other hand, the human ESC-derived cortical structure keeps a continuous NE structure for a long time (up to 90 days) (Eiraku et al. 2008; Kadoshima et al. 2013). Interestingly, such continuous human ESC-derived NE frequently shows graded

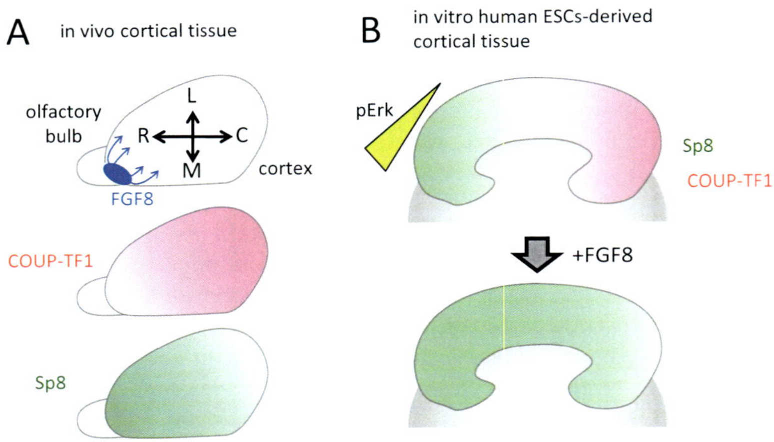

Fig. 1.4 Cortical arealization in vivo and in vitro. (**a**) The transcription factors COUP-TF1 and Sp8 are expressed in gradients along the rostrocaudal axis of the fetal cortex. FGF8 is secreted from the anterior patterning center, the anterior neural ridge, which later becomes the commissural plate. (**b**) Graded expression of COUP-TF1 and Sp8 in human ESC-derived cortical neuroepithelium. Phosphorylated ERK is localized in the cortical neuroepithelium on the side with strong Sp8 expression. Treatment with exogenous FGF8 causes broad expression of Sp8 at the expense of COUP-TF1 expression

expression of Coup-TF1 and also reverse expression of SP8 as seen in early fetal cortex (Kadoshima et al. 2013) (Fig. 1.4b). Moreover, a higher level of phospho-ERK signals is observed on the side opposite to Coup-TF1 expression (Fig. 1.4b), suggesting the involvement of upstream signals of ERK activation at the region. Indeed, treatment with extrinsic FGF8 causes broad expression of Sp8 at the expense of Coup-TF1 expression (Kadoshima et al. 2013) (Fig. 1.4b). These data indicate that exogenous FGF8 signaling induces rostralization of early cortical tissues in vitro, mimicking the in vivo situation. Besides arealization along the AP axis, subregionalization of the cortical lobe, such as the occipital lobe, frontal lobe, and parietal lobe, was also observed in cerebral organoids (Lancaster et al. 2013). In addition to FGF signals, artificial manipulation of cortical arealization by exogenous signals in 3D PSC culture will be an interesting topic for future study. These advances move us closer to being able to efficiently generate an intended cortical area.

Lastly, induction of 3D cortical tissues allows the examination of human oRG-like cells that requires markers and a morphological character to detect proper differentiation. In human ESC-derived cortical tissue in SFEBq culture, a large number of $Pax6^{+}/Sox2^{+}/Tbr2^{-}$ progenitors appear outside the VZ on day 91, whereas they are hardly present on day 70 (Fig. 1.3b). Phospho-vimentin (M phase-specific phosphorylation) staining reveals that they have a basal process but not an apical process during mitosis, like in vivo oRG cells. Moreover, the cleavage plane of the human

ESC-derived oRG-like cells tended to be horizontal, as observed in oRG cells in vivo (Fietz et al. 2010; Hansen et al. 2010). These data indicate that human ESC-derived cortical tissues recapitulate oRG characteristics, such as similar timing in their appearance and similar molecular marker expression, cellular morphology, and location. Therefore, 3D cortical tissue generated from human PSCs will be a useful tool in studying these species-specific progenitors, which likely play a significant role in cortical evolution.

1.3 Ventralizing Telencephalic Tissues

1.3.1 Development of the Ventral Telencephalon In Vivo

The ventral telencephalon (subpallium) is subdivided into the lateral ganglionic eminence (LGE), medial ganglionic eminence (MGE), caudal ganglionic eminence (CGE), the septum, and the ventral-most telencephalic stalk regions (e.g., the preoptic and anterior entopeduncular areas) (Sousa and Fishell 2010). The LGE is the origin of striatal projection neurons (Marín et al. 2000), whereas MGE and CGE appear to be the primary source of cortical and striatal GABAergic interneurons (Anderson et al. 1997, Nóbrega-Pereira et al. 2010). Previous studies have demonstrated the importance of patterning signaling via sonic hedgehog (Shh) in subpallial development (Shimamura et al. 1995; Echelard et al. 1993; Fuccillo et al. 2004, 2006), and the signals trigger cascades that lead to generation of cortical interneurons by inducing key transcription factors, such as Gsh2 in LGE, Nkx2.1 in MGE, and CoupTFII in CGE (Butt et al. 2008; Fuccillo et al. 2006; Corbin et al. 2003; Wonders and Anderson 2006). In the mouse embryo, Shh is first observed at an early developmental phase (E8–E9.5) in the diencephalon and mesendoderm, adjacent to the ventral telencephalon. This Shh expression induces Nkx2.1 in the MGE area (Sousa and Fishell 2010). Then Shh is expressed by the MGE and preoptic area by E12.5, and Gsh2 is expressed between the Nkx2.1 and Pax6 domains. The $Gsh2^{+}$/$Nkx2.1^{-}$ domain possesses LGE identity (Flandin et al. 2010; Waclaw et al. 2009) (Fig. 1.5a).

There are temporal time windows in the development of ventral telencephalon. Removal of Shh responsiveness from the ventral telencephalon by conditional deletion of smoothened by E9.5 results in strong abnormalities of patterning, whereas removal by E12.5 causes only minor patterning abnormalities (Dessaud et al. 2007). In humans, some studies report that most cortical interneurons undergo their terminal mitosis in the cortical subventricular zone (Letinic et al. 2002), yet recent studies show that cortical interneurons are actually generated from subpallium and then tangentially migrate into the cortical layer. This suggests a preserved developmental patterning of ventral telencephalon for generating GABAergic interneurons, as in mice (Anderson et al. 1997; Ma et al. 2013; Hansen et al. 2013).

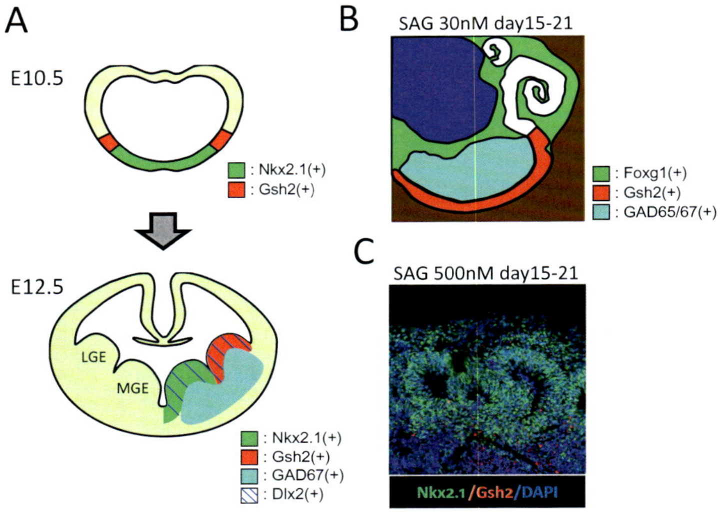

Fig. 1.5 Induction of ventral telencephalon. (**a**) Developmental process of ventral telencephalon in mice. Previous studies have demonstrated that Shh is first observed at an early developmental phase (E8-E9.5) in the diencephalon and mesendoderm adjacent to the ventral telencephalon, and Shh induces Nkx2.1 in the MGE area. Then, Shh is expressed in the MGE and preoptic area by E12.5, and Gsh2 is expressed between Nkx2.1 and Pax6 domain, and Gsh2$^+$/Nkx2.1$^-$ domain possesses LGE identity. Also the eminence structure emerges during this period. (**b**) Schematic of human ES cell-derived cortex–LGE tissues induced by a moderate treatment with SAG. Continuous tissue including cortical (Pax6$^+$) and LGE (Gsh2$^+$) domains was generated in sequential order, as seen in vivo. A mass of GAD65$^+$ GABAergic neurons was generated underneath the Gsh2$^+$ LGE NE, whereas the rest of the telencephalic NE was largely positive for the cortical NE marker Pax6. (**c**) Higher concentrations of SAG (500 nM, days 15–21) induced the medial ganglionic eminence (MGE) marker Nkx2.1 at the cost of Pax6 and Gsh2 expression. Scale bar: 100 μm

1.3.2 Development of the Ventral Telencephalon In Vitro

Once the telencephalic fate is determined, the DV pattern within the telencephalic region can be modified by patterning signals (Xu et al. 2010). Thus, for hPSC-derived cortical tissues, treatment with Shh signals can generate more ventral regions (LGE and MGE), depending on concentration and duration. Consistent with this idea, subpallium differentiation by treating ESC-derived telencephalic tissues with Shh has been achieved (Aubry et al. 2008; Danjo et al. 2011; Kadoshima et al. 2013; Eiraku and Sasai 2012; Li et al. 2009). In addition, different levels of Shh signal can selectively induce LGE- and MGE-type differentiation using SFEBq culture (Fig. 1.1c).

The NE of the cortex is Pax6$^+$/Gsh2$^-$, whereas subpallium NE cells are Pax6$^-$/Gsh2$^+$, and Nkx2.1 is expressed in the MGE and preoptic area, but not in the LGE or CGE (Corbin et al. 2003; Nery et al. 2002). The majority of subpallium-derived neurons are GABAergic and express GAD65/67 (Marin et al. 2000; Nóbrega-Pereira et al. 2010). In a mouse ES cell study, SFEBq-cultured mouse ESC aggregates differentiated preferentially into cortical progenitors (Eiraku et al. 2008) in the absence of exogenous Shh signals. When Shh signaling was blocked by the hedgehog (Hh) inhibitor cyclopamine, the majority of aggregates expressed the cortical markers Pax6 or Tbr1, whereas few expressed ventral markers such as Gsh2, Nkx2.1, Dlx2, and GAD67 (Danjo et al. 2011). However, a moderate dose of Shh (10 nM) strongly inhibited Pax6 or Tbr1 expression, and the ESC-derived Foxg1$^+$ domain expressed subpallial markers Gsh2, Dlx2, or GAD67. Although the MGE marker Nkx2.1 was rarely seen in cells cultured with a moderate dose of Shh (10 nM), the aggregates treated with a higher concentration (30 nM) contained a substantial number of Nkx2.1$^+$ cells (Danjo et al. 2011). When induced under these conditions and purified by a fluorescence-activated cell sorter, telencephalic cells efficiently differentiated into Nolz1$^+$/Ctip2$^+$ LGE neuronal precursors and subsequently, both in culture and after in vivo grafting, into DARPP32$^+$ medium-sized spiny neurons. Purified telencephalic progenitors treated with high doses of the hedgehog (Hh) agonist SAG (smoothened agonist) differentiated into MGE- and CGE-like tissues. Interestingly, in addition of Shh signaling, the efficient specification of MGE cells requires Fgf8 signaling but is inhibited by treatment with Fgf15/19. In contrast, CGE differentiation is promoted by Fgf15/19 but suppressed by Fgf8, suggesting that specific Fgf signals play different, critical roles in the positional specification of ESC-derived ventral subpallial tissues.

As in mouse ES cells, human ES cell-derived cortical tissues can also acquire a ventral telencephalic fate after Shh treatment (Kadoshima et al. 2013). When the human ESC-derived telencephalic NE was partially ventralized by a moderate level of hedgehog agonist (30 nM smoothened agonist or SAG for days 15–21), a substantial portion of Foxg1$^+$ NE expressed Gsh2 (Fig. 1.5b). A mass of GAD65$^+$ GABAergic neurons was generated underneath this LGE NE, as seen in vivo, whereas the rest of the telencephalic NE was largely positive for the cortical NE marker Pax6. Higher concentrations of SAG induced the MGE marker Nkx2.1 at the cost of Pax6 and Gsh2 expression (Fig. 1.5c). Interestingly, in the human aggregates, continuous epithelial tissues were retained, resulting in the generation of cortical (Pax6$^+$) and LGE (Gsh2$^+$) domains as seen in vivo, suggesting that our improved culture method allows self-formation of pallial–subpallial structures in sequential order (Fig. 1.5b). Generation of pallium–subpallium in one sphere will offer the analysis of tangential migration of GABAergic cortical interneurons using the SFEBq method; thus, SFEBq culture relating to ventral telencephalon is an efficient in vitro culture system for cortical interneurons and is also very interesting for not only stem cell biology but also human developmental biology. Besides, analysis of Shh signaling cascades on the subpallial patterning of ESC-derived telencephalic tissues is also an interesting topic.

Deficiencies in subtypes of interneurons, and the resulting excitatory–inhibitory imbalances, have been associated with many neurodevelopmental and neurodegenerative diseases, such as epilepsy, schizophrenia, autism, and Huntington's disease (Sigurdsson and Duvarci 2015; Kumar et al. 2015), so the recapitulation of ventral telencephalic tissues using human PSCs will model not only human ventral forebrain development but also human disease pathology. Actually, transplantation studies using hPSC-derived cortical interneurons serve as a foundation for examining cortical interneuron involvement in disease pathology (Maroof et al. 2013; Nicholas et al. 2013). As a future aim, the induction of subtypes of human interneurons would be useful for developing treatments for diseases resulting from interneuron dysfunction.

1.4 Dorsalizing Telencephalic Tissues

1.4.1 Development of Dorsomedial Telencephalic Structure In Vivo

In the more dorsomedial portion of the cerebral cortex, there are three distinct regions: the choroid plexus, cortical hem, and medial pallium (Furuta et al. 1997; Monuki et al. 2001; Arnold and Trojanowski 1996) (Fig. 1.6a). The most dorsomedial portion of dorsal telencephalon is the choroid plexus (Hébert et al. 2002). In early development, choroid plexus is induced via BMP signals from the roof plate (E9.0) (Hébert et al. 2002). Previous studies have showed that many BMP proteins (such as BMP2, 4, 6, and 7) are expressed at the dorsal midline of telencephalon with a signaling gradient and disrupting the BMP signaling led to a lack of choroid plexus tissues, suggesting the necessity of BMP signaling in specifying dorsal telencephalic cell fate (Cheng et al. 2006; Hébert et al. 2002) (Fig. 1.6a). In addition, during the embryonic to adult phase, choroid plexus has essential developmental and homeostatic roles regarding the generation of cerebrospinal fluid (CSF) and the formation of the blood–CSF barrier (Lehtinen et al. 2013). The cortical hem is the structure located next to choroid plexus, and it expresses several Wnt (Wnt 2b, 3a, and 5a) and BMP (BMP 6 and 7) proteins that induce dorsomedial patterning of dorsal telencephalon (Lee et al. 2000; Failli et al. 2002).

During development, it is the medial pallium that generates the hippocampus, and it lies between the neocortex and the cortical hem (Arnold and Trojanowski 1996; Grove et al. 1998; Lee et al. 2000; Monuki et al. 2001; Tole et al. 1997; Zhao et al. 1999) (Fig. 1.6a). The ability to form memories, during long-term tasks such as learning and short-term tasks such as remembering a telephone number, relies on the neuronal activity of the hippocampus (Geuze et al. 2005). During the embryonic phase, Prox1$^+$ hippocampal dentate granule (DG) neurons are first induced in the dorsomedial edge of medial pallium (E12.5-), and they migrate along a curved trajectory (E12.5–15.5). Then the cell fate of hippocampal CA neurons is determined (CA3 (KA1$^+$): E14.5- and CA1 (SCID$^+$): E15.5-), resulting in the formation of a typical hippocampal structure (Grove and Tole 1999) (Fig. 1.6a). Importantly, the cortical hem is also known as the hippocampal organizer (Mangale et al. 2008). When expression of Wnt 3a in the cortical hem was reduced using conditional

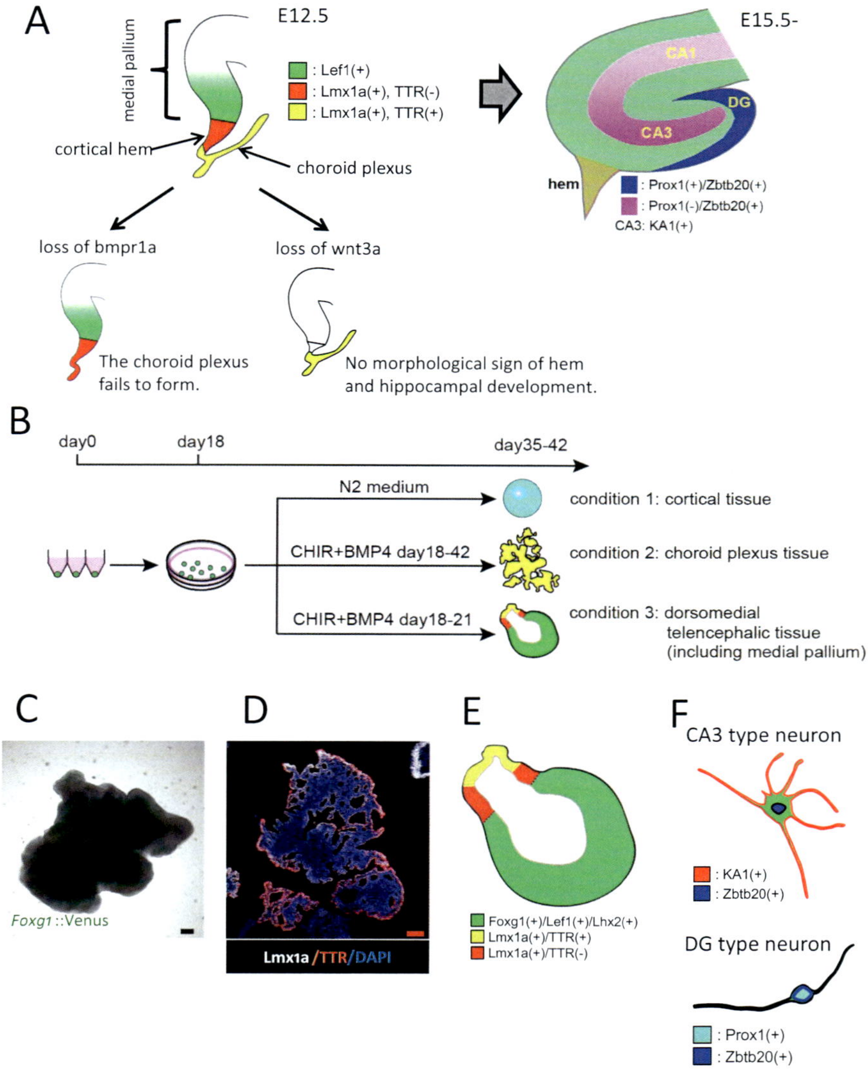

Fig. 1.6 Induction of dorsomedial telencephalon. (**a**) Schematic of the segmental development of the dorsomedial telencephalon. There are three distinct regions: the choroid plexus, cortical hem, and medial pallium. During the embryonic phase, Prox1+ hippocampal dentate granule neurons are first induced in the dorsomedial edge of medial pallium (E12.5-) and migrate along a curved trajectory (E12.5–15.5). Then the cell fate of CA neurons is determined (CA3 ($KA1^+$): E14.5- and CA1 ($SCID^+$): E15.5-) with migration like that of DG neurons, resulting in the formation of a typical hippocampal structure. Disrupting the BMP signaling caused lack of choroid plexus tissues, whereas reduction of Wnt 3a in cortical hem severely affected the formation of hippocampal structure. (**b**) Schematic of the induction method for dorsomedial telencephalic tissues. (**c**) Bright-field view of one aggregate cultured using choroid plexus induction method on day 42. (**d**) Immunohistochemistry of choroid plexus tissues derived from hESCs. The pleated epithelia are positive for Lmx1a and TTR. TTR mainly stained apically. (**e**) Schematic of hESC-derived dorsomedial telencephalic tissues. Choroid plexus, hem, and medial pallium regions are continuously generated as seen in vivo. (**f**) Schematic of hESC-derived hippocampal neurons. KA1-positive neurons (CA type) tended to have large cell bodies; in contrast, Prox1-positive neurons (DG type) mostly had small somata. Scale bars: 500 μm in C and 200 μm in D

knock-out mice, the formation of hippocampal structure was severely affected (Lee et al. 2000; Galceran et al. 2000). In contrast, the formation of ectopic cortical hem by chimera mouse of conditional Lhx2 knock-out showed ectopic generation of hippocampal structure next to additionally induced hem (Mangale et al. 2008). Taken together, these in vivo results suggest the necessity and sufficiency of cortical hem and Wnt signaling for induction of hippocampus and the necessity of BMP signaling for the patterning of dorsomedial telencephalic tissues.

1.4.2 Generation of Dorsomedial Telencephalic Tissues In Vitro

Based on in vivo developmental cues, the more dorsal regions (cortical hem, choroid plexus, and medial pallium) can be generated by treatment with Wnt and BMP signals (Eiraku et al. 2008; Sakaguchi et al. 2015). Regarding choroid plexus induction in vitro, there have been three reports using mouse and human ES cells (Eiraku et al. 2008; Watanabe et al. 2012; Sakaguchi et al. 2015). In mouse ES cells, treatment with BMP4 promoted the generation of transthyretin (TTR)-positive cells, and TTR is a marker specific for choroid plexus in nervous tissues (Eiraku et al. 2008; Watanabe et al. 2012). The induced aggregates contained a thin epithelial or vesicle-like choroid plexus structure, with tight junctions confirmed by ZO-1 expression and electron microscopy (Watanabe et al. 2012). In addition, the structure could integrate into endogenous choroid plexus epithelium following intraventricular transplantation. This dorsalizing strategy can also be applied to a human ES cell system, and treatment with BMP4 alone could sufficiently induce TTR^+ choroid plexus tissues as in the mouse study (Watanabe et al. 2012; Sakaguchi et al. 2015). Furthermore, treatment with BMP4 combined with Wnt signaling further enhanced choroid plexus induction (Sakaguchi et al. 2015) (Fig. 1.6b). The treatment of SFEBq-induced neocortical tissues with BMP4 ligand plus CHIR 99021 (GSK3 inhibitor, also known to accelerate canonical Wnt signaling), a condition designed to mimic the in vivo dorsal midline and cortical hem signaling center, significantly increased *lmx1a*, *otx2*, and *ttr* mRNA expression. Immunohistochemistry showed that Lmx1a (Failli et al. 2002), water channel Aqp1 (Praetorius and Nielsen 2006), and TTR signal were mainly seen apically in the thin NE portion and tight junction marker ZO-1 was on the surface of epithelia (Fig. 1.6c, d). Interestingly, these aggregates have a pleated NE structure as seen in human embryonic choroid plexus tissues. This morphological character does not occur without Wnt treatment (Sakaguchi et al. 2015). These characteristics in morphology and marker distribution are reminiscent of those in human fetal choroid plexus epithelia. Collectively, treatment with CHIR + BMP signals to SFEBq-induced neocortical tissues could dominantly induce choroid plexus tissues from human ESCs in 3D culture. This culture system might be useful for exploring choroid plexus development with regard to structure formation, CSF generation, barrier formation, and so on.

To date, there have been only two reports regarding hippocampal induction from PSCs (Yu et al. 2014a; Sakaguchi et al. 2015). Yu et al. first reported a protocol for selective differentiation of DG granule neurons using human ES and iPS cells (Yu et al. 2014a). They differentiated Prox1$^+$ DG granule neurons by Wnt3a/BDNF treatment; these neurons were electrically functional and were able to integrate into endogenous DG circuitry. A recent paper further showed that titrating BMP and Wnt exposure to neocortical tissues allowed the self-organization of dorsomedial telencephalic tissues that include Foxg1$^+$/Lef1$^+$/Lhx2$^+$ medial pallium tissues (Sakaguchi et al. 2015). Interestingly, the sequential expression pattern of genetic markers of choroid plexus (Foxg1$^-$/Lmx1a$^+$/TTR$^+$), cortical hem (Foxg1$^-$/Lmx1a$^+$/TTR$^-$), and medial pallium (Foxg1$^+$/Lhx2$^+$/Lef1$^+$) suggested that a self-organized form of differentiation was occurring in the same chronological order as in in vivo development (Fig. 1.6e). These SFEBq-induced dorsomedial telencephalic tissues give rise to hippocampal granule (DG)- and pyramidal (CA)-like neurons through further long-term culture. On day 61, Lef1$^+$/Foxg1$^+$ medial pallium-like continuous NE portions clearly formed adjacent to the Lmx1a$^+$ choroid plexus- and hem-like domain, and a more basal layer of the Lef1$^+$ NE was positive for neuropillin2 and Zbtb20 (BTB/POZ zinc-finger family), both of which are known as entire hippocampal markers in the embryonic hippocampus (Chen et al. 1997; Xie et al. 2010). IHC showed clear Zbtb20 expression in cells beneath the Lef1$^+$ NE and weak Zbtb20 expression in the NE. Collectively, these results demonstrate the successful in vitro generation of hippocampal primordium-like tissue.

Furthermore, using dissociation culture, these dorsomedial telencephalic tissues gave rise to Zbtb20$^+$/Prox1$^+$ granule neurons (corresponding to DG neurons) and Zbtb20$^+$/KA1$^+$ pyramidal neurons (corresponding to CA3 neurons) (Fig. 1.6f), both of which were electrically functional and positive for the maturation marker CaMKII. Notably, by examining intracellular calcium dynamics, calcium transients at 8 weeks after dissociation showed a synchronized pattern of network activity that was not seen at 4 weeks after dissociation, suggesting the development of synaptic connections and a neuronal network formation. Thus, it is suggested that the hippocampal primordium tissues induced by SFEBq culture could recapitulate human hippocampus development, allowing the generation of functional hippocampal granule- and pyramidal-like neurons (Sakaguchi et al. 2015).

As mentioned above, the complex process of hippocampal development involves cell fate determination, cell movement, dentate gyrogenesis, and circuit formations (Yu et al. 2014b); this complexity has made it difficult to induce hippocampal tissues in vitro. Thus, the discovery of a method for inducing continuous dorsomedial structures in vitro is very intriguing. A speculated mechanism of hippocampal tissue generation using SFEBq is as follows: (i) a small portion of hESC-derived cortical tissue differentiates into dorsomedial tissue by culturing in the medium for choroid plexus induction (days 18–21); (ii) despite culturing in medium based on the conditions for neocortex induction from day 21, the induced dorsomedial tissue domain functions as a dorsalizing organizer; and (iii) with the effect of this organizer, the remaining cortical domain can express Lef1 and Lhx2, which results in dorsomedial pattern formation as seen in vivo (days 21–42). Our technology now gives researchers new means to investigate hippocampal biology, disease, and potential therapies.

1.5 Closing Remarks and Future Perspective

Here we have reviewed recent advances in 3D organogenesis with regard to telencephalic tissue induction. Generation of 3D tissue will allow the morphogenesis as seen in human development in vivo. The human PSC-derived cortical tissues in 3D culture are thought to have a very similar transcriptional profile to human fetal cortex (Mariani et al. 2012; Camp et al. 2015). These studies reveal that human PSC-derived 3D human cortical tissues recapitulate not only the cortical architecture and cellular behavior but also genetic features. Therefore human PSC-derived cortical tissues can be a powerful tool for understanding human corticogenesis and for neurodevelopmental disease modeling. Indeed, the cortical tissues derived from PSCs from a patient with microcephaly showed premature neuronal differentiation (Lancaster et al. 2013). However, to generate fully functional cortical tissues in vitro, we have to overcome some technical limitations. In particular, an insufficient supply of oxygen and nutrients is a critical problem for long-term culture. Vascular network formation in the in vitro human cortical tissues could be one of the promising solutions. In the mouse fetal cortex, vascularization starts around E12.5, and it supplies nutrients and provides niches for neural stem/progenitor cells (Stubbs et al. 2009). A recent study showed the generation of vascularized liver from human PSCs in 3D culture (Takebe et al. 2013), and this method is also applicable to brain tissues (Takebe et al. 2015). Such attempts would allow better recapitulation of human corticogenesis in vitro. Moreover, induction of more complex telencephalic tissues will open the next generation of tissue formation and medical applications with regard to the complex functions of the human brain. Generation of a broad area of telencephalon including pallium–subpallium or neo-to-archi cortical tissue in sequential order may allow the induction of more complex tissue such as amygdala, hippocampus, or claustrum in vitro. If so, this would enable a deeper exploration of higher brain functions such as the formation of episodic memory, personality, intention, or emotion.

References

Anderson SA, Eisenstat DD, Shi L, Rubenstein JLR (1997) Interneuron migration from basal forebrain to neocortex: dependence on Dlx genes. Science 278(5337):474–476

Angevine JB, Sidman RL (1961) Autoradiographic study of cell migration during histogenesis of cerebral cortex in the mouse. Nature 192:766–768

Armentano M, Chou SJ, Tomassy GS, Leingärtner A, O'Leary DD, Studer M (2007) COUP-TFI regulates the balance of cortical patterning between frontal/motor and sensory areas. Nat Neurosci 10(10):1277–1286

Arnold SE, Trojanowski JQ (1996) Human fetal hippocampal development: I. Cytoarchitecture, myeloarchitecture, and neuronal morphologic features. J Comp Neurol 367(2):274–292
Aubry L, Bugi A, Lefort N, Rousseau F, Peschanski M, Perrier AL (2008) Striatal progenitors derived from human ES cells mature into DARPP32 neurons in vitro and in quinolinic acid-lesioned rats. Proc Natl Acad Sci USA 105(43):16707–16712
Butt SJ, Sousa VH, Fuccillo MV, Hjerling-Leffler J, Miyoshi G, Kimura S, Fishell G (2008) The requirement of Nkx2-1 in the temporal specification of cortical interneuron subtypes. Neuron 59(5):722–732
Camp JG, Badsha F, Florio M, Kanton S, Gerber T, Wilsch-Bräuninger M, Lewitus E, Sykes A, Hevers W, Lancaster M, Knoblich JA, Lachmann R, Pääbo S, Huttner WB, Treutlein B (2015) Human cerebral organoids recapitulate gene expression programs of fetal neocortex development. Proc Natl Acad Sci USA 112(51):15672–15677
Chen H, Chédotal A, He Z, Goodman CS, Tessier-Lavigne M (1997) Neuropilin-2, a novel member of the neuropilin family, is a high affinity receptor for the semaphorins Sema E and Sema IV but not Sema III. Neuron 19(3):547–559
Cheng X, Hsu CM, Currle DS, Hu JS, Barkovich AJ, Monuki ES (2006) Central roles of the roof plate in telencephalic development and holoprosencephaly. J Neurosci 26(29):7640–7649
Clancy B, Finlay BL, Darlington RB, Anand KJS (2007) Extrapolating brain development from experimental species to humans. Neurotoxicology 28(5):931–937
Corbin JG, Rutlin M, Gaiano N, Fishell G (2003) Combinatorial function of the homeodomain proteins Nkx2. 1 and Gsh2 in ventral telencephalic patterning. Development 130(20):4895–4906
Cox WG, Hemmati-Brivanlou A (1995) Caudalization of neural fate by tissue recombination and bFGF. Development 121(12):4349–4358
Crossley PH, Martin GR (1995) The mouse Fgf8 gene encodes a family of polypeptides and is expressed in regions that direct outgrowth and patterning in the developing embryo. Development 121(2):439–451
Danjo T, Eiraku M, Muguruma K, Watanabe K, Kawada M, Yanagawa Y, Rubenstein JLR, Sasai Y (2011) Subregional specification of embryonic stem cell-derived ventral telencephalic tissues by timed and combinatory treatment with extrinsic signals. J Neurosci 31(5):1919–1933
DeFelipe J, Alonso-Nanclares L, Arellano JI (2002) Microstructure of the neocortex: comparative aspects. J Neurocytol 31(3–5):299–316
Dessaud E, Yang LL, Hill K, Cox B, Ulloa F, Ribeiro A, Mynett A, Novitch BG, Briscoe J (2007) Interpretation of the sonic hedgehog morphogen gradient by a temporal adaptation mechanism. Nature 450(7170):717–720
Durston AJ, Van Der Wees J, Pijnappel WWM, Godsave SF (1998) Retinoids and related signals in early development of the vertebrate central nervous system. Curr Top Dev Biol 40:112–177
Echelard Y, Epstein DJ, St-Jacques B, Shen L, Mohler J, McMahon JA, McMahon AP (1993) Sonic hedgehog, a member of a family of putative signaling molecules, is implicated in the regulation of CNS polarity. Cell 75(7):1417–1430
Eiraku M, Sasai Y (2012) Self-formation of layered neural structures in three-dimensional culture of ES cells. Curr Opin Neurobiol 22(5):768–777
Eiraku M, Watanabe K, Matsuo-Takasaki M, Kawada M, Yonemura S, Matsumura M, Wataya T, Nishiyama A, Muguruma K, Sasai Y (2008) Self-organized formation of polarized cortical tissues from ESCs and its active manipulation by extrinsic signals. Cell Stem Cell 3(5):519–532
Espuny-Camacho I, Michelsen KA, Gall D, Linaro D, Hasche A, Bonnefont J, Bali C, Orduz D, Bilheu A, Herpoel A, Lambert N, Gaspard N, Péron S, Schiffmann SN, Giugliano M, Gaillard A, Vanderhaeghen P (2013) Pyramidal neurons derived from human pluripotent stem cells integrate efficiently into mouse brain circuits in vivo. Neuron 77(3):440–456
Failli V, Bachy I, Rétaux S (2002) Expression of the LIM-homeodomain gene Lmx1a (dreher) during development of the mouse nervous system. Mech Dev 118(1):225–228
Fietz SA, Kelava I, Vogt J, Wilsch-Brauninger M, Stenzel D, Fish JL, Corbeil D, Riehn A, Distler W, Nitsch R, Huttner WB (2010) OSVZ progenitors of human and ferret neocortex are epithelial-like and expand by integrin signaling. Nat Neurosci 13(6):690–699

Flandin P, Kimura S, Rubenstein JL (2010) The progenitor zone of the ventral medial ganglionic eminence requires Nkx2-1 to generate most of the globus pallidus but few neocortical interneurons. J Neurosci 30(8):2812–2823
Fuccillo M, Rallu M, McMahon AP, Fishell G (2004) Temporal requirement for hedgehog signaling in ventral telencephalic patterning. Development 131(20):5031–5040
Fuccillo M, Joyner AL, Fishell G (2006) Morphogen to mitogen: the multiple roles of hedgehog signalling in vertebrate neural development. Nat Rev Neurosci 7(10):772–783
Fukuchi-Shimogori T, Grove EA (2001) Neocortex patterning by the secreted signaling molecule FGF8. Science 294(5544):1071–1074
Fukuchi-Shimogori T, Grove EA (2003) Emx2 patterns the neocortex by regulating FGF positional signaling. Nat Neurosci 6(8):825–831
Furuta Y, Piston DW, Hogan BL (1997) Bone morphogenetic proteins (BMPs) as regulators of dorsal forebrain development. Development 124(11):2203–2212
Galceran J, Miyashita-Lin EM, Devaney E, Rubenstein JL, Grosschedl R (2000) Hippocampus development and generation of dentate gyrus granule cells is regulated by LEF1. Development 127(3):469–482
Garel S, Huffman KJ, Rubenstein JL (2003) Molecular regionalization of the neocortex is disrupted in Fgf8 hypomorphic mutants. Development 130(9):1903–1914
Gaspard N, Bouschet T, Hourez R, Dimidschstein J, Naeije G, van den Ameele J, Espuny-Camacho I, Herpoel A, Passante L, Schiffmann SN, Gaillard A, Vanderhaeghen P (2008) An intrinsic mechanism of corticogenesis from embryonic stem cells. Nature 455(7211):351–357
Geuze E, Vermetten E, Bremner JD (2005) MR-based in vivo hippocampal volumetrics: 2. Findings in neuropsychiatric disorders. Mol Psychiatry 10(2):160–184
Godsave SF, Slack JMW (1989) Clonal analysis of mesoderm induction in *Xenopus laevis*. Dev Biol 134(2):486–490
Grove EA, Tole S (1999) Patterning events and specification signals in the developing hippocampus. Cereb Cortex 9(6):551–561
Grove EA, Tole S, Limon J, Yip LW, Ragsdale CW (1998) The hem of the embryonic cerebral cortex is defined by the expression of multiple Wnt genes and is compromised in Gli3-deficient mice. Development 125(12):2315–2325
Grunz H, Tacke L (1989) Neural differentiation of *Xenopus laevis* ectoderm takes place after disaggregation and delayed reaggregation without inducer. Cell Differ Dev 28(3):211–217
Hansen DV, Lui JH, Parker PR, Kriegstein AR (2010) Neurogenic radial glia in the outer subventricular zone of human neocortex. Nature 464(7288):554–561
Hansen DV, Lui JH, Flandin P, Yoshikawa K, Rubenstein JL, Alvarez-Buylla A, Kriegstein AR (2013) Non-epithelial stem cells and cortical interneuron production in the human ganglionic eminences. Nat Neurosci 16:1576–1587
Hébert JM, Mishina Y, McConnell SK (2002) BMP signaling is required locally to pattern the dorsal telencephalic midline. Neuron 35(6):1029–1041
Hevner RF, Daza RA, Rubenstein JL, Stunnenberg H, Olavarria JF, Englund C (2003) Beyond laminar fate: toward a molecular classification of cortical projection/pyramidal neurons. Dev Neurosci 25(2–4):139–151
Huttner WB, Brand M (1997) Asymmetric division and polarity of neuroepithelial cells. Curr Opin Neurobiol 7(1):29–39
Kadoshima T, Sakaguchi H, Nakano T, Soen M, Ando S, Eiraku M, Sasai Y (2013) Self-organization of axial polarity, inside-out layer pattern, and species-specific progenitor dynamics in human ES cell-derived neocortex. Proc Natl Acad Sci USA 110:20284–20289
Kamiya D, Banno S, Sasai N, Watanabe K, Kawada M, Yakura R, Jakt LM, Nishikawa SI, Sasai Y (2011) Intrinsic transition of embryonic stem-cell differentiation into neural progenitors. Nature 470(7335):503–509
Kriegstein AR, Noctor SC (2004) Patterns of neuronal migration in the embryonic cortex. Trends Neurosci 27(7):392–399
Kriegstein A, Noctor S, Martínez-Cerdeño V (2006) Patterns of neural stem and progenitor cell division may underlie evolutionary cortical expansion. Nat Rev Neurosci 7(11):883–890

Kumar A, Singh SK, Kumar V, Kumar D, Agarwal S, Rana MK (2015) Huntington's disease: an update of therapeutic strategies. Gene 556(2):91–97
Lancaster MA, Renner M, Martin CA, Wenzel D, Bicknell LS, Hurles ME, Homfray T, Penninger JM, Jackson AP, Knoblich JA (2013) Cerebral organoids model human brain development and microcephaly. Nature 501(7467):373–379
Lee KJ, Jessell TM (1999) The specification of dorsal cell fates in the vertebrate central nervous system. Annu Rev Neurosci 22(1):261–294
Lee SM, Tole S, Grove E, McMahon AP (2000) A local Wnt-3a signal is required for development of the mammalian hippocampus. Development 127(3):457–467
Lehtinen MK, Bjornsson CS, Dymecki SM, Gilbertson RJ, Holtzman DM, Monuki ES (2013) The choroid plexus and cerebrospinal fluid: emerging roles in development, disease, and therapy. J Neurosci 33(45):17553–17559
Letinic K, Zoncu R, Rakic P (2002) Origin of GABAergic neurons in the human neocortex. Nature 417(6889):645–649
Levine AJ, Brivanlou AH (2007) Proposal of a model of mammalian neural induction. Dev Biol 308(2):247–256
Li XJ, Zhang X, Johnson MA, Wang ZB, LaVaute T, Zhang SC (2009) Coordination of sonic hedgehog and Wnt signaling determines ventral and dorsal telencephalic neuron types from human embryonic stem cells. Development 136(23):4055–4063
Lui JH, Hansen DV, Kriegstein AR (2011) Development and evolution of the human neocortex. Cell 146(1):18–36
Lumsden A, Krumlauf R (1996) Patterning the vertebrate neuraxis. Science 274(5290):1109–1115
Ma T, Wang C, Wang L, Zhou X, Tian M, Zhang Q, Zhang Y, Li J, Liu Z, Cai Y, Liu F, You Y, Chen C, Campbell K, Song H, Ma L, Rubenstein JL, Yang Z (2013) Subcortical origins of human and monkey neocortical interneurons. Nat Neurosci 16:1588–1597
Mangale VS, Hirokawa KE, Satyaki PR, Gokulchandran N, Chikbire S, Subramanian L, Shetty AS, Martynoga B, Paul J, Mai MV, Li Y, Flanagan LA, Tole S, Monuki ES (2008) Lhx2 selector activity specifies cortical identity and suppresses hippocampal organizer fate. Science 319(5861):304–309
Mariani J, Simonini MV, Palejev D, Tomasini L, Coppola G, Szekely AM, Horvath TL, Vaccarino FM (2012) Modeling human cortical development in vitro using induced pluripotent stem cells. Proc Natl Acad Sci USA 109(31):12770–12775
Marín O, Anderson SA, Rubenstein JL (2000) Origin and molecular specification of striatal interneurons. J Neurosci 20(16):6063–6076
Maroof AM, Keros S, Tyson JA, Ying SW, Ganat YM, Merkle FT, Liu B, Goulburn A, Stanley EG, Elefanty AG, Widmer HR, Eggan K, Goldstein PA, Anderson SA, Studer L (2013) Directed differentiation and functional maturation of cortical interneurons from human embryonic stem cells. Cell Stem Cell 12(5):559–572
Miyata T, Kawaguchi A, Okano H, Ogawa M (2001) Asymmetric inheritance of radial glial fibers by cortical neurons. Neuron 31(5):727–741
Molyneaux BJ, Arlotta P, Menezes JR, Macklis JD (2007) Neuronal subtype specification in the cerebral cortex. Nat Rev Neurosci 8(6):427–437
Monuki ES, Porter FD, Walsh CA (2001) Patterning of the dorsal telencephalon and cerebral cortex by a roof plate-Lhx2 pathway. Neuron 32(4):591–604
Muhr J, Graziano E, Wilson S, Jessell TM, Edlund T (1999) Convergent inductive signals specify midbrain, hindbrain, and spinal cord identity in gastrula stage chick embryos. Neuron 23(4):689–702
Nasu M, Takata N, Danjo T, Sakaguchi H, Kadoshima T, Futaki S, Sekiguchi K, Eiraku M, Sasai Y (2012) Robust formation and maintenance of continuous stratified cortical neuroepithelium by lamininContaining matrix in mouse ES cell culture. PLoS One 7:e53024
Nery S, Fishell G, Corbin JG (2002) The caudal ganglionic eminence is a source of distinct cortical and subcortical cell populations. Nat Neurosci 5(12):1279–1287

Nicholas CR, Chen J, Tang Y, Southwell DG, Chalmers N, Vogt D, Arnold CM, Chen YJ, Stanley EG, Elefanty AG, Sasai Y, Alvarez-Buylla A, Rubenstein JL, Kriegstein AR (2013) Functional maturation of hPSC-derived forebrain interneurons requires an extended timeline and mimics human neural development. Cell Stem Cell 12(5):573–586
Nóbrega-Pereira S, Gelman D, Bartolini G, Pla R, Pierani A, Marín O (2010) Origin and molecular specification of globus pallidus neurons. J Neurosci 30(8):2824–2834
Noctor SC, Flint AC, Weissman TA, Dammerman RS, Kriegstein AR (2001) Neurons derived from radial glial cells establish radial units in neocortex. Nature 409(6821):714–720
Noctor SC, Martínez-Cerdeño V, Ivic L, Kriegstein AR (2004) Cortical neurons arise in symmetric and asymmetric division zones and migrate through specific phases. Nat Neurosci 7(2):136–144
Nordström U, Jessell TM, Edlund T (2002) Progressive induction of caudal neural character by graded Wnt signaling. Nat Neurosci 5:525–532
Northcutt RG, Kaas JH (1995) The emergence and evolution of mammalian neocortex. Trends Neurosci 18(9):373–379
O'Leary DD (1989) Do cortical areas emerge from a protocortex? Trends Neurosci 12(10):400–406
O'Leary DD, Chou SJ, Sahara S (2007) Area patterning of the mammalian cortex. Neuron 56:252–269
Piccolo S, Sasai Y, Lu B, De Robertis EM (1996) Dorsoventral patterning in Xenopus: inhibition of ventral signals by direct binding of chordin to BMP-4. Cell 86(4):589–598
Praetorius J, Nielsen S (2006) Distribution of sodium transporters and aquaporin-1 in the human choroid plexus. Am J Phys Cell Phys 291(1):C59–C67
Rakic P (1988) Specification of cerebral cortical areas. Science 241(4862):170–176
Rubenstein JL, Shimamura K, Martinez S, Puelles L (1998) Regionalization of the prosencephalic neural plate. Annu Rev Neurosci 21(1):445–477
Sahara S, Kawakami Y, Belmonte JCI, O'Leary DD (2007) Sp8 exhibits reciprocal induction with Fgf8 but has an opposing effect on anterior-posterior cortical area patterning. Neural Dev 2(1):10
Sakaguchi H, Kadoshima T, Soen M, Narii N, Ishida Y, Ohgushi M, Takahashi J, Eiraku M, Sasai Y (2015) Generation of functional hippocampal neurons from self-organizing human embryonic stem cell-derived dorsomedial telencephalic tissue. Nat Commun 6:8896
Sansom SN, Hébert JM, Thammongkol U, Smith J, Nisbet G, Surani MA, McConnell SK, Livesey FJ (2005) Genomic characterisation of a Fgf-regulated gradient-based neocortical protomap. Development 132(17):3947–3961
Sasai Y (2013a) Next-generation regenerative medicine: organogenesis from stem cells in 3D culture. Cell Stem Cell 12(5):520–530
Sasai Y (2013b) Cytosystems dynamics in self-organization of tissue architecture. Nature 493(7432):318–326
Sasai Y, De Robertis EM (1997) Ectodermal patterning in vertebrate embryos. Dev Biol 182(1):5–20
Sasai Y, Lu B, Steinbeisser H, De Robertis EM (1995) Regulation of neural induction by the Chd and Bmp-4 antagonistic patterning signals in Xenopus. Nature 376(6538):333–336
Sato SM, Sargent TD (1989) Development of neural inducing capacity in dissociated Xenopus embryos. Dev Biol 134(1):263–266
Shen Q, Wang Y, Dimos JT, Fasano CA, Phoenix TN, Lemischka IR, Ivanova NB, Stifani S, Morrisey EE, Temple S (2006) The timing of cortical neurogenesis is encoded within lineages of individual progenitor cells. Nat Neurosci 9(6):743–751
Shimamura K, Hartigan DJ, Martinez S, Puelles L, Rubenstein JL (1995) Longitudinal organization of the anterior neural plate and neural tube. Development 121(12):3923–3933
Shitamukai A, Konno D, Matsuzaki F (2011) Oblique radial glial divisions in the developing mouse neocortex induce self-renewing progenitors outside the germinal zone that resemble primate outer subventricular zone progenitors. J Neurosci 31:3683–3695

Sigurdsson T, Duvarci S (2015) Hippocampal-prefrontal interactions in cognition, behavior and psychiatric disease. Front Syst Neurosci 9:190
Smart IH, Dehay C, Giroud P, Berland M, Kennedy H (2002) Unique morphological features of the proliferative zones and postmitotic compartments of the neural epithelium giving rise to striate and extrastriate cortex in the monkey. Cereb Cortex 12(1):37–53
Smukler SR, Runciman SB, Xu S, van der Kooy D (2006) Embryonic stem cells assume a primitive neural stem cell fate in the absence of extrinsic influences. J Cell Biol 172(1):79–90
Sousa VH, Fishell G (2010) Sonic hedgehog functions through dynamic changes in temporal competence in the developing forebrain. Curr Opin Genet Dev 20(4):391–399
Stubbs D, DeProto J, Nie K, Englund C, Mahmud I, Hevner R, Molnár Z (2009) Neurovascular congruence during cerebral cortical development. Cereb Cortex 19:bhp040
Sur M, Rubenstein JL (2005) Patterning and plasticity of the cerebral cortex. Science 310(5749):805–810
Suzuki IK, Vanderhaeghen P (2015) Is this a brain which I see before me? Modeling human neural development with pluripotent stem cells. Development 142(18):3138–3150
Takebe T, Sekine K, Enomura M, Koike H, Kimura M, Ogaeri T, Zhang RR, Ueno Y, Zheng YW, Koike N, Aoyama S, Adachi Y, Taniguchi H (2013) Vascularized and functional human liver from an iPSC-derived organ bud transplant. Nature 499(7459):481–484
Takebe T, Enomura M, Yoshizawa E, Kimura M, Koike H, Ueno Y, Matsuzaki T, Yamazaki T, Toyohara T, Osafune K, Nakauchi H, Yoshikawa HY, Taniguchi H (2015) Vascularized and complex organ buds from diverse tissues via mesenchymal cell-driven condensation. Cell Stem Cell 16(5):556–565
Tiberi L, Vanderhaeghen P, Van Den Ameele J (2012) Cortical neurogenesis and morphogens: diversity of cues, sources and functions. Curr Opin Cell Biol 24(2):269–276
Tole S, Christian C, Grove EA (1997) Early specification and autonomous development of cortical fields in the mouse hippocampus. Development 124(24):4959–4970
Tropepe V, Hitoshi S, Sirard C, Mak TW, Rossant J, van der Kooy D (2001) Direct neural fate specification from embryonic stem cells: a primitive mammalian neural stem cell stage acquired through a default mechanism. Neuron 30:65–78
Waclaw RR, Wang B, Pei Z, Ehrman LA, Campbell K (2009) Distinct temporal requirements for the homeobox gene Gsx2 in specifying striatal and olfactory bulb neuronal fates. Neuron 63(4):451–465
Wang X, Tsai JW, LaMonica B, Kriegstein AR (2011) A new subtype of progenitor cell in the mouse embryonic neocortex. Nat Neurosci 14:555–561
Watanabe K, Kamiya D, Nishiyama A, Katayama T, Nozaki S, Kawasaki H, Watanabe Y, Mizuseki K, Sasai Y (2005) Directed differentiation of telencephalic precursors from embryonic stem cells. Nat Neurosci 8:288–296
Watanabe K, Ueno M, Kamiya D, Nishiyama A, Matsumura M, Wataya T, Takahashi J, Nishikawa S, Nishikawa S, Muguruma K, Sasai Y (2007) A ROCK inhibitor permits survival of dissociated human embryonic stem cells. Nat Biotechnol 25:681–686
Watanabe M, Kang YJ, Davies LM, Meghpara S, Lau K, Chung CY, Kathiriya J, Hadjantonakis AK, Monuki ES (2012) BMP4 sufficiency to induce choroid plexus epithelial fate from embryonic stem cell-derived neuroepithelial progenitors. J Neurosci 32(45):15934–15945
Wataya T, Ando S, Muguruma K, Ikeda H, Watanabe K, Eiraku M, Kawada M, Takahashi J, Hashimoto N, Sasai Y (2008) Minimization of exogenous signals in ES cell culture induces rostral hypothalamic differentiation. Proc Natl Acad Sci USA 105(33):11796–11801
Wilson SW, Houart C (2004) Early steps in the development of the forebrain. Dev Cell 6(2):167–181
Wonders CP, Anderson SA (2006) The origin and specification of cortical interneurons. Nat Rev Neurosci 7(9):687–696
Wree A, Zilles K, Schleicher A (1983) A quantitative approach to cytoarchitectonics. Anat Embryol 166(3):333–353

Xie Z, Ma X, Ji W, Zhou G, Lu Y, Xiang Z, Wang YX, Zhang L, Hu Y, Ding YQ, Zhang WJ (2010) Zbtb20 is essential for the specification of CA1 field identity in the developing hippocampus. Proc Natl Acad Sci USA 107(14):6510–6515
Xu Q, Guo L, Moore H, Waclaw RR, Campbell K, Anderson SA (2010) Sonic hedgehog signaling confers ventral telencephalic progenitors with distinct cortical interneuron fates. Neuron 65(3):328–340
Ying QL, Stavridis M, Griffiths D, Li M, Smith A (2003) Conversion of embryonic stem cells into neuroectodermal precursors in adherent monoculture. Nat Biotechnol 21:183–186
Yoshihara SI, Omichi K, Yanazawa M, Kitamura K, Yoshihara Y (2005) Arx homeobox gene is essential for development of mouse olfactory system. Development 132(4):751–762
Yu DX, Di Giorgio FP, Yao J, Marchetto MC, Brennand K, Wright R, Mei A, McHenry L, Lisuk D, Grasmick JM, Silberman P, Silberman G, Jappelli R, Gage FH (2014a) Modeling hippocampal neurogenesis using human pluripotent stem cells. Stem Cell Rep 2(3):295–310
Yu DX, Marchetto MC, Gage FH (2014b) How to make a hippocampal dentate gyrus granule neuron. Development 141(12):2366–2375
Zembrzycki A, Griesel G, Stoykova A, Mansouri A (2007) Genetic interplay between the transcription factors Sp8 and Emx2 in the patterning of the forebrain. Neural Dev 2(1):8
Zhao Y, Sheng HZ, Amini R, Grinberg A, Lee E, Huang S, Taira M, Westphal H (1999) Control of hippocampal morphogenesis and neuronal differentiation by the LIM homeobox gene Lhx5. Science 284(5417):1155–1158
Zhou C, Tsai SY, Tsai MJ (2001) COUP-TFI: an intrinsic factor for early regionalization of the neocortex. Genes Dev 15(16):2054–2059

Chapter 2
Self-Organized Cerebellar Tissue from Human Pluripotent Stem Cells and Its Application to Clinical Medicine

Keiko Muguruma

Abstract The cerebellum is a highly ordered brain structure with several well-defined types of cells, which are indispensable for motor functions. It is known that the damage of the cerebellum leads to impairments in motor and postural control. The damage is often caused by cerebellar neurodegenerative diseases that arise sporadically or from inherited genetic defects. The investigation of the pathophysiology and the discovery of drugs for these diseases require the generation of accurate disease models. However, it has been difficult to generate them due to limited information on the human cerebellar development and unavailability of living human cerebellar neurons in experiments.

We recently reported that cerebellar neurons are generated from human pluripotent stem cells (hPSCs) such as embryonic stem cells (hESCs) and induced pluripotent stem cells (iPSCs). We have developed a three-dimensional (3D) culture system of hPSCs that recapitulates the in vivo microenvironments associated with the isthmic organizer, the source of self-inductive signaling. In this chapter, I explain the cerebellar neurogenesis from hPSCs, focusing on the notion of the positional information and self-formation of polarized neuroepithelium. I also discuss how in vitro culture systems using hPSCs facilitate the mechanistic understanding of cerebellar development in humans and the establishment of accurate neurodegenerative disease models.

Keywords Purkinje cells • Cerebellum • Pluripotent stem cells • Spinocerebellar ataxia • SCA • Self-organization • Neural differentiation • Cerebellar development

---

K. Muguruma (✉)
Laboratory for Cell Asymmetry, RIKEN Center for Developmental Biology,
2-2-3 Minatojima-minamimachi, Chuo, Kobe 650-0047, Hyogo, Japan
e-mail: muguruma@cdb.riken.jp

T. Tsuji (ed.), *Organ Regeneration Based on Developmental Biology*,
DOI 10.1007/978-981-10-3768-9_2

2.1 Introduction

Over the last decade, stem cell research has revolutionized the way to investigate the intractable neuronal diseases. iPSCs have emerged as important tools for regenerative medicine against brain damage. Recent advances in the techniques that differentiate iPSCs into specific types of cells shed light on a novel replacement therapy. In addition, these techniques enabled us to establish in vitro disease models from the patient-derived iPSCs that allow us to understand the pathophysiology related to disease mechanisms.

We have developed three-dimensional (3D) culture systems of hPSCs for neural induction with high efficiency. In serum-free floating culture of embryoid body-like aggregates with quick reaggregation (SFEBq) method, robust neural differentiation is observed (Eiraku et al. 2008; Wataya et al. 2008). As observed in the development in vivo, neural induction is the "default pathway" of PSC differentiation (Hemmati-Brivanlou and Melton 1997; Muñoz-Sanjuán and Brivanlou 2002; Sasai and De Robertis 1997; Stern 2006; Weinstein and Hemmati-Brivanlou 1999). It occurs in PSCs when they are cultured in the absence of external inductive signals for mesoderm or endoderm (Chambers et al. 2009; Kamiya et al. 2011; Kawasaki et al. 2000; Smukler et al. 2006; Tropepe et al. 2001; Watanabe et al. 2005; Ying et al. 2003). In SFEBq culture, PSCs are dissociated (to minimize possible effects of the culture substrate or matrix) and subsequently reaggregated using a low-cell-adhesion V-bottomed 96-well culture plate. PSCs selectively differentiate into neural progenitors (>90% in total cells). Inhibition of endogenous SMAD signaling with TGF-β antagonists can promote neural induction, particularly in hPSCs (Chambers et al. 2009). Within few hours in this culture, hPSCs start to form reaggregates and to express pluripotent stem cell markers (e.g., Oct3, Nanog), which gradually disappear by day 7. Around 14 days, the majority of hPSC-derived tissue becomes N-cadherin$^+$, Nestin$^+$, Pax6$^+$, and Sox2$^+$ neural progenitors, and they form neuroepithelial-like structures with apicobasal polarity (Fig. 2.1).

As in vivo, the most anterior region of the brain can be easily induced in PSC culture, due to default differentiation tendency of naive neuroectoderm (Eiraku et al. 2008; Gaspard et al. 2008; Muguruma and Sasai 2012; Sasai 2013; Vallier et al. 2004; Wataya et al. 2008). As for more caudal regions, the character and identity of PSC-derived neural progenitors are specified by a combination of embryonic positional signals, including Fgfs, Wnts, retinoic acid (RA), and Shh, along the anteroposterior (AP) and dorsoventral (DV) axes in neural tube (Anderson and Vanderhaeghen 2014; Elkabetz et al. 2008; Gaspard and Vanderhaeghen 2010; Kirkeby et al. 2012; Lee et al. 2000; Li et al. 2005; Lu et al. 2016; Mizuseki et al. 2003; Su et al. 2006; Watanabe et al. 2005; Wichterle et al. 2002; Zeng et al. 2010) (Fig. 2.2). Ample evidence suggests that PSC-derived neural progenitors have the ability to sense positional information through patterning signals. Manipulation of the positional information by addition or inhibition of secreted signals led to establishment of reproducible differentiation protocols for cortical, striatal, midbrain dopaminergic, spinal motor, and autonomic nervous system from hPSCs.

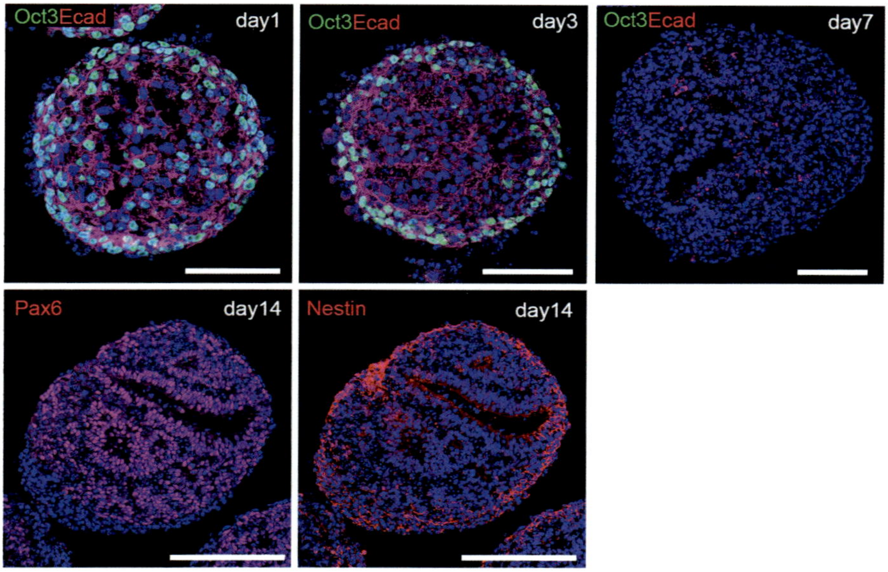

Fig. 2.1 Self-induction of neural progenitors from hESCs in SFEBq culture. (*Upper panel*) hESCs form an aggregate within few hours. The expressions of Oct3/4 and E-cadherin on day 1 gradually disappear by day 7. (*Lower panel*) Around 14 days, the majority of hESC-derived tissue expresses Nestin and Pax6, which are markers for neural progenitors. *Ecad* E-cadherin. The *scale bars* represent 100 μm (*upper panel*) and 200 μm (*lower panel*) (Modified from Muguruma 2016)

The cerebellum has been considered as a major component of the motor system. It sends outputs to a variety of brain regions that are related to motor functions and receives feedback inputs from them. It is well known that the damage in the cerebellum leads to impairments in motor and postural control. In addition to the well-known motor functions, recent studies, especially those in humans, have revealed an involvement of the cerebellum in cognitive functions, such as attention, language processing, and supervised learning. Several labs had attempted cerebellar induction from PSCs, but the efficiency of differentiation was low (Erceg et al. 2010; Salero and Hatten 2007; Su et al. 2006; Tao et al. 2010). In addition, these protocols have not yet been widely used by other researchers, and their reproducibility remains to be confirmed. Such current circumstances hamper the research of cerebellar diseases with the iPSC technologies, which have already been extensively used for investigation of other neurological disorders (Wiethoff et al. 2015).

We recently demonstrated that cerebellar neurons are generated from PSCs (Morino et al. 2015; Muguruma et al. 2010, 2015). To effectively generate cerebellar tissues from PSCs, we took an indirect strategy to induce the isthmic organizer tissue that secondarily self-induces the cerebellar tissues. In this culture using hPSCs, we found that hPSC self-organizes a polarized cerebellar structure in 3D culture.

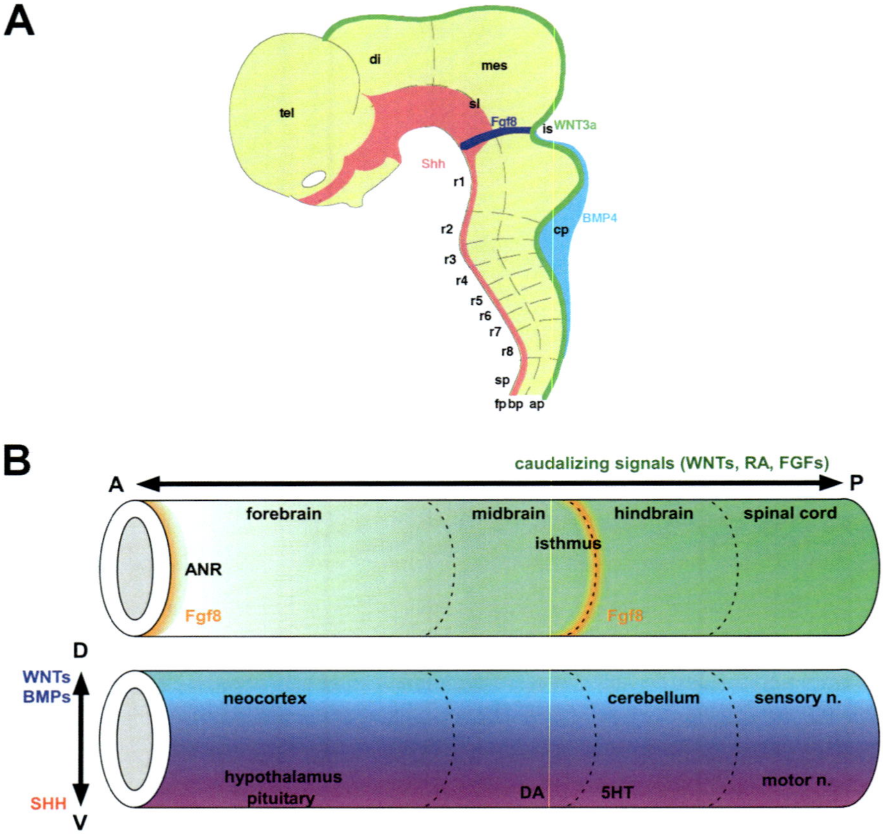

Fig. 2.2 Schematic diagram showing early patterning of the neural tube. (**a**) Lateral view of embryonic brain showing the distribution of secreting signaling molecules. *Tel* telencephalon, *di* diencephalon, *mes* mesencephalon, *sl* sulcus limitans, *is* isthmus, *cp* choroid plexus, *fp* floor plate, *bp* basal plate, *ap* alar plate, and *r1-8* rhombomere 1-8. (**b**) Simplified diagram showing the axial structure of the neural tube. The neural tube, which initially has an anterior identity, is subsequently patterned into discrete domains by secreted factors along the anteroposterior and the dorsoventral axes. *ANR* anterior neural ridge, *DA* dopaminergic neurons, and *5HT* serotonergic neurons

Stepwise addition of components that recapitulate human cerebellar development induced the generation of continuous cerebellar-like NE that are stratified as seen in the early cerebellar plate. Here I discuss the self-organizing nature of hPSC-derived cerebellar tissues with regard to spontaneous polarity formation in 3D culture. I also discuss applications to intractable cerebellar disease utilizing patient-specific iPSCs.

2.2 Self-Organization of Cerebellar Tissue from hPSCs in 3D Culture

2.2.1 Development of Mammalian Cerebellum

In humans, the cerebellum develops over a long period of time, from the early embryonic days to the first postnatal year. Such protracted development could be the source of vulnerability and the cause of broad spectrum of developmental disorders. Cellular and molecular studies on the cerebellar development have been done mostly in model animals. However, it is not known whether the mechanisms revealed in model animals are simply applicable to human or not. Thus, the cellular and molecular studies in humans have been long-awaited for elucidation of functions and dysfunctions of the human cerebellum.

The cerebellum is a highly ordered brain structure with several well-defined types of cells (Fig. 2.3). The early development of the cerebellum is conserved among amniotes including human. The initial phase of cerebellar development is the formation of the isthmic organizer, which lies at the midbrain-hindbrain boundary (MHB). The isthmic organizer secretes positional signals such as Fgf8 and Wnt1 and controls the AP pattern of the caudal mesencephalon (midbrain)-rostral hindbrain. Fgf8 is essential for the induction of cerebellar development, while transplantation of Fgf8-soaked beads in the rostral midbrain can cause ectopic formation of cerebellar structures (Sato et al. 2001). The isthmic organizer is formed and maintained by the intricate regulatory functions of region-specific transcription factors such as En1/En2, Pax2/5/8, Otx2, and Gbx2. The expression of these factors is under the control of Fgf8 secreted by the isthmic organizer itself (Wurst and Bally-Cuif 2001), and the format of a positive feedback loop involves Fgf8 contributing to the maintenance of the tissue's identity. Under its inductive influence, the cerebellar anlage arises in the dorsal region of rostral hindbrain (alar plate of rhombomere 1) (Nakamura et al. 2005; Zervas et al. 2005).

Cerebellar cells are generated in the two distinct germinal zones in the rhombomere 1. One is the ventricular zone (VZ) of the cerebellar plate, which expresses the basic helix-loop-helix (bHLH) transcription factor Ptf1a. The Ptf1a^{+} progenitors produce GABAergic neurons of the cerebellar cortex (Purkinje, basket, Golgi, and stellate cells) and of the deep cerebellar nuclei (DCN). The other one is the upper rhombic lip (RL), which expresses another bHLH factor, Atoh1. The Atoh1^{+} progenitors generate cerebellar glutamatergic neurons, including granule cells and large DCN projection neurons (Butts et al. 2014; Hoshino 2006; Martinez et al. 2013; Millen and Gleeson 2008) (Fig. 2.4). The most prominent neuron type of the cerebellum is the Purkinje cell, with its large soma arranged in a plane (known as the Purkinje cell layer) and an extensive, flattened dendritic tree projecting into the molecular layer (Fig. 2.3). Purkinje cells have a central role in integrating heterogeneous neural inputs from the mossy and climbing fibers. The dendrites of Purkinje cells receive synaptic input from parallel fibers of granule cells, the main excitatory interneurons of the cerebral cortex. In humans, each Purkinje cell dendritic tree may

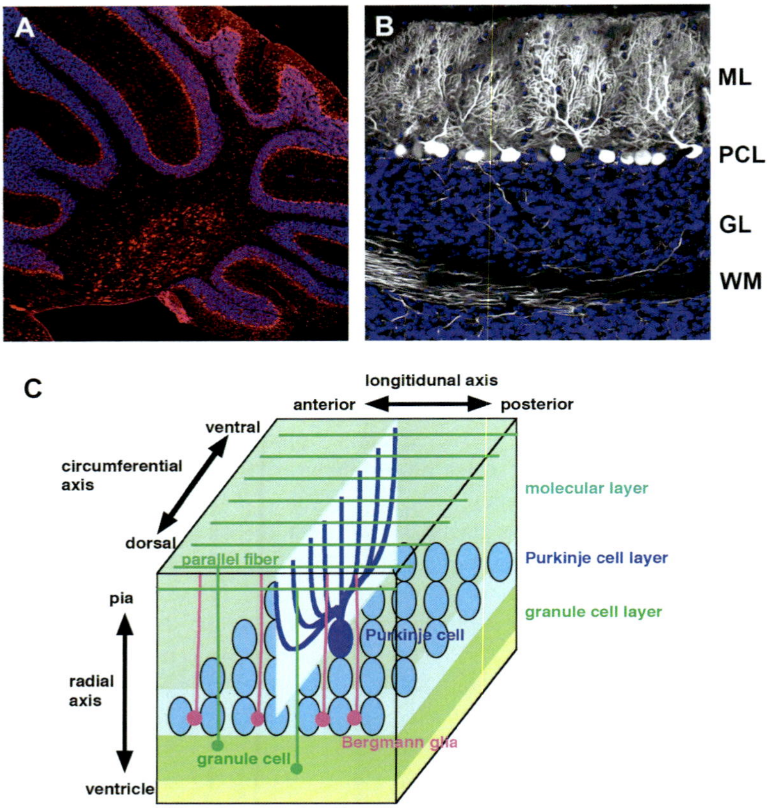

Fig. 2.3 Histological organization of the cerebellum. (**a**) Sagittal section of adult mouse cerebellum processed with fluorescent Nissl (*red*) and nuclear (*blue*) staining. (**b**) High-power view of the mouse cerebellar cortex showing the layer structure. Purkinje cells are visualized with L7/PCP2 antibody (*white*) (Modified from Muguruma 2016). (**c**) Diagram of cerebellar cortical neural circuits. Purkinje cells extend elaborate flat dendrites into the molecular layer within a plane perpendicular to the parallel fibers. *ML* molecular layer, *PCL* Purkinje cell layer, *GL* granule cell layer, and *WM* white matter

be intersected by as many as 200,000 parallel fibers (Braitenberg and Atwood 1958). Purkinje cell dysfunction (hereditary or acquired) leads to severe motor discoordination.

2.2.2 Attempt to Generate Cerebellar Neurons from PSCs

The developmental complexity of the cerebellum, arising from the existence of the influential isthmic organizer, the ordered cell migration from two different germinal zones, and the specialized cell morphology and properties, has hampered establishment of effective differentiation culture system.

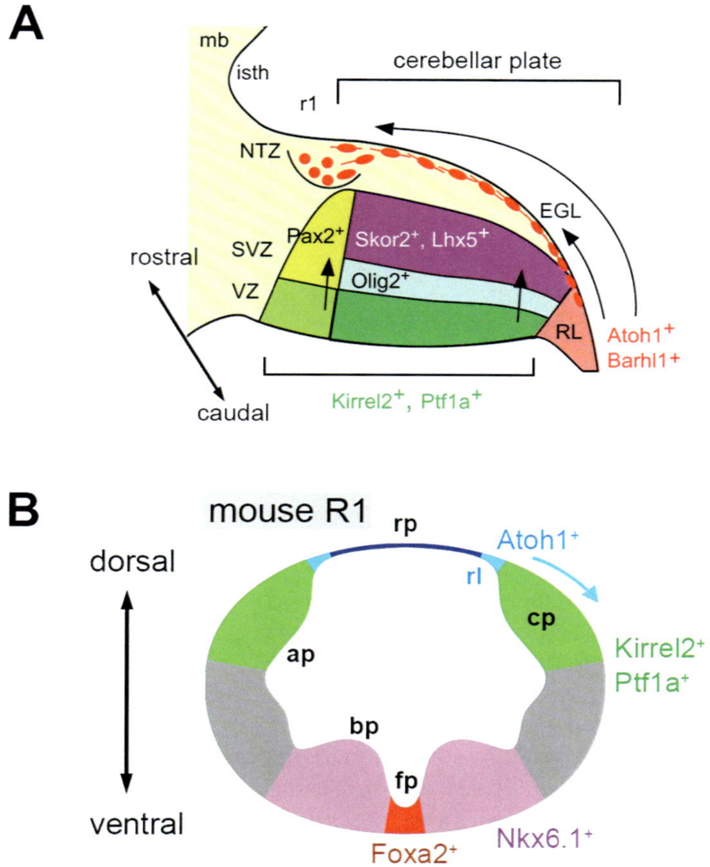

Fig. 2.4 Schematic of the organization the embryonic cerebellar plate. Sagittal view of cerebellar plate (**a**) and coronal view of neural tube (**b**). *EGL* external granular layer, *is* isthmus. *mb* midbrain, *NTZ* nuclear transitory zone, *r1* rhombomere 1, *RL* rhombic lip, *SVZ* subventricular zone, *VZ* ventricular zone, *rp* roof plate, *ap* alar plate, *bp* basal plate, *fp* floor plate, and *cp* cerebellar plate (Modified from Muguruma et al. 2015)

At first, cerebellar differentiation from mouse ESCs (mESC) has accomplished a combinatory treatment of positional signals (Salero and Hatten 2007; Su et al. 2006). In this culture, mESCs were treated sequentially with secreted factors (Fgf8, Wnt1 and RA) known to initiate cerebellar primordium; BMPs and Gdf7, which induce granule cell progenitors; and mitogens (Shh and Jagged 1). While Atoh1$^+$ granule cell progenitors were differentiated from mESCs at relatively high efficiency (>15%), the generation of Purkinje cells was extremely low (less than 1% of the total). A similar approach, exposing a series of inductive signals, was applied in hESC culture (Erceg et al. 2010). Although this protocol somewhat improves the efficiency of Purkinje cell differentiation, it requires a large amount of growth factors, and it does not appear to induce the characteristic morphology, such as dendritic extension and large soma. This protocol has not been widely used by other groups thus far. Improved differentiation protocol of mESC-derived Purkinje cells utilized coculture system with feeder cells to promote generation, survival, and maturation (Tao et al. 2010). The efficiency was a twofold to ninefold increase in numbers of Purkinje cells. They showed somewhat morphological and electrophysiological features of Purkinje cells, but reproduced results using hPSCs have not been shown.

2.2.3 *Self-Induction of Cerebellar Neurons from hPSCs*

Several labs had attempted cerebellar induction from PSCs (Erceg et al. 2010; Salero and Hatten 2007; Su et al. 2006; Tao et al. 2010), but the efficiency of differentiation has been low. Therefore, we took a different indirect strategy that focuses on mimicking the innate program of initial cerebellar development associated with isthmic organizer (Muguruma et al. 2010). In SFEBq culture of mESCs, robust differentiation (>80% efficacy) of $En2^+$ progenitors, representing the caudal midbrain to rostral hindbrain regions, is observed when the cell aggregate is transiently treated with moderate caudalizing factors (Fgf2 and insulin). A majority of the induced progenitors express rostral hindbrain markers. Further treatment with a hedgehog inhibitor, which promotes dorsalization indirectly by inhibiting the ventralizing effect of endogenous Shh, causes expression of Ptf1a and Kirrel2 (a cell surface marker upregulated by Ptf1a, a transcription factor essential for Purkinje cell generation) (Mizuhara et al. 2010), suggesting efficient generation of cerebellar neuroepithelium. The induced $Kirrel2^+$ progenitors subsequently differentiate into Purkinje cells at high efficiency when cocultured with mouse RL-derived granule cells. It is not exactly known why the combination of Fgf2 and insulin is so effective for induction of cerebellar neurons in SFEBq culture. An intriguing finding in this context is that insulin, a commonly used additive in various culture media, has moderate but substantial caudalizing effects on ESC-derived neural progenitors. Indeed, the addition of insulin suppresses the expression of Six3 and Rax (rostral forebrain markers) and increases the expression Fgf8 and En2 (expressed in MHB) (Muguruma et al. 2010; Wataya et al. 2008), although the exact in vivo role of insulin (or IGF) signaling in mammalian neural patterning remains to be elucidated.

These two factors do not seem to play instructive roles in rostral hindbrain induction; rather, these moderate signals may be permissive for differentiation of broad MHB regions in the mESC aggregate. Importantly, the combination of insulin and Fgf2 can reproducibly support the self-formation, albeit small in area, of around isthmic organizer tissue ($En2^+/Otx2^+$) in culture, which secondarily promotes the specification of cerebellar anlage in the aggregate by emanating the organizer signals Fgf8 and Wnt1. The self-formation of isthmic organizer tissue in this culture conditions is also dependent on Fgf8 and Wnt1 and suppressed by their inhibitors (Muguruma et al. 2010), suggesting that an autoregulatory feedback loop operates in the maintenance of this tissue in small area. Direct application of Fgf8 and Wnt1 might destabilize this autoregulatory loop (Sasai 2013).

This self-formation principle is applicable to the generation of human cerebellar neurons from hPSC with some minor modification. This includes the addition of TGFß inhibitor, which inhibits mesenchymal differentiation and promotes neuroectodermal differentiation (Muguruma et al. 2015). These hPSC-derived neural progenitors express En2 and Gbx2, midbrain-hindbrain markers, after 21 days in culture. Around 35 days in culture, substantial populations of hPSC-derived cells express Purkinje cell progenitor markers, Lhx5, Kirrel2, Ptf1a, and Skor2. Following long-term coculture with mouse RL-derived granule cells, hPSC-derived cerebellar progenitors become to L7$^+$, Calbindin$^+$, GABA$^+$, and Aldolase C$^+$ (specific markers for mature Purkinje cells). L7$^+$/Calbindin$^+$ hPSC-derived Purkinje cells developed elaborate dendritic branches and spines that are positive for Purkinje cell-specific glutamate receptor GRID2 (GluRδ2). GRID2 was associated with CBLN1, which is expressed at the presynaptic termini of the parallel fibers. The electrophysiological properties of hPSC-derived Purkinje cells showed typical characters as seen in rodents (Muguruma et al. 2015) (Fig. 2.5).

It is known that BMP signals from the roof plate promote RL formation (Alder et al. 1999). We previously showed that BMP signals are necessary for the induction of RL-derived Atoh1$^+$ progenitors in mESC culture (Muguruma et al. 2010). In contrast, Atoh1$^+$ cells, Atoh1$^+$/Barhl1$^+$ granule cells, and Tbr1$^+$/SMI-32$^+$ DCN projection neurons were generated from hPSCs even in the absence of BMP4 (Muguruma et al. 2015).

2.2.4 Self-Formation of Polarized Cerebellar Structure

The formation of polarized cerebellar neuroepithelium is more or less stochastic. Interestingly, sequential application of Fgf19 and SDF1 (CXCL2) promoted self-formation of polarized cerebellar tissue consisting of a RL-like region at the edge of the sheet and the rest of the sheet displaying an identity reminiscent of the cerebellar field. Remarkably, in this condition, hPSC-derived cerebellar tissue displays an apicobasally layered arrangement of multiple cell types, including progenitors similar to the cerebellar VZ, an intermediate layer containing precursors of Purkinje neurons, and an outermost layer occupied by derivatives of the RL, which have the characteristics of granule cell precursors (Fig. 2.6). hPSC-derived neural progenitors appear to have more potential self-organizing ability to differentiate into cerebellar tissues than those from mouse PSCs.

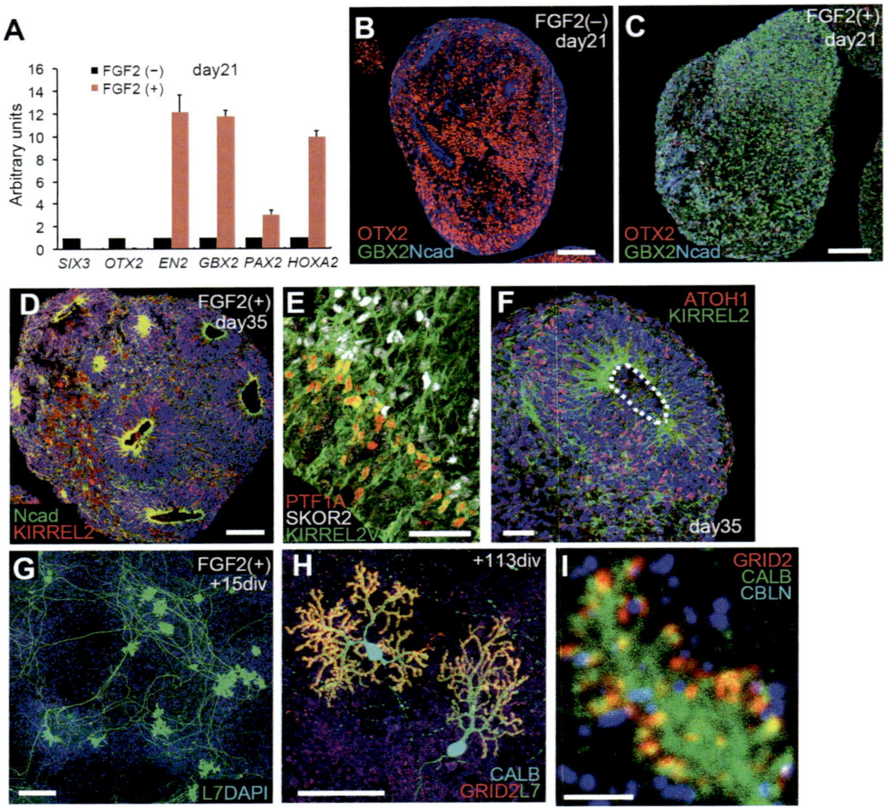

Fig. 2.5 Self-organization of the cerebellar plate neuroepithelium from hESCs. (**a**) qPCR analysis for region-specific genes in hESC aggregate cultured for 21 days with or without Fgf2. Each bar represents mean ± SD. (**b** and **c**) Immunostaining for Otx2, Gbx2, and N-cadherin of hESC aggregates (day 21) cultured with or without Fgf2. (**d**) Expression of Kirrel2 and N-cadherin in the aggregate on day 35. (**e**) Expression of cerebellar progenitor-specific markers in Fgf2-treated aggregate on day 35. (**f**) Immunostaining of the aggregates on day 35. Atoh1 was expressed within and in the periphery of Kirrel2$^+$ cerebellar plate neuroepithelium. (**g**) Immunostaining of dissociated coculture of hESC-derived progenitors with mouse granule cells. L7$^+$ neurons appeared on day 15. (**h**, **i**) Immunostaining of hESC-derived Purkinje cells on culture day 113. The *scale bars* represent 100 μm (**b**–**d**, **g**, **h**), 50 μm (**f**), 25 μm (**e**), and 5 μm (**i**) (Modified from Muguruma et al. 2015)

2.3 Construction of Human Cerebellar Disease Models with Patient-Derived iPSCs

Construction of neuronal disease models with hPSCs has been extensively studied by many researchers. It depends primarily on the presence of efficient differentiation protocols (Aurbry et al. 2008; Chambers et al. 2012; Corti et al. 2012; Delli Carri et al. 2013; Devlin et al. 2015; Egawa et al. 2012; Hu and Zhang 2009; Israel et al. 2012; Jeon et al. 2012; Kriks et al. 2011; Kondo et al. 2013; Kondo et al. 2014;

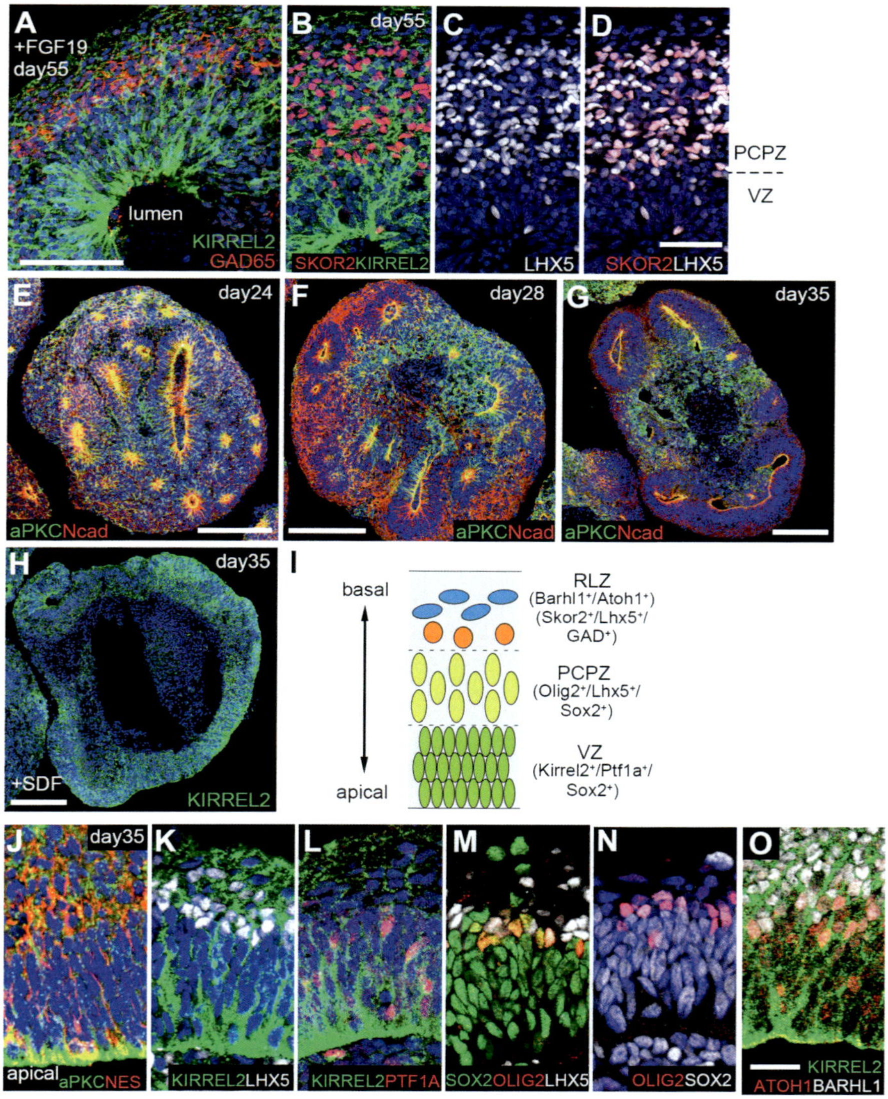

Fig. 2.6 Self-formation of stratified cerebellar neuroepithelial structure. (**a**–**d**) Immunostaining of Fgf19-treated aggregates on day 55. PCPZ, Purkinje cell precursor zone; VZ, ventricular zone. (**e**–**g**) Self-formation of neuroepithelial structure in hESC aggregate. Immunostaining for PKCζ (aPKC) and N-cadherin. (**h**) Immunostaining of SDF-1-treated aggregates on day 35. (**i**) Laminar structures found in hESC-derived cerebellar neuroepithelium. RLZ, RL-derivative zone. (**j**–**o**) Immunostaining on day 35 for cerebellar precursor-specific markers. The scale bars represent 200 μm (**e**–**g**), 100 μm (**a** and **h**) and 50 μm (**d** and apply to **b** and **c**), and 25 μm (**o** and apply to **j**–**n**) (Modified from Muguruma et al. 2015)

Li et al. 2005, 2008; Patani et al. 2011; Perrier et al. 2004; Perrier and Peschanski 2012; Sánchez-Danés et al. 2012; Shi et al. 2012; Yan et al. 2005). Among the cerebellar diseases, the cerebellar ataxia is a diverse group of neurological disorders, defined by a loss of motor coordination that results from the degeneration of specific populations of neurons in the cerebellar cortex (predominantly Purkinje cell), brainstem, and spinocerebellar tracts (Klockgether 2011; Manto and Marmolino 2009). Purkinje cells, whose large soma, extensive fan-shaped dendritic tree, and characteristic electrophysiological properties (as described herein below) render them particularly vulnerable to proteostatic insults and disturbances in ion channel function caused by mutations in spinocerebellar ataxia disease genes (Hekman and Gomez 2014). Disease models using patient-derived iPSCs for neurodegenerative disease, including amyotrophic lateral sclerosis (ALS), Parkinson's disease (PD), spinal muscular atrophy (SMA), Alzheimer's disease (AD), and Huntington's disease (HD), have extensively studied for understanding disease mechanisms and developing therapies. However few studies on iPSC-derived models of cerebellar diseases have been reported (Bird et al. 2014; Eigentler et al. 2013; Hick et al. 2013; Koch et al. 2011; Ku et al. 2010). To date, these studies still remain at the level of the generation of disease-specific iPSCs or, at most, generation of iPSC-derived ßIII-tubulin positive neurons. While the analyses with patient-derived iPSCs or neurons differentiated from them may provide some insights into the pathogenesis of the cerebellar diseases, they cannot directly assay the functions and dysfunctions that are specific to mature cerebellar neurons. Recapitulation of accurate phenotypes of patients in vitro with patient-derived cerebellar neurons would provide much more information and be crucial for understanding the cerebellar diseases. It is important to generate specific neuronal subtypes for construction of the disease models (Inoue et al. 2014; Watson et al. 2015; Wiethoff et al. 2015). We are currently establishing the culture systems for generating the various cerebellar cell types from human iPSCs. Recently, we have succeeded in the generation of Purkinje cells from disease-specific iPSC (Muguruma et al. 2015; Morino et al. 2015). Our efficient culture protocol is going to be widely used by other groups as it was used in a recent report (Wang et al. 2015). Thus, the combination of the generation of patient-derived iPSCs and the differentiation of them into mature cerebellar neurons will lead to the establishment of disease models that accelerate the development of effective therapies against the human cerebellar diseases. Although only a few therapies are effective to the cerebellar diseases, disease-specific iPSC research should provide us new knowledge to overcome these diseases.

2.4 Conclusion

Disease modeling using patient-derived iPSCs is likely to provide significant insights to study of neurodegenerative disease, including the inherited and sporadic cerebellar ataxia. In particular, human iPSCs-derived Purkinje cells should provide us useful information on the developmental mechanisms, pathological processes, and selective vulnerability. The self-organizing culture technique that differentiates hPSCs into cerebellar neurons will become a powerful tool for investigating intractable cerebellar disease.

Acknowledgment This work is supported by grants-in-aid from Ministry of Health, Labour, and Welfare; grant-in-aid for scientific research (C) from Japan Society for the Promotion of Science (JSPS); and the Core Program for Disease Modeling using iPS cells from JST and AMED.

References

Alder J, Lee KJ, Jessell TM et al (1999) Generation of cerebellar granule neurons in vivo by transplantation of BMP-treated neural progenitor cells. Nat Neurosci 2:535–540

Anderson S, Vanderhaeghen P (2014) Cortical neurogenesis from pluripotent stem cells: complexity emerging from simplicity. Curr Opin Neurobiol 27:151–157

Aubry L, Bugi A, Lefort N et al (2008) Striatal progenitors derived from human ES cells mature into DARPP32 neurons *in vitro* and in quinolinic acid-lesioned rats. Proc Natl Acad Sci U S A 43:16707–16712

Bird MJ, Needham K, Frazier AE et al (2014) Functional characterization of Friedreich ataxia iPS derived neuronal progenitors and their integration on the adult brain. PLoS One 9:e101718

Braitenberg V, Atwood RP (1958) Morphological observations on the cerebellar cortex. J Comp Neurol 109:1–33

Butts T, Green MJ, Wingate RJT (2014) Development of the cerebellum: simple steps to make a 'little brain'. Development 141:4031–4041

Chambers SM, Fasano CA, Papapetrou EP et al (2009) Highly efficient neural conversion of human ES and iPS cells by dual inhibition of SMAD signaling. Nat Biotechnol 27:275–280

Chambers SM, Qi Y, Mica Y et al (2012) Combined small-molecule inhibition accelerates developmental timing and converts human pluripotent stem cells into nociceptors. Nat Biotechnol 30:715–720

Corti S, Nizzardo M, Simone C et al (2012) Genetic correction of human induced pluripotent stem cells from patients with spinal muscular atrophy. Sci Transl Med 4:165ra2

Delli Carri A, Onorati M, Lelos MJ et al (2013) Developmentally coordinated extrinsic signals drive human pluripotent stem cells differentiation toward authentic DARPP-32^{+} medium-sized spiny neurons. Development 140:301–312

Devlin AC, Burr K, Borooah S et al (2015) Human iPSC-derived motoneurons harbouring TARDBP or C9ORF ALS mutations are dysfunctional despite maintaining viability. Nat Commun 6:5999

Egawa N, Kitaoka S, Tsukita K et al (2012) Drug Screening for ALS using patient-specific induced pluripotent stem cells. Sci Transl Med 4:145ra104

Eigentler A, Boesch S, Schneider R et al (2013) Induced pluripotent stem cells from Friedreich ataxia patients fail to upregulate frataxin during in vitro differentiation to peripheral sensory neurons. Stem Cells Dev 22:3271–3282

Eiraku M, Watanabe K, Matsuo-Takasaki M et al (2008) Self-organized formation of polarized cortical tissues from ESCs and its active manipulation by extrinsic signals. Cell Stem Cell 3:519–532

Elkabetz Y, Panagiotakos G, Shamy GA et al (2008) Human ES cell-derived neural rosettes reveal a functionally distinct early neural stem cell stage. Gene Dev 22:152–165

Erceg S, Ronaghi M, Zipancic I et al (2010) Efficient differentiation of human embryonic stem cells into functional cerebellar-like cells. Stem Cells Dev 19:1745–1756

Gaspard N, Vanderhaeghen P (2010) Mechanisms of neural specification from embryonic stem cells. Curr Opin Neurobiol 20:37–43

Gaspard N, Bouschet T, Hourez R et al (2008) An intrinsic mechanism of corticogenesis from embryonic stem cells. Nature 455:351–357
Hekman KE, Gomez CM (2014) The autosomal dominant spinocerebellar ataxias: emerging mechanistic themes suggest pervasive Purkinje cell vulnerability. J Neurol Neurosurg Psychiatry 86:554–561
Hemmati-Brivanlou A, Melton D (1997) Vertebrate embryonic cells will become nerve cells unless told otherwise. Cell 88:13–17
Hick A, Wattenhofer-Donze M, Chintawar S et al (2013) Neurons and cardiomyocytes derived from induced pluripotent stem cells as a model for mitochondrial defects in Friedreich's ataxia. Dis Model Mech 6:608–621
Hoshino M (2006) Molecular machinery governing GABAergic neuron specification in the cerebellum. Cerebellum 5:193–198
Hu BY, Zhang SC (2009) Differentiation of spinal motor neurons from pluripotent human stem cells. Nat Protoc 4:1295–1304
Inoue H, Nagata N, Kurokawa H et al (2014) iPS cells: a game changer for future medicine. EMBO J 33:409–417
Israel MA, Yuan SH, Bardy C et al (2012) Probing sporadic and familial Alzheimer's disease using induced pluripotent stem cells. Nature 482:216–220
Jeon I, Lee N, Li JY et al (2012) Neuronal properties, in vivo effects, and pathology of Huntington's disease patient-derived induced pluripotent stem cells. Stem Cells 30:2054–2062
Kamiya D, Banno S, Sasai N et al (2011) Intrinsic transition of embryonic stem-cell differentiation into neural progenitors. Nature 470:503–509
Kawasaki H, Mizuseki K, Nishikawa S et al (2000) Induction of midbrain dopaminergic neurons from ES cells by stromal cell-derived inducing activity. Neuron 28:31–40
Kirkeby A, Grealish S, Wolf DA et al (2012) Generation of regionally specified neural progenitors and functional neurons from human embryonic stem cells under defined conditions. Cell Rep 1:703–714
Kriks S, Shim JW, Piao J et al (2011) Dopamine neurons derived human ES cells efficiently engraft in animal models of Parkinson's disease. Nature 480:547–551
Klockgether T (2011) Update on degenerative ataxias. Curr Opin Neurol 24:339–245
Koch P, Breuer P, Peitz M et al (2011) Excitation-induced ataxin-3 aggregation in neurons from patients with Machado-Joseph disease. Nature 480:543–546
Kondo T, Asai M, Tsukita K et al (2013) Modeling Alzheimer's disease with iPSC reveals stress phenotypes associated with intracellular Aß and differential drug responsiveness. Cell Stem Cell 12:487–496
Kondo T, Funayama M, Tsukita K et al (2014) Focal transplantation of human iPSC-derived glial-rich neural progenitors improves lifespan of ALS mice. Stem Cell Rep 3:242–249
Ku S, Soragni E, Campau E et al (2010) Friedreich's ataxia induced pluripotent stem cells model intergenerational GAA•TTC triplet repeat instability. Cell Stem Cell 7:631–637
Lee SH, Lumelsky N, Studer L et al (2000) Efficient generation of midbrain and hindbrain neurons from mouse embryonic stem cells. Nat Biotechnol 18:675–679
Li XJ, Du ZW, Zarnowska WD et al (2005) Specification of motoneurons from human embryonic stem cells. Nat Biotech 23:215–221
Li XJ, Hu BY, Jones SA et al (2008) Directed differentiation of ventral spinal progenitors and motor neurons from human embryonic stem cells by small molecules. Stem Cells 26:886–893
Lu J, Zhong X, Liu H et al (2016) Generation of serotonin neurons from human pluripotent stem cells. Nat Biotech 34:89–94
Manto M, Marmolino D (2009) Cerebellar ataxias. Curr Opin Neurol 22:419–429
Martinez S, Andreu A, Mecklenburg N et al (2013) Cellular and molecular basis of cerebellar development. Front Neuroanat 7:1–12
Millen KJ, Gleeson JG (2008) Cerebellar development and disease. Curr Opin Neurobiol 18:12–19
Mizuhara E, Minaki Y, Nakatani T et al (2010) Purkinje cells originate from cerebellar ventricular zone progenitors positive for Neph3 and E-cadherin. Dev Biol 338:202–214

Mizuseki K, Sakamoto T, Watanabe K et al (2003) Generation of neural crest-derived peripheral neurons and floor plate cells from mouse and primate embryonic stem cells. Proc Natl Acad Sci U S A 100:5823–5833
Morino H, Matsuda Y, Muguruma K et al (2015) A mutation in the low voltage-gated calcium channel *CACNA1G* alters the physiological properties of the channel, causing spinocerebellar ataxia. Mol Brain 8:89
Muguruma K (2016) Construction of iPS-derived functional brain tissues for investigation of disease mechanisms (Japanese). Exp Med 34:551–556
Muguruma K, Sasai Y (2012) In vitro recapitulation of neural development using embryonic stem cells: From neurogenesis to histogenesis. Develop Growth Differ 54:349–357
Muguruma K, Nishiyama A, Ono Y et al (2010) Ontogeny-recapitulating generation and tissue integration of ES cell-derived Purkinje cells. Nat Neurosci 13:1171–1180
Muguruma K, Nishiyama A, Kawakami H et al (2015) Self-organization of polarized cerebellar tissue in 3D culture of human pluripotent stem cells. Cell Rep 10:537–550
Muñoz-Sanjuán I, Brivanlou AH (2002) Default model and embryonic stem cells. Nat Rev Neurosci 3:271–280
Nakamura H, Katahira T, Matsunaga E et al (2005) Isthmus organizer for midbrain and hindbrain development. Brain Res Brain Res Rev 49:12–126
Patani R, Hollins AJ, Wishart TM et al (2011) Retinoid-independent motor neurogenesis from human embryonic stem cells reveals a medial columnar ground state. Nat Commun 2:214
Perrier A, Peschanski M (2012) How can human pluripotent stem cells help decipher and cure Huntington's disease? Cell Stem Cell 11:153–161
Perrier AL, Tabar V, Barberi T et al (2004) Derivation of midbrain dopamine neurons from human embryonic stem cells. Proc Natl Acad Sci U S A 101:12543–12548
Salero E, Hatten ME (2007) Differentiation of ES cells into cerebellar neurons. Proc Natl Acad Sci U S A 104:2997–3002
Sánchez-Danés A, Richaud-Patin Y, Carballo-Carbajal I et al (2012) Disease-specific phenotypes in dopamine neurons from human iPS-based models of genetic and sporadic Parkinson's disease. EMBO Mol Med 4:380–395
Sasai Y (2013) Next-generation regenerative medicine: organogenesis from stem cells in 3D culture. Cell Stem Cell 12:520–530
Sasai Y, De Robertis EM (1997) ectodermal patterning in vertebrate embryos. Dev Biol 182:5–20
Sato T, Araki I, Nakamura H (2001) Inductive signal and tissue responsiveness defining the tectum and the cerebellum. Development 128:2461–2469
Shi Y, Kirwan P, Livesey FJ (2012) Directed differentiation of human pluripotent stem cells to cerebral cortex neurons and neural networks. Nat Protoc 10:1836–1846
Smukler SR, Runciman SB, Xu S et al (2006) Embryonic stem cells assume a primitive neural stem cell fate in the absence of extrinsic influences. J Cell Biol 172:79–90
Stern CD (2006) Neural induction: 10 years on since the 'default model'. Curr Opin Cell Biol 18:692–697
Su HL, Muguruma K, Matsuo-Takasaki M et al (2006) Generation of cerebellar neuron precursors from embryonic stem cells. Dev Biol 290:287–296
Tao O, Shimazaki T, Okada Y et al (2010) Efficient generation of mature cerebellar Purkinje cells from mouse embryonic stem cells. J Neurosci Res 88:234–247
Tropepe V, Hitoshi S, Sirard C et al (2001) Direct neural fate specification from embryonic stem cells: A primitive mammalian neural stem cell stage acquired through a default mechanism. Neuron 30:65–78
Vallier L, Reynolds D, Pedersen RA (2004) Nodal inhibits differentiation of human embryonic stem cells along the neuroectodermal default pathway. Dev Biol 275:403–421
Wang S, Wang B, Pan N et al (2015) Differentiation of human induced pluripotent stem cells to mature functional Purkinje neurons. Sci Rep 5:9232
Watanabe K, Kamiya D, Nishiyama A et al (2005) Directed differentiation of telencephalic precursors from embryonic stem cells. Nat Neurosci 8:288–296

Wataya T, Ando S, Muguruma K et al (2008) Minimization of exogenous signals in ES cell culture induces rostral hypothalamic differentiation. Proc Natl Acad Sci U S A 105:11796–11801
Watson LM, Wong MMK, Becker EBE (2015) Induced pluripotent stem cell technology for modeling and therapy of cerebellar ataxia. Open Biol 5:150056
Weinstein DC, Hemmati-Brivanlou A (1999) Neural induction. Annu Rev Cell Dev Biol 15:411–433
Wichterle H, Lieberam I, Porter JA et al (2002) Directed differentiation of embryonic stem cells into motor neurons. Cell 110:385–397
Wiethoff S, Arber C, Li A et al (2015) Using human induced pluripotent stem cells to model cerebellar disease: hope and hype. J Neurogenetics 29:95–102
Wurst W, Bally-Cuif L (2001) Neural plate patterning: upstream and downstream of the isthmic organizer. Nat Rev Neurosci 2:99–108
Yan Y, Yand D, Zarnowska ED et al (2005) Directed differentiation of dopaminergic neuronal subtypes from human embryonic stem cells. Stem Cell 23:781–790
Ying QL, Stavridis M, Griffiths D et al (2003) Conversion of embryonic stem cells into neuroectodermal precursors in adherent monoculture. Nat Biotechnol 21:183–186
Zeng H, Guo M, Martins-Taylor K et al (2010) Specification of region-specific neurons including forebrain glutamatergic neurons from human induced pluripotent stem cells. PLoS One 5:e11853
Zervas M, Bllaess S, Joyner AL (2005) Classical embryological studies and modern genetic analysis of midbrain and cerebellum development. Curr Top Dev Biol 69:101–138

Chapter 3
Functional Pituitary Tissue Formation Recapitulating Hypothalamus and Pituitary Development Using ES/iPS Cells

Hidetaka Suga and Chikafumi Ozone

Abstract The hypothalamic-pituitary system is essential for maintaining homeostasis by controlling systemic hormones. However, it can be disrupted by various diseases, resulting in lifelong serious symptoms.

Pluripotent stem cells, such as embryonic stem (ES) cells and induced pluripotent stem (iPS) cells, differentiate into neuroectodermal progenitors when cultured as floating aggregates under serum-free conditions. Recent results have shown that strict removal of exogenous patterning factors induces efficient generation of rostral hypothalamic progenitors from mouse ES cells. Those hypothalamic-like progenitors generated rostral-dorsal hypothalamic neurons, in particular magnocellular vasopressinergic neurons, which release hormones upon stimulation.

We subsequently reported self-formation of adenohypophysis using mouse ES cell aggregates. The aggregates were stimulated to differentiate into both non-neural head ectoderm and hypothalamic neuroectoderm in adjacent layers, followed by treatment with a Sonic Hedgehog agonist. Self-organization of Rathke's pouch-like structures occurred at the interface of the two epithelia in vitro. Subsequently, various endocrine cells including corticotrophs were produced from Rathke's pouch-like structures. The induced corticotrophs efficiently secreted adrenocorticotropic hormone in response to corticotropin-releasing hormone. Furthermore, when engrafted in vivo, these cells rescued systemic glucocorticoid levels in hypopituitary mice.

H. Suga (✉)
Department of Endocrinology and Diabetes, Graduate School of Medicine, Nagoya University, 65 Tsurumai-cho, Showa-ku, Nagoya, Aich 466-8550, Japan
e-mail: sugahide@med.nagoya-u.ac.jp

C. Ozone
Department of Endocrinology and Diabetes, Graduate School of Medicine, Nagoya University, 65 Tsurumai-cho, Showa-ku, Nagoya, Aich 466-8550, Japan

RIKEN Center for Developmental Biology, Kobe 650-0047, Hyogo, Japan

T. Tsuji (ed.), *Organ Regeneration Based on Developmental Biology*,
DOI 10.1007/978-981-10-3768-9_3

Our latest study aimed to prepare hypothalamic and pituitary tissues from human pluripotent stem cells and establish effective transplantation techniques for future clinical applications. We succeeded in establishing the differentiation method using human ES cells. The culture method is characterized by replication of stepwise embryonic differentiation. Therefore, these methods could potentially be used as developmental and disease models, as well as for future regenerative medicine.

Keywords Hypothalamus • Oral ectoderm • Pituitary • Embryonic stem cells • Recapitulation • Differentiation • SFEBq

3.1 Introduction

The hypothalamus and adenohypophysis maintain physiological homeostasis by controlling the endocrine system. A collection of studies exploring their development and function has shown that they are essential for the regulation of vital functions. However, their regeneration remains largely unclear.

Recently, somatic stem cells have been recognized as a major source for tissue maintenance and regeneration. Also in the adenohypophysis, the existence of somatic stem cells was reported (Chen et al. 2005). Subsequent studies have discussed their functions during early postnatal pituitary maturation (Fauquier et al. 2001; Kikuchi et al. 2007; Chen et al. 2009; Gremeaux et al. 2012; Mollard et al. 2012), after pituitary damage (Luque et al. 2011; Fu et al. 2012; Langlais et al. 2013), and in pituitary tumorigenesis (Gaston-Massuet et al. 2011; Andoniadou et al. 2012; Garcia-Lavanderia et al. 2012; Li et al. 2012).

In addition to somatic stem cells, studies have focused on embryonic stem (ES) cells and induced pluripotent stem (iPS) cells. These pluripotent stem cells exhibit self-renewal properties and pluripotent differentiation abilities. Therefore, they have attracted attention as a cell source for differentiated tissues in clinical applications.

3.2 A Need for Hypothalamus and Adenohypophysis Regenerative Medicine

The hypothalamus and adenohypophysis are located in adjacent regions, connected with portal vein, and coordinated as the center for the endocrine systems. In case of their dysfunction, patients suffer from various systemic symptoms. Current treatment consists of hormone replacement therapy, but various factors can complicate its proper dose. Drug administration cannot precisely mimic the circadian or stress-induced changes of hormone requirements. For example, we have reported that some patients with central diabetes insipidus show unstable serum Na levels, resulting in a poor prognosis (Arima et al. 2014). This instability seems to be caused by the lack of positive and negative control systems, which is characteristic of hormone-producing cells. As for hypopituitarism, it has been reported that adrenal crisis occurs in a substantial proportion of hypopituitary patients and adrenal crisis-associated mortality is not negligible, even in educated patients (Hahner et al. 2015). Furthermore, adrenocorticotropic hormone (ACTH)-dependent adrenal insufficiency, as well as high-dose hydrocortisone treatment, serves as a predictor for acromegaly-associated mortality (Sherlock et al. 2009; Ben-Shlomo 2010). Taken together, there are many prospects for pituitary regenerative medicine.

3.3 Mouse Embryonic Stem Cells

The establishment of mouse pluripotent ES cells significantly contributed to the advancement of biology and medicine. In 1981, Evans and Kaufman successfully established mouse ES cells from the inner cell mass of mouse blastocyst-stage embryos (Evance and Kaufman 1981). Various knockout and knock-in mice have been established using genetically modified ES cells, which have contributed to our understanding of gene functions (Bernstein and Breitman 1989; Babinet and Cohen-Tannoudji 2001).

There are two reasons for the use of mouse ES cells as a first step, rather than human ES cells, in our recent studies. One reason is the short developmental period; the duration of mouse fetal development is about 20 days, which is much shorter than the 300 days of human development. Therefore, mouse ES cells are suitable for establishing novel differentiation methods with numerous trial-and-error processes. Another reason is the similarity between mouse and human cells. For example, the retinal differentiation method from human ES cells (Nakano et al. 2012) was established based on a previous report using mouse ES cells (Eiraku et al. 2011). Although the two inducing culture methods considerably differ, their key principles are similar. The fundamental processes of mouse ES cells appear to be applicable to human ES cells.

3.4 Pituitary Gland Embryology

The adenohypophysis, which corresponds to the anterior pituitary gland, contains several types of endocrine cells that secrete factors including adrenocorticotropic hormone (ACTH), growth hormone (GH), prolactin (PRL), thyroid-stimulating hormone, luteinizing hormone, and follicle-stimulating hormone. The posterior pituitary gland consists of axons and terminals of hypothalamic neurons, i.e., vasopressin and oxytocin neurons. The development of the adenohypophysis is a complex

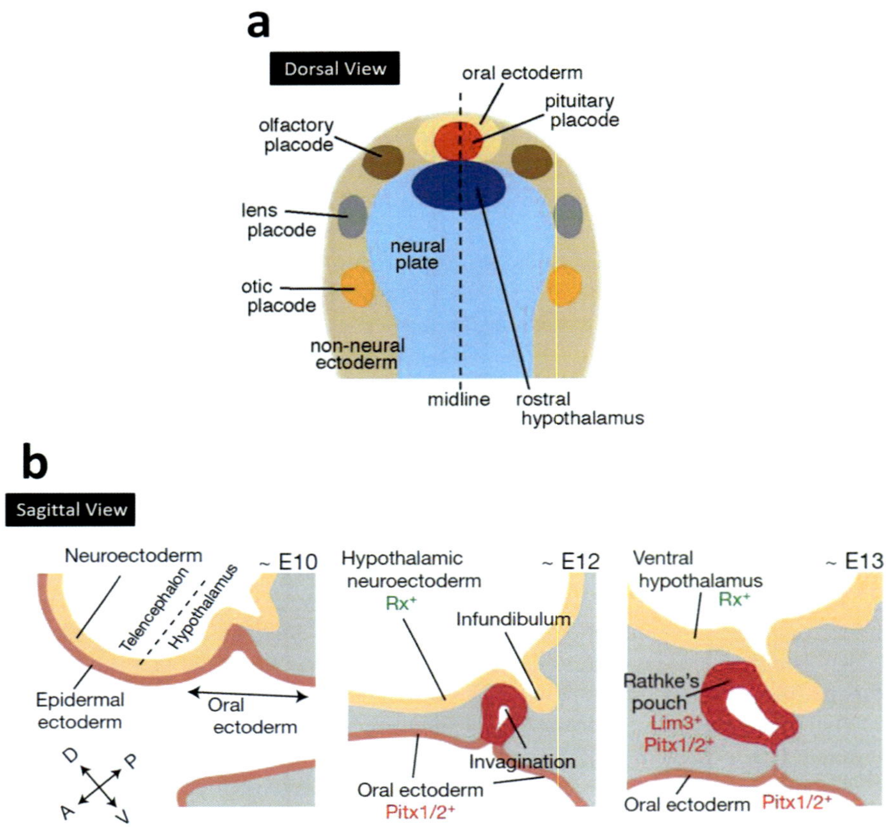

Fig. 3.1 Diagram of mouse pituitary development. (**a**) Dorsal view of neural plate and placodes. (**b**) Sagittal view of pituitary embryogenesis

process. During the early development, the adenohypophysis anlage originates as a placode in the nonneural ectoderm adjacent to the anterior neural plate (Fig. 3.1a). Both the adenohypophysis placode and hypothalamic anlage interact with each other. Accordingly, the thickened placode invaginates and subsequently detaches from the oral ectoderm to form a hollowed vesicle termed “Rathke’s pouch” (Zhu et al. 2007) (Fig. 3.1b). The molecular nature of this local inductive interaction during this initial phase of pituitary formation has been intensively investigated, but still remains elusive. Among them, FGF and BMP signals appear to be involved as important factors (Takuma et al. 1998; Brinkmeier et al. 2007).

3.5 Three-Dimensional ES Cell Culture

Organ formation during embryogenesis consists of complicated processes that involve various local interactions between different tissues or cells. Despite this complexity, organogenesis can be modeled in vivo. Our colleagues established a three-dimensional culture method for ES cells called "serum-free culture of embryoid body-like aggregates with quick reaggregation (SFEBq)" (Watanabe et al. 2005; Eiraku et al. 2008). The culture method is quite simple. First of all, the quality of maintenance for undifferentiated ES cells is quite important. For SFEBq culture, maintained ES cells are dissociated to single cells in trypsin or something similar. The cells are then quickly aggregated using low-cell adhesion 96-well plates in differentiation medium suitable for each differentiation purpose.

This culture method is appropriate for induction of various ectodermal derivatives from ES cells. In SFEBq cultures, the ES cell aggregates exhibit self-organization (Sasai et al. 2012) and spontaneous formation of a highly ordered structure or patterning. This floating culture has revealed intrinsic programs that drive locally autonomous modes of organogenesis and homeostasis. Using the SFEBq method, mesencephalic dopamine neurons (Kawasaki et al. 2002; Morizane et al. 2006), cortex neurons (Eiraku et al. 2008; Danjo et al. 2011; Kadoshima et al. 2013), the optic cup (Ikeda et al. 2005; Osakada et al. 2008; Eiraku et al. 2011), cerebellar neurons (Muguruma et al. 2010), and hippocampal neurons (Sakaguchi et al. 2015) have been generated from mouse and human ES cells.

3.6 Induction of Hypothalamic Neurons from Mouse ES Cells

Using SFEBq cultures, hypothalamic neurons, such as vasopressin-positive neurons, have been induced from mouse ES cells (Wataya et al. 2008). The differentiation occurs efficiently when the ES cell aggregates are cultured in growth factor-free, chemically defined medium (gfCDM). Strict removal of exogenous patterning factors during early differentiation steps induces efficient generation of rostral hypothalamic-like progenitors (Rax+/Six3+/Vax1+; these combinations are characteristic for hypothalamic precursors) in mouse ES cell aggregates. The use of gfCDM is critical. For example, even the presence of exogenous insulin, which is commonly used in cell culture, strongly inhibits differentiation *via* the Akt-dependent pathway. The ES cell-derived hypothalamic progenitors generate Otp+/Brn2+ neuronal precursors (characteristic of rostral-dorsal hypothalamic neurons) and subsequent magnocellular vasopressinergic neurons that release vasopressin upon stimulation. Additionally, differentiation markers of rostral-"ventral" hypothalamic precursors and neurons have been induced from ES cell-derived Rax + progenitors by treatment with Sonic Hedgehog (Shh).

Thus, in the absence of exogenous growth factors in the medium, ES cell-derived neuroectodermal cells spontaneously differentiated into rostral (particularly rostral-dorsal) hypothalamic-like progenitors, which generate characteristic hypothalamic neuroendocrine neurons in a stepwise fashion, as observed in vivo. These findings indicated that, instead of the addition of inductive signals, minimization of exogenous patterning signaling played a key role in rostral hypothalamic specification of neural progenitors derived from pluripotent cells. This work also showed that the default fate of mouse ES cells is the rostral hypothalamus (Wataya et al. 2008).

3.7 Two-Layer Formation In Vitro Is the First Step of Adenohypophysis Differentiation

We next established an in vitro differentiation method for the anterior pituitary (Suga et al. 2011). It is known that Rathke's pouch is formed as a result of interactions between the hypothalamus and neighboring oral ectoderm (Zhu et al. 2007). To recapitulate these embryonic pituitary developmental processes, we co-induced these two tissues within one ES cell aggregate.

Previous results have shown hypothalamic differentiation from mouse ES cells (Wataya et al. 2008). Mouse ES cells can be induced to differentiate into hypothalamic cells when cultured as floating aggregates using the SFEBq method with gfCDM. Therefore, the present study used some technical modifications to co-induce oral ectodermal differentiation in addition to hypothalamic differentiation.

We attempted to slightly shift the positional information so that the oral ectoderm coexisted with hypothalamic tissues (Suga et al. 2011). As shown in Fig. 3.1a, the oral ectoderm is generated from the rostral and midline region adjacent to the hypothalamic region in the mouse embryo. Therefore, the rostral and midline shifting information was theoretically relevant for mouse ES cell aggregates in the SFEBq culture. We tested many culture conditions known to affect early ectodermal patterning. We ultimately identified two conditions that efficiently induced oral ectoderm. One condition was the addition of bone morphogenetic protein 4 (BMP4). However, treatment with 0.5 μM BMP4 strongly inhibited hypothalamic neuron differentiation, instead of inducing oral ectodermal differentiation. The other condition was high-density cell aggregation (10,000 cells per aggregate instead of 3000 in SFEBq culture), which we refer to as large cell aggregation (LCA) (Fig. 3.2a). In the LCA culture, both the oral ectoderm (Pitx1/2+) and hypothalamic tissues coexisted within one aggregate (Fig. 3.2b).

LCA culture allows for the formation of oral ectoderm epithelium on the surface of mouse ES cell aggregates, as well as hypothalamic neural tissue in the inner layer adjacent to the oral ectoderm (Fig. 3.2b). Treatment with a BMP4 antagonist, dorsomorphin, has been shown to suppress the generation of oral ectoderm (Suga et al. 2011). Quantitative polymerase chain reaction analyses revealed significantly higher internal BMP4 expression in LCA aggregates (Suga et al. 2011). Moreover,

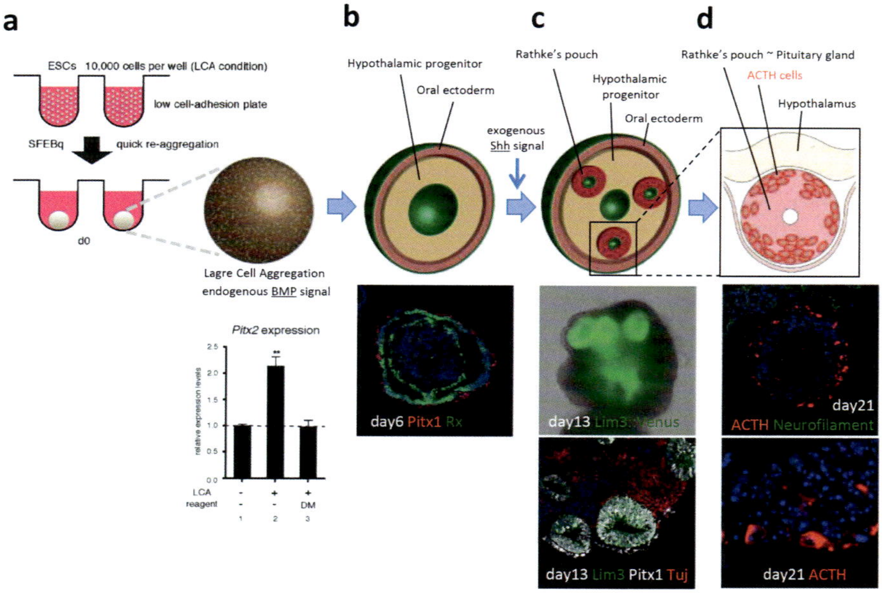

Fig. 3.2 In vitro differentiation into anterior pituitary from mouse ES cells. (**a**) Diagram of SFEBq. (**b**) Two-layer formation in LCA aggregates. (**c**) Self-formation of Rathke's pouches. (**d**) Subsequent generation of ACTH$^+$ cells

Koehler et al. succeeded in differentiating the otic placode (Fig. 3.1a) (Koehler et al. 2013), which belongs to the head and oral ectoderm, following BMP treatment of mouse ES cells, which supports the reliability and robustness of this strategy. Our recent study showed that very low concentrations (picomolar level) of exogenous BMP4 treatment facilitated differentiation into nonneural ectoderms, which contained not only pituitary primordium but also dental germs (Ochiai et al. 2015). Taken together, appropriate BMP4 signal appears to be important for head ectoderm induction (Wilson and Hemmati-Brivanlou 1995; Basch and Bronner-Fraser 2006; Davis and Camper 2007).

3.8 Self-Formation of Rathke's Pouch

In the developing embryo, Rathke's pouch forms at the midline of the head ectoderm. Shh is expressed in the ventral diencephalon and oral ectoderm but is excluded from the invaginating Rathke's pouch (Zhu et al. 2007; Wang et al. 2010). Rathke's pouch receives Shh signals from neighboring tissues in vivo, and Shh is known to provide positional information to adjust toward the midline (Zhu et al. 2007). Therefore, we added smoothened agonist (SAG) as a strong Shh signal to the differentiation medium of mouse ES cell aggregates in vitro. On day 13, multiple oval structures formed in the SAG-treated LCA SFEBq aggregates (Fig. 3.2c). The vesicles were situated between the oral ectoderm and hypothalamic neurons. Lim3 (formal gene name is Lhx3) expression indicated that the vesicles had similar characteristics to Rathke's pouch. These Lim3⁺ tissues appeared as a thick epithelium on the surface, which then invaginated and finally formed hollowed vesicles. The length of the major axis was about 200 μm, which is almost equal to the size of the embryonic Rathke's pouch. The size of Rathke's pouch seems to be prescribed.

Interactions between oral ectoderm and hypothalamic neurons appear to be critically important for in vitro induction of Rathke's pouch. Neither isolated surface ectoderm alone nor isolated hypothalamic tissues alone formed $Lim3^+$ pouches. Only in cases where the two divided components are reassembled does $Lim3^+$ expression recover to some extent (Suga et al. 2011).

These findings demonstrate self-formation of Rathke's pouch in mouse ES cell aggregates. It has also been shown that Rathke's pouch forms even without mesenchymal cells, because this model contains only ectodermal cells.

Interestingly, a single aggregate often contains several pouches, whereas there is usually only one pouch in the embryo (Suga et al. 2011). This finding suggests that several morphogenetic fields for pituitary placodes can be independently generated within the oral ectoderm epithelium on the surface of the ES cell aggregate, which is reminiscent of the *Vax1* knockout mouse (Bharti et al. 2011). A second Rathke's pouch develops in addition to the orthotopic anlage in the *Vax1* knockout mouse. Ectopic expression of FGF10, which is expressed in the infundibulum and implicated in pituitary induction, is also detected in the hypothalamic neuroepithelium overlying the second pouch. Thus, Vax1 likely limits the hypothalamic neuroepithelium area that generates pituitary-inducing signals. Indeed, *Vax1* expression in vivo is eliminated near the infundibulum, which has inducing activity for pituitary development. In the mouse ES aggregates used for pituitary differentiation in the present study, Vax1-positive cells did not exist in the hypothalamic area. Conversely, Wataya's aggregate for hypothalamic differentiation (Wataya et al. 2008) has been shown to contain Vax1-positive cells. We speculate that precise positioning in the hypothalamus slightly shifts as a result of BMP4 and Shh signals.

3.9 Differentiation into Hormone-Producing Endocrine Cells

During pituitary development in the embryo, $Lim3^+$ pituitary progenitors commit to several lineages (Davis et al. 2011), i.e., corticotroph, somatotroph, lactotroph, thyrotroph, gonadotroph, and melanotroph lineages. Among them, the ACTH-producing corticotroph lineage expresses the transcription factor Tbx19 prior to ACTH expression. It is known that Notch signaling inhibits Tbx19 expression (Lamolet et al. 2001; Zhu et al. 2006; Kita et al. 2007). Therefore, we evaluated the effect of the Notch inhibitor DAPT. As a result, DAPT treatment increased Tbx19 expression in SAG-treated LCA SFEBq aggregates. A substantial number of $ACTH^+$ cells appeared in the $Tbx19^+$ lesion (Fig. 3.2d). Without DAPT treatment, corticotroph differentiation efficiency was decreased, and other lineages were not detected.

Previous reports have shown that canonical Wnt signaling promotes Pit1 expression (DiMattia et al. 1997; Olson et al. 2006; Sornson et al. 1996). Consistent with this finding, treatment with the Wnt agonist BIO increased Pit1 expression, resulting in subsequent GH^+ and PRL^+ cell differentiation.

Head mesenchyme has been suggested to promote pituitary development in vivo (Gleiberman et al. 1999). Therefore, we applied conditioned medium from PA6 stromal cells to SAG-treated LCA SFEBq aggregates. As a result, we successfully induced luteinizing hormone-positive, follicle-stimulating hormone-positive, and thyroid-stimulating hormone-positive cells. Further investigation is necessary to identify factors in the PA6-conditioned medium.

Lim3 is essential for these hormone-producing lineages. To suppress Lim3 expression in differentiating mouse ES cells, we used the Tet-inducible shRNA expression lentivirus vector system (kindly gifted from Hiroyuki Miyoshi at RIKEN BioResource Center). Knockdown of Lim3 inhibited subsequent differentiation into hormone-producing cells, which supports altered pituitary development in Lim3 knockout mice (Sheng et al. 1996).

These results demonstrate the competence of ES cell-derived pituitary progenitors to generate multiple endocrine lineages in vitro.

3.10 Functionality of Induced $ACTH^+$ Cells

Positive and negative regulations by exogenous stimuli are characteristic for endocrine cells. To investigate in vitro functionality, we induced $ACTH^+$ cells for evaluation because they are most efficiently generated using the SAG-treated LCA SFEBq method.

After 10 min of stimulation by corticotropin-releasing hormone (CRH), substantial amounts of ACTH were secreted from SAG-treated LCA SFEBq aggregates in vitro (Fig. 3.3a). The secreted ACTH concentration was similar to levels in mouse peripheral blood. ACTH secretion from the pituitary gland is negatively regulated by the downstream glucocorticoid hormone in vivo. Consistent with this control principle, in vitro ACTH secretion as a result of CRH stimulation was suppressed by glucocorticoid pretreatment (Fig. 3.3b).

Similar to in vivo endocrine systems, these data demonstrate that mouse ES cell-derived $ACTH^+$ cells respond to both positive and negative regulators. These hormonal responses to surrounding regulators are indispensable for homeostasis. For this reason, the generation of anterior pituitary tissue that retains regulatory hormonal control in vitro is an important step for the development of cell transplantation therapies for pituitary diseases. Furthermore, we suggest that the endocrine organoid formed in this three-dimensional culture condition might better reflect the in vivo microenvironment. Such approaches may be beneficial for producing other functionally mature endocrine tissues.

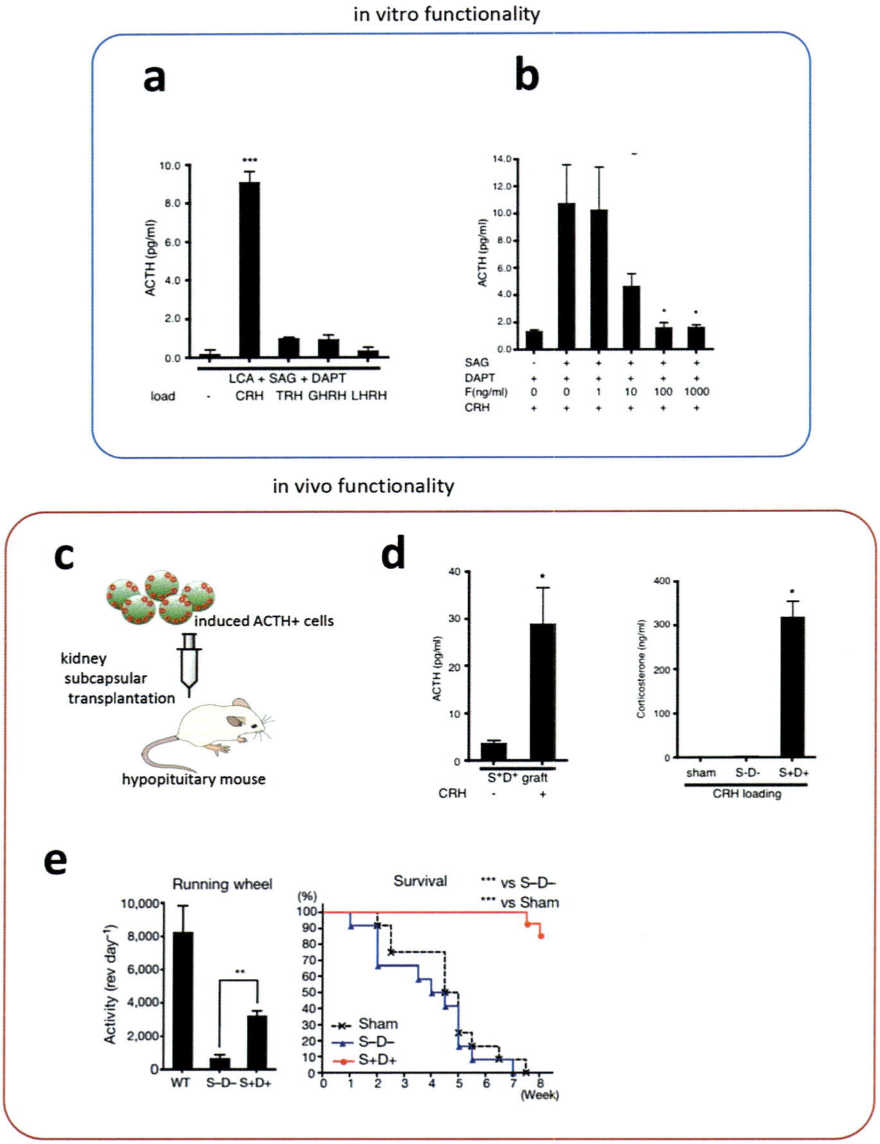

Fig. 3.3 Functional tests of mouse ES-derived ACTH⁺ cells. (**a**) In vitro release from mouse ES-derived ACTH⁺ cells. "F" = glucocorticoid pretreatment. Among the releasing factors, CRH most efficiently induces ACTH secretion. (**b**) Negative feedback test. Pretreatment with hydrocortisone suppresses CRH-stimulated ACTH secretion from aggregates. (**c**) In vivo functional test by ectopic transplantation. All mice, except for the WT mice, received a hypophysectomy; hypopituitarism was confirmed by CRH loading. "S + D+" =SAG- and DAPT-treated aggregates. "S−D−" = no SAG or DAPT treatment. The values shown on graphs represent mean ± s.e.m. $*P < 0.05$, $**P < 0.01$, and $***P < 0.001$ (from Suga et al., 2011, modified). (**d**) Blood ACTH and subsequent release of corticosterone. (**e**) Improved activity and survival

3.11 Effect of Transplantation into Hypophysectomized Model Animals

Finally, we evaluated the transplantation effect of the induced $ACTH^+$ cells. Because of technical difficulties, we chose ectopic transplantation into the kidney subcapsule (Fig. 3.3c), instead of orthotopic transplantation into the sella turcica. At 1 week after transplantation, blood ACTH levels were slightly, but significantly, increased. CRH loading induced a substantial elevation in blood ACTH levels (Fig. 3.3d). The downstream glucocorticoid hormone corticosterone was also significantly increased, indicating that ACTH from the graft sufficiently induced the downstream hormone (Fig. 3.3d).

Even without CRH loading, the basal levels of ACTH were higher. Importantly, corticosterone levels were also increased, suggesting that partial recovery of blood ACTH has a moderate, but biologically significant, effect (note that ED50 of the ACTH receptor MC2R for glucocorticoid production is around 9 pg/mL) (Wang and Majzoub 2011). In accordance with this, the treated hypophysectomized mice displayed higher spontaneous locomotor activities and survived significantly longer (Fig. 3.3e). Although CRH, which is secreted from the hypothalamus, should be diluted in the peripheral site, mESC-derived pituitary tissues rescued survival and spontaneous activities, suggesting that basal secretion from these tissues was sufficient for those effects.

These findings showed that induced $ACTH^+$ cells derived from mouse ES cells acted as endocrine tissues and that regenerative medicine for pituitary dysfunction is feasible.

3.12 Adaptation to Human ES/iPS Cell Culture

The recovery of lost pituitary function is an important issue for medical studies because the anterior pituitary has poor potential for regeneration. Because some pituitary dysfunctions cannot be solely treated by drugs (Arima et al. 2014; Hahner et al. 2015; Sherlock et al. 2009), regenerative therapy employing stem cells should be considered as a new form of therapeutic intervention. Our SFEBq method (Suga et al. 2011) induces pituitary cells that can autoregulate hormonal secretion and respond to changing circumstances. The application of this culture method to human ES cells is necessary for clinical purposes. However, poor survival of human ES cells in SFEB culture might limit the use of these cells for future medical applications. Our colleagues found that a selective Rho-associated kinase (ROCK) inhibitor, Y-27632, markedly diminished dissociation-induced apoptosis of human ES cells and enabled the cells to form aggregates in SFEB culture (Watanabe et al. 2007). Using this fundamentally important discovery, we attempted to adapt our pituitary-differentiating culture method for human ES cell culture. We were able to obtain corticotrophs and somatotrophs from the human ES cells (Ozone et al. 2016).

The characteristics for differentiating pluripotent stem cells into pituitary cells were as follows:

– Simultaneous induction of neighboring hypothalamic neuroectoderm and oral ectodermal tissue, similar to embryo.
– Self-formation of pituitary anlage (Rathke's pouch) as a result of interaction between those two layers.
– Generation of multiple endocrine lineages from Lim3+ pituitary progenitors.
– Functionality confirmation as endocrine tissue.

Combining these approaches, we designed a cell culture scheme for human ES cells.

Our results demonstrated that the anterior pituitary self-forms in vitro following co-induction of the hypothalamic and oral ectoderm (Fig. 3.4a). The juxtaposition of these tissues facilitated the formation of the pituitary placode, and their features were consistent with characteristics of Rathke's pouch in vivo. The human ES cell-derived Rathke's pouch was much larger than the pouch formed by mouse ES cells, which was in accordance with the size difference between human and mouse embryonic Rathke's pouches. These pituitary placodes subsequently differentiated into pituitary hormone-producing cells. All six types of pituitary hormone-producing cells were identified (Fig. 3.4b). Among them, we confirmed that the human ES-derived corticotroph responded normally to releasing and feedback signals. Electron microscopy revealed secretory granules stored in the cytoplasm of these cells (Fig. 3.4c).

For both mouse and human ES cells, SFEB culture is a favorable method that can generate functional pituitary cells. Future studies will confirm whether human iPS cells can differentiate into pituitary cells using the same culture methods.

3.13 Future Perspectives

There are two primary uses for human ES/iPS cell-derived pituitary cells. One is the human model of development or disease. Results from our study showed that the present culture methods recapitulated embryogenesis, suggesting that it could be used in the area of developmental biology. In terms of diseases due to gene mutations, tissues derived from disease-specific iPS cells can be used for therapy screenings in a human disease model.

The second major use for human ES/iPS cell-derived pituitary cells is for regenerative medicine. Although stem cell-based therapeutics provide high expectations for the treatment of diabetes mellitus, the use of regenerative medicine for hypothalamus-hypophyseal dysfunctions has received little attention.

ES cell-derived ACTH-producing cells function even after ectopic transplantation. This finding raises the possibility of relatively simple grafting of artificial ES/iPS cell-derived pituitary tissues into a peripheral site. These cells can function effectively if hormone secretion can be extrinsically controlled by releasing factors

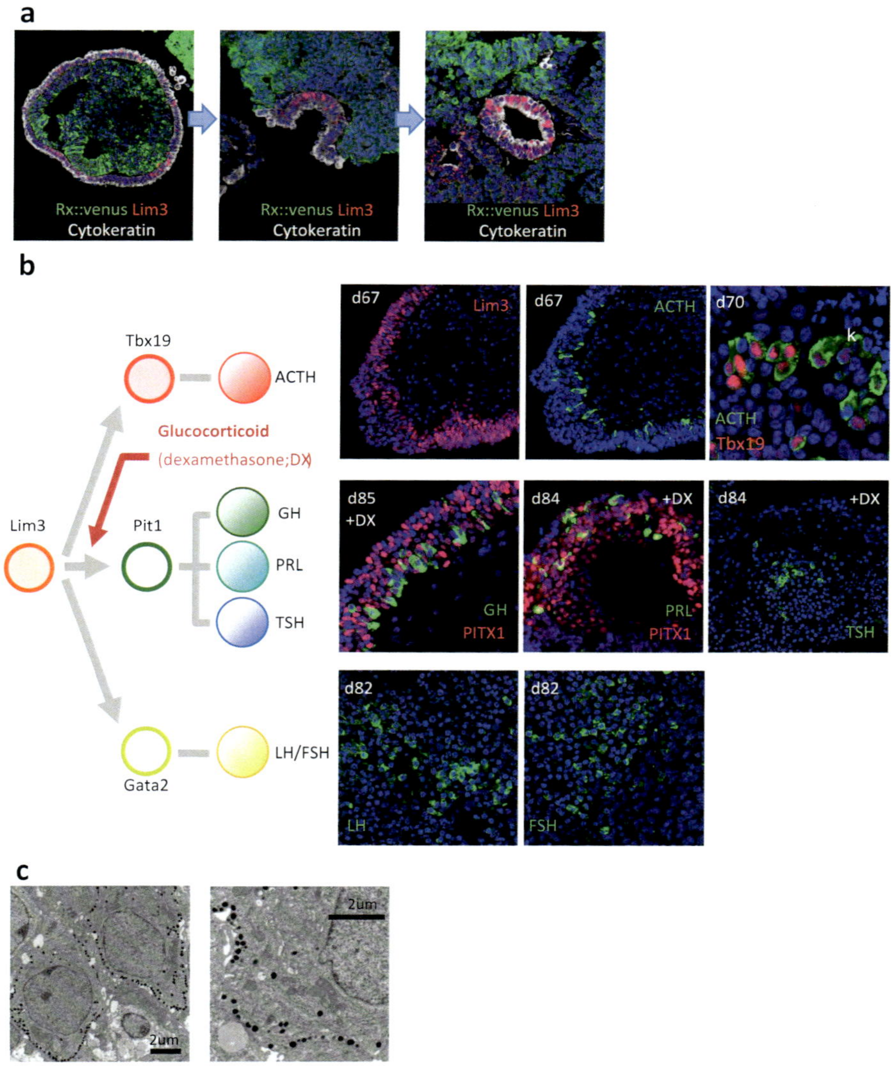

Fig. 3.4 Human ES culture (**a**) Recapitulation of Rathke's pouch formation. (**b**) Differentiation into multiple lineages. (**c**) Secretory granules characteristic of endocrine cells

or small molecule agonists. However, ectopic transplantation is not perfect, because physiological CRH released from the hypothalamus does not directly affect these grafts. Orthotopic transplantation of hormone-producing cells that are controlled by positive and negative regulators is one of the future candidates for complete therapy.

In future studies, it will be challenging to recapitulate an entire anterior pituitary gland that contains all endocrine components in three-dimensional cultures of human ES or iPS cells and to use such artificial pituitary tissues for orthotopic transplantation into the sella of a large mammal. To achieve this long-term goal, further studies are needed before pituitary regenerative medicine can be directly transferred to clinical use.

References

Andoniadou CL, Gaston-Massuet C, Reddy R et al (2012) Identification of novel pathways involved in the pathogenesis of human adamantinomatous craniopharyngioma. Acta Neuropathol 124:259–271

Arima H, Wakabayashi T, Nagatani T et al (2014) Adipsia increases risk of death in patients with central diabetes insipidus. Endocr J 61:143–148

Babinet C, Cohen-Tannoudji M (2001) Genome engineering via homologous recombination in mouse embryonic stem (ES) cells: an amazingly versatile tool for the study of mammalian biology. An Acad Bras Cienc 73:365–383

Basch ML, Bronner-Fraser M (2006) Neural crest inducing signals. Adv Exp Med Biol 589:24–31

Ben-Shlomo A (2010) Pituitary gland: predictors of acromegaly-associated mortality. Nat Rev Endocrinol 6:67–69

Bernstein A, Breitman M (1989) Genetic ablation in transgenic mice. Mol Biol Med 6:523–530

Bharti K, Gasper M, Bertuzzi S et al (2011) Lack of the ventral anterior homeodomain transcription factor VAX1 leads to induction of a second pituitary. Development 138:873–878

Brinkmeier ML, Potok MA, Davis SW et al (2007) Camper SA: TCF4 deficiency expands ventral diencephalon signaling and increases induction of pituitary progenitors. Dev Biol 311:396–407

Chen J, Hersmus N, Van Duppen V, Caesens P, Denef C, Vankelecom H et al (2005) The adult pituitary contains a cell population displaying stem/progenitor cell and early embryonic characteristics. Endocrinology 146:3985–3998

Chen J, Gremeaux L, Fu Q et al (2009) Pituitary progenitor cells tracked down by side population dissection. Stem Cells 27:1182–1195

Danjo T, Eiraku M, Muguruma K et al (2011) Subregional specification of embryonic stem cell-derived ventral telencephalic tissues by timed and combinatory treatment with extrinsic signals. J Neurosci 31:1919–1933

Davis SW, Camper SA (2007) Noggin regulates Bmp4 activity during pituitary induction. Dev Biol 305:145–160

Davis SW, Mortensen AH, Camper SA (2011) Birthdating studies reshape models for pituitary gland cell specification. Dev Biol 352:215–227

DiMattia GE, Rhodes SJ, Krones A et al (1997) The Pit-1 gene is regulated by distinct early and late pituitary-specific enhancers. Dev Biol 182:180–190

Eiraku M, Watanabe K, Matsuo-Takasaki M et al (2008) Self-organized formation of polarized cortical tissues from ESCs and its active manipulation by extrinsic signals. Cell Stem Cell 3:519–532

Eiraku M, Takata N, Ishibashi H et al (2011) Self-organizing optic-cup morphogenesis in three-dimensional culture. Nature 472:51–56

Evans MJ, Kaufman MH (1981) Establishment in culture of pluripotential cells from mouse embryos. Nature 292:154–156

Fauquier T, Guérineau NC, McKinney RA et al (2001) Folliculostellate cell network: a route for long-distance communication in the anterior pituitary. Proc Natl Acad Sci USA 98:8891–8896

Fu Q, Gremeaux L, Luque RM et al (2012) The adult pituitary shows stem/progenitor cell activation in response to injury and is capable of regeneration. Endocrinology 153:3224–3235

Garcia-Lavandeira M, Saez C, Diaz-Rodriguez E et al (2012) Craniopharyngiomas express embryonic stem cell markers (SOX2, OCT4, KLF4, and SOX9) as pituitary stem cells but do not coexpress RET/GFRA3 receptors. J Clin Endocrinol Metab 97:E80–E87

Gaston-Massuet C, Andoniadou CL, Signore M et al (2011) Increased Wingless (Wnt) signaling in pituitary progenitor/stem cells gives rise to pituitary tumors in mice and humans. Proc Natl Acad Sci USA 108:11482–11487

Gleiberman AS, Fedtsova NG, Rosenfeld MG (1999) Tissue interactions in the induction of anterior pituitary: role of the ventral diencephalon, mesenchyme, and notochord. Dev Biol 213:340–353

Gremeaux L, Fu Q, Chen J et al (2012) Activated phenotype of the pituitary stem/progenitor cell compartment during the early-postnatal maturation phase of the gland. Stem Cells Dev 21:801–813

Hahner S, Spinnler C, Fassnacht M et al (2015) High incidence of adrenal crisis in educated patients with chronic adrenal insufficiency: a prospective study. J Clin Endocrinol Metab 100:407–416

Ikeda H, Osakada F, Watanabe K et al (2005) Generation of Rx+/Pax6+ neural retinal precursors from embryonic stem cells. Proc Natl Acad Sci USA 102:11331–11336

Kadoshima T, Sakaguchi H, Nakano T et al (2013) Self-organization of axial polarity, inside-out layer pattern, and species-specific progenitor dynamics in human ES cell-derived neocortex. Proc Natl Acad Sci USA 110:20284–20289

Kawasaki H, Suemori H, Mizuseki K et al (2002) Generation of dopaminergic neurons and pigmented epithelia from primate ES cells by stromal cell-derived inducing activity. Proc Natl Acad Sci USA 99:1580–1585

Kikuchi M, Yatabe M, Kouki T et al (2007) Changes in E- and N-cadherin expression in developing rat adenohypophysis. Anat Rec (Hoboken) 290:486–490

Kita A, Imayoshi I, Hojo M et al (2007) Hes1 and Hes5 control the progenitor pool, intermediate lobe specification, and posterior lobe formation in the pituitary development. Mol Endocrinol 21:1458–1466

Koehler KR, Mikosz AM, Molosh AI et al (2013) Generation of inner ear sensory epithelia from pluripotent stem cells in 3D culture. Nature 500:217–221

Lamolet B, Pulichino AM, Lamonerie T et al (2001) A pituitary cell-restricted T box factor, Tpit, activates POMC transcription in cooperation with Pitx homeoproteins. Cell 104:849–859

Langlais D, Couture C, Kmita M et al (2013) Adult pituitary cell maintenance: lineage-specific contribution of self-duplication. Mol Endocrinol 27:1103–1112

Li H, Collado M, Villasante A et al (2012) p27(Kip1) directly represses Sox2 during embryonic stem cell differentiation. Cell Stem Cell 11:845–852

Luque RM, Lin Q, Córdoba-Chacón J et al (2011) Metabolic impact of adult-onset, isolated, growth hormone deficiency (AOiGHD) due to destruction of pituitary somatotropes. PLoS One 6:e15767

Mollard P, Hodson DJ, Lafont C et al (2012) A tridimensional view of pituitary development and function. Trends Endocrinol Metab 23:261–269

Morizane A, Takahashi J, Shinoyama M, Ideguchi M, Takagi Y, Fukuda H, Koyanagi M, Sasai Y, Hashimoto N et al (2006) Generation of graftable dopaminergic neuron progenitors from mouse ES cells by a combination of coculture and neurosphere methods. J Neurosci Res 83:1015–1027

Muguruma K, Nishiyama A, Ono Y et al (2010) Ontogeny-recapitulating generation and tissue integration of ES cell-derived Purkinje cells. Nat Neurosci 13:1171–1180

Nakano T, Ando S, Takata N et al (2012) Self-formation of optic cups and storable stratified neural retina from human ESCs. Cell Stem Cell 10:771–785
Ochiai H, Suga H, Yamada T et al (2015) BMP4 and FGF strongly induce differentiation of mouse ES cells into oral ectoderm. Stem Cell Res 15:290–298
Olson LE, Tollkuhn J, Scafoglio C et al (2006) Homeodomain-mediated beta-catenin-dependent switching events dictate cell-lineage determination. Cell 125:593–605
Osakada F, Ikeda H, Mandai M et al (2008) Toward the generation of rod and cone photoreceptors from mouse, monkey and human embryonic stem cells. Nat Biotechnol 26:215–224
Ozone C, Suga H, Eiraku M et al (2016) Functional anterior pituitary generated in self-organizing culture of human embryonic stem cells. Nat Commun 7:10351
Sakaguchi H, Kadoshima T, Soen M et al (2015) Generation of functional hippocampal neurons from self-organizing human embryonic stem cell-derived dorsomedial telencephalic tissue. Nat Commun 6:8896
Sasai Y, Eiraku M, Suga H (2012) In vitro organogenesis in three dimensions: self-organising stem cells. Development 139:4111–4121
Sheng HZ, Zhadanov AB, Mosinger B Jr et al (1996) Specification of pituitary cell lineages by the LIM homeobox gene Lhx3. Science 272:1004–1007
Sherlock M, Reulen RC, Alonso AA et al (2009) ACTH deficiency, higher doses of hydrocortisone replacement, and radiotherapy are independent predictors of mortality in patients with acromegaly. J Clin Endocrinol Metab 94:4216–4223
Sornson MW, Wu W, Dasen JS et al (1996) Pituitary lineage determination by the prophet of Pit-1 homeodomain factor defective in Ames dwarfism. Nature 384:327–333
Suga H, Kadoshima T, Minaguchi M et al (2011) Self-formation of functional adenohypophysis in three-dimensional culture. Nature 480:57–62
Takuma N, Sheng HZ, Furuta Y et al (1998) Formation of Rathke's pouch requires dual induction from the diencephalon. Development 125:4835–4840
Wang O, Majzoub JA (2011) Adrenocorticotropin. In: Melmed S (ed) The pituitary, 3rd edn. Academic, London, pp 47–81
Wang Y, Martin JF, Bai CB (2010) Direct and indirect requirements of Shh/Gli signaling in early pituitary development. Dev Biol 348:199–209
Watanabe K, Kamiya D, Nishiyama A et al (2005) Directed differentiation of telencephalic precursors from embryonic stem cells. Nat Neurosci 8:288–296
Watanabe K, Ueno M, Kamiya D et al (2007) A ROCK inhibitor permits survival of dissociated human embryonic stem cells. Nat Biotechnol 25:681–686
Wataya T, Ando S, Muguruma K et al (2008) Minimization of exogenous signals in ES cell culture induces rostral hypothalamic differentiation. Proc Natl Acad Sci USA 105:11796–11801
Wilson PA, Hemmati-Brivanlou A (1995) Induction of epidermis and inhibition of neural fate by Bmp-4. Nature 376:331–333
Zhu X, Zhang J, Tollkuhn J et al (2006) Sustained Notch signaling in progenitors is required for sequential emergence of distinct cell lineages during organogenesis. Genes Dev 20:2739–2753
Zhu X, Gleiberman AS, Rosenfeld MG (2007) Molecular physiology of pituitary development: signaling and transcriptional networks. Physiol Rev 87:933–963

Chapter 4
Inner Ear Organoids: Recapitulating Inner Ear Development in 3D Culture

Alhasan N. Elghouche, Rick F. Nelson, and Eri Hashino

Abstract The inner ear contains sensory epithelia composed of mechanosensitive hair cells, supporting cells, and sensory neurons that work in concert to detect sound and positional information and transmit those signals to the brain. Within the backdrop of embryogenesis, inner ear development follows an intricate pathway of signaling cues and morphological changes, leading to its complex final three-dimensional (3D) structure. Application of various small molecules and recombinant proteins to mouse embryonic stem cells at specific time points in vitro has enabled recapitulation of developmental cues with subsequent formation of inner ear organoids. This has resulted in a model system of inner ear development that is easily derived, manipulated, and analyzed. These organoids contain functional mechanosensitive hair cells, supporting cells, and sensory neurons, which phenocopy functional components of the inner ear responsible for detection of positional information. The potential applications of this system include investigation of inner ear development, disease modeling, drug screening, and therapy development. This chapter highlights the process of in vivo inner ear development, the rationale and process behind inner ear organoid formation, and potential applications and limitations of this in vitro model system.

Keywords Inner ear • Hearing • Balance • Embryonic development • Stem cells • 3D culture • Organogenesis

4.1 Introduction

The mammalian inner ear is a complex, labyrinthine structure containing two sensory organs—the cochlea and the vestibular apparatus. The cochlea detects sound with the auditory sensory epithelium in the organ of Corti, whereas the vestibular apparatus detects positional information with sensory epithelia in the cristae of the

A.N. Elghouche • R.F. Nelson • E. Hashino (✉)
Department of Otolaryngology-Head & Neck Surgery, Indiana University School of Medicine, Indianapolis, IN 46202, USA
e-mail: ehashino@iupui.edu

T. Tsuji (ed.), *Organ Regeneration Based on Developmental Biology*,
DOI 10.1007/978-981-10-3768-9_4

semicircular canals and the maculae of the utricle and saccule (Fig. 4.1). Key to the function of both the auditory and vestibular organs is the mechanosensitive hair cell. There are four types of hair cells: inner and outer cochlear hair cells and type I and type II vestibular hair cells. The common feature of all hair cells is the presence of a hair bundle, an organized array of stereocilia, and a kinocilium, on the apical surface. Mechanical stimuli from sound or positional changes cause deflection of hair bundles, which is transduced into neural impulses through synaptic contacts at the basal side of the hair cells (Breneman et al. 2009). This electrical signal is transmitted to the brain through primary afferent neurons. In addition to sensory hair cells and neural components, the inner ear also contains a diverse group of supporting cells. Supporting cells facilitate ion homeostasis and structural support and have a potential to help repair injury to sensory hair cells.

The second key to the mammalian inner ear is the inability of mature hair cells to regenerate when damaged. Thus, inner ear injury generally causes permanent hearing and/or equilibrium dysfunction. There is a wide spectrum of etiologies which compromise hair cell function and cause inner ear pathology including congenital syndromes, ototoxic medications, noise-induced trauma, and aging. Thus, establishment of an accurate in vitro model will allow effective investigation of various gene and pharmacologic therapies for hearing loss and vestibular dysfunction (Park 2015). Other potential, long-term applications include cell-based therapies for deafness; for example, introduction of partially differentiated human fetal auditory stem cells into the cochlea of a gerbil model of neuropathic deafness demonstrated implantation and survival of transplanted cells in addition to improvements in auditory function (Chen et al. 2012).

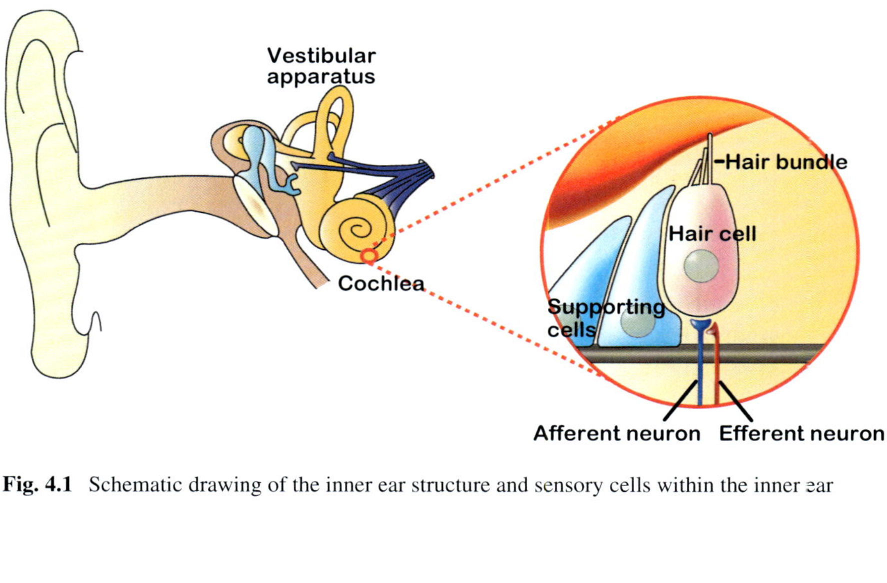

Fig. 4.1 Schematic drawing of the inner ear structure and sensory cells within the inner ear

The differentiation of embryonic stem cells (ESCs) into inner ear organoids comprising sensory epithelia, including hair cells, has emerged through translation of normal developmental cues by way of in vitro application of proteins and small molecules at precise time points in a three-dimensional (3D) cell culture system (Koehler et al. 2013; Koehler and Hashino 2014).

4.2 Inner Ear Development

The anatomic arrangement of inner ear components reflects the immense intricacy in the embryologic transition from ectoderm to the fully developed and functional inner ear. The relevant, sequential stages of inner ear development may be generally ordered as definitive ectoderm, nonneural ectoderm (NNE), preplacodal region/ectoderm (PPR), otic epibranchial placode domain (OEPD), and otic placode (Fig. 4.2). Further development of the otic placode gives rise to hair cells with structural and functional properties of native mechanosensitive hair cells, sensory neurons, and supporting cells of the inner ear.

During the process of gastrulation, in which the blastula is rearranged into three germ layers, the definitive ectoderm is divided into neural and nonneural components (Litsiou et al. 2005; Saint-Jeannet and Moody 2014). The neuroectoderm is destined to form the central nervous system and neural crest derivatives, whereas the nonneural ectoderm gives rise to the inner ear, other placodal derivatives, and the epidermis. The lineage choice between neuroectoderm and nonneural ectoderm is mediated by bone morphogenetic protein (BMP) signaling—in which high levels of BMP in the lateral ectoderm induce NNE formation (Kwon et al. 2010; Leung et al. 2013). BMP-mediated induction of NNE may be facilitated by transcription factors from the Dlx and GATA families (Pieper et al. 2012). Analysis of zebrafish

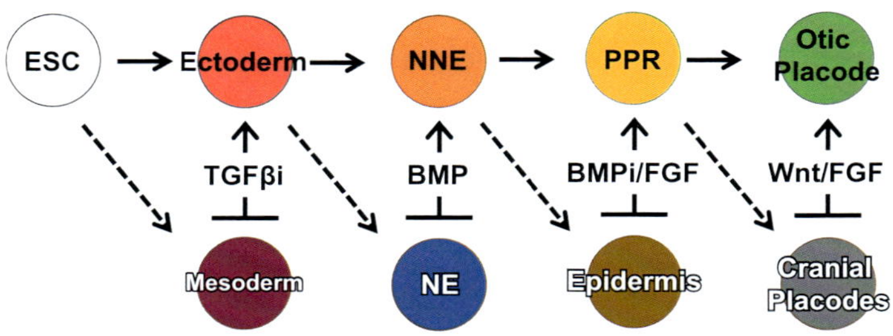

Fig. 4.2 Temporal sequence of embryological steps and pertinent signals involved in the induction of the inner ear within described 3D stem cell differentiation protocol. *ESC* embryonic stem cell, *NNE* nonneural ectoderm, *NE* neural ectoderm, *PPR* preplacodal region, *TGFβi* inhibition of TGFβ signaling, *BMPi* inhibition of BMP signaling

embryos demonstrates that early BMP signaling during gastrulation induces four transcription factors (Tfap2a, Tfap2c, Foxi1, and Gata3) that subsequently enable PPR formation within the NNE (Saint-Jeannet and Moody 2014).

4.2.1 Formation of the Preplacodal Region

The PPR is a horseshoe-shaped portion of ectoderm around the anterior neural plate and neural crest (Schlosser 2014). The PPR is named for the various sensory placodes residing within it. These placodes give rise to a myriad of cell types and neurons that form the cranial sensory ganglia and tissues (i.e., the lens, inner ear, and olfactory epithelium) (Bailey and Streit 2005; Streit 2007). The PPR emerges through attenuation of both BMP and Wnt signaling, in conjunction with increased fibroblast growth factor (FGF) signaling, at the interface of the neuroectoderm and NNE (Ahrens and Schlosser 2005; Litsiou et al. 2005; Esterberg and Fritz 2009; Saint-Jeannet and Moody 2014). The role of reduced BMP signaling is supported by studies which demonstrate that local decreases in BMP signaling expand the PPR in both chick and *Xenopus* (Glavic et al. 2004; Ahrens and Schlosser 2005; Litsiou et al. 2005). In zebrafish, this dissipation of BMP signaling may be facilitated by Dlx3-mediated expression of BMP antagonists, cv2 and Bambi-b (Esterberg and Fritz 2009; Reichert et al. 2013). This BMP antagonism with concurrent FGF signaling is sufficient for PPR emergence in zebrafish, chick, and frog (Ahrens and Schlosser 2005; Esterberg and Fritz 2009; Saint-Jeannet and Moody 2014).

The changes in BMP and FGF signaling lead to the induction of the transcription factor Six1 and its cofactor Eya1, the well-established markers for PPR formation (Schlosser and Ahrens 2004; Ahrens and Schlosser 2005; Sato et al. 2010; Schlosser 2014). Confirming this embryologic role, overexpression of Six1 in *Xenopus* increases the PPR at the expense of the neural crest and epidermis (Brugmann et al. 2004). In humans, mutations in Six1 and Eya1 cause the branchio-oto-renal syndrome, clinical features of which include ear anomalies and sensorineural hearing loss (Ruf et al. 2004). Aside from members of the Six family, studies in zebrafish and *Xenopus* implicate a network of additional transcription factors involved in *PPR* specification including Tfap2a/c, Foxi1/3, and Gata2/3 (Litsiou et al. 2005; Pieper et al. 2012; Bhat et al. 2013; Saint-Jeannet and Moody 2014).

4.2.2 Formation of the Otic Epibranchial Placode Domain

The PPR becomes further subdivided into thickened patches of ectoderm which represent the various cranial placodes (Pieper et al. 2012). It is from these cranial placodes that various cranial sensory organs, such as the inner ear, emerge (Litsiou et al. 2005; Pieper et al. 2012; Saint-Jeannet and Moody 2014). The cranial placodes are arranged anteroposteriorly, and the development of each placode is dependent upon the expression of various Pax proteins. Pax proteins are transcription factors that modulate organogenesis through regulation of cell proliferation, apoptosis, migration, and maintenance of stem cell pluripotency (Chi and Epstein 2002; Lang et al. 2007). In vertebrates, Six1 is necessary for Pax protein expression and placode development (Christophorou et al. 2009). Pax2 and Pax8 are central to inner ear development and are markers of the otic epibranchial placode domain (OEPD) (Baker and Bronner-Fraser 2001; Bouchard et al. 2004; Ohyama et al. 2006; Schlosser 2006; Pieper et al. 2012). Induction of the OEPD and subsequent Pax2/Pax8 expression is mediated by FGF signaling. Despite the lack of uniformity of FGF signaling among species, Fgf3/8/10/19 mediates induction of OEPD in mammals. (Wright and Mansour 2003; Schimmang 2007; Ladher et al. 2010; Urness et al. 2010).

4.2.3 Formation of the Otic Placode

After induction of the OEPD, the functional segregation of the otic and epibranchial placodes is mediated by Wnt signaling and further upregulation of Pax2 within the otic placode (Groves and Bronner-Fraser 2000; Wright and Mansour 2003). In mice, specification of the otic placode is mediated by expression of Wnt ligands including Wnt8a from the caudal hindbrain (Ladher et al. 2000; Freter et al. 2008; Vendrell et al. 2013). In addition, the role of Wnt signaling in otic specification is further supported by mouse studies in which deletion of β-catenin reduces the size of the otic placode (Wright and Mansour 2003). It has been postulated that Pax2 contributes to the thickened pseudostratified morphology of the otic placode as compared to the surrounding epithelium (Christophorou et al. 2010). This thickening may enable actions on the basal and apical aspects of placodal cells, such as expression of cellular adhesion molecules, facilitating subsequent placode invagination, and formation of the otic vesicle (Christophorou et al. 2010; Sai and Ladher 2015).

From its superficial position in the embryo, the otic placode undergoes morphogenetic changes to attain the relatively insulated position of the fully developed inner ear. Invagination into the head mesenchyme may be mediated by expansion of the basal region of the placodal epithelium followed by apical constriction (Sai et al. 2014). This biphasic process promotes depression of the otic placode to form the otic pit, followed by complete encapsulation as a hollow sphere known as the otic vesicle or otocyst wherein differentiation of sensory cells and neuronal components takes place (Freter et al. 2008).

4.3 Generation of Inner Ear Organoids

One of the earliest protocols for differentiation of mouse embryonic stem cells (ESCs) into hair cell-like cells involved placement of stem cell aggregates onto an adhesive culture surface in the presence of epidermal growth factor (EGF) and insulin-like growth factor 1 (IGF-1) for a specified time period followed by an addition of bFGF to form inner ear progenitor cells. These progenitors were isolated and induced toward further differentiation by removal of growth factors, which resulted in detection of hair cell markers including Myo7a and espin (Li et al. 2003).

After confirmation that ESCs can be differentiated into hair cell-like cells, attempts to generate functional/mechanosensitive hair cells followed. Cells generated in a protocol using induced pluripotent stem (iPS) cells, which included addition of IGF-1 and FGF2, displayed stereocilia and a kinocilium mimicking endogenous vestibular hair cells. This protocol was dependent on plating stem cell-derived otic progenitors over inactivated, fibroblast-like cells from embryonic chicken utricles. Despite the relatively low yield of successfully derived hair cell-like cells and the need for exogenous tissue, mechanical stimulation of these cells elicited adaptation properties and transduction currents consistent with immature hair cells (Oshima et al. 2010).

Recent insights and advances have enabled in vitro generation of inner ear sensory epithelia and corresponding neural structures from mouse ESC aggregates. This organogenesis was facilitated through the use of three-dimensional (3D) cell culture techniques and precisely timed administration of recombinant proteins and small molecules, with recognition of the importance of the NNE and the PPR as precursors to otic differentiation. These aggregates subsequently develop and organize in a self-directed manner to form both mechanosensitive hair cells and corresponding sensory neural components (Koehler et al. 2013; Koehler and Hashino 2014).

Organogenesis is increasingly being studied in the context of 3D floating cell culture. In vitro organ models have been developed with 3D culture techniques because they facilitate self-organization of cells into complex organ-like structures, (Eiraku et al. 2008, 2011; Sato et al. 2009; Suga 2011) as opposed to traditional monolayer culture which restricts free movements of cells (Sasai et al. 2012). As evidenced above, an array of intricate signaling gradients and morphological relationships takes place during organ development; 3D culture more closely models physiologic embryogenesis and appears to facilitate self-organization of cells into recognizable tissues (Sasai et al. 2012; Xinaris et al. 2015). This is due to the increased freedom for morphologic changes (e.g., invagination and vesicle formation) and cellular interactions within the developing tissue (Sasai et al. 2012).

Organoids of neuroectodermal fate, namely, cerebral cortex and retina, have been formed in vitro (Eiraku et al. 2008; Eiraku et al. 2011; Li et al. 2013). We postulated that inner ear organoids may be generated in a similar manner by driving the ectoderm toward NNE as opposed to neuroectoderm. Detailing self-guided differentiation within a 3D culture system of the functional anterior pituitary gland tissue (derived from that anterior aspect of the PPR) has been documented. After NNE formation, treatment with the medium that did not contain any factors to affect tissue organization and patterning facilitated self-guided differentiation. Furthermore, the authors demonstrated that treatment of smaller stem cell aggregates with exogenous BMP also led to induction of NNE markers (Suga 2011).

4.3.1 Formation of Nonneural Ectoderm (NNE)

We recognized that epithelium similar to the definitive ectoderm (Nanog$^-$ Laminin$^+$ E-cadherin$^+$) is present on day 3 of ESC aggregates after an addition of Matrigel (Fig. 4.3a–d). It was postulated that the cultured aggregates could be directed to form tissue arising from the NNE, such as the inner ear, via induction of NNE from this definitive ectoderm-like epithelium. As outlined above, the differentiation of the definitive ectoderm into the neuroectoderm and NNE is mediated by BMP

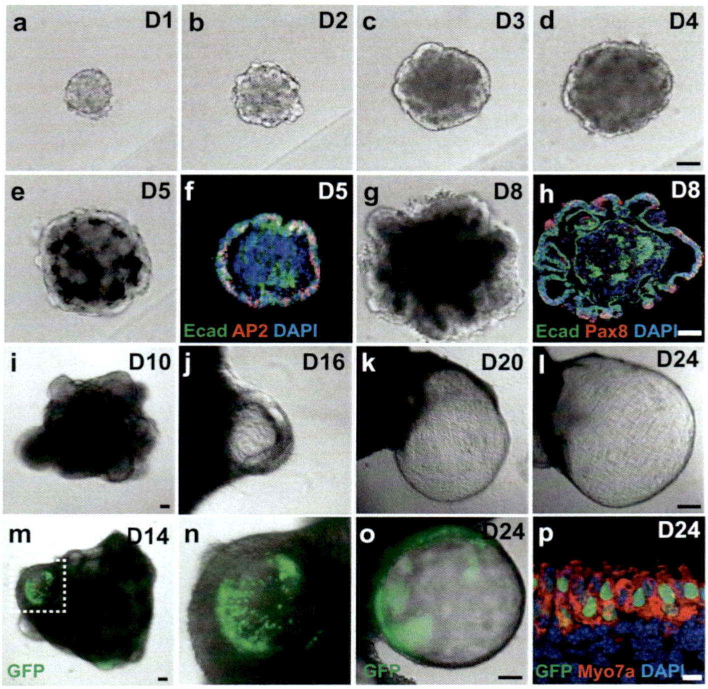

Fig. 4.3 Development of inner ear organoids in 3D culture. Morphological progression of an ESC aggregate from day 1 to day 24 (**a–e**, **g**, **i–l**). Immunofluorescence for Ecad/AP2 and Ecad/Pax8 in days 5 and 8 aggregates, respectively (**f**, **h**). *Atoh1*-GFP expression marks hair cells arising in organoids (**m–o**). Immunofluorescence for Myosin 7a (Myo7a) confirms that *Atoh1*-GFP-positive cells are indeed hair cells (**p**). Scale bars: 100 μm (**d**, **h**, **i**, **l**, **m**, **o**), 10 μm (**p**)

signaling—this was recapitulated in vitro in the experiments to induce anterior pituitary gland tissue from floating ESC aggregates (Suga 2011). Consistent with these findings, we observed that addition of human BMP4 to the aggregates on day 3 upregulated the NNE marker Dlx3 and the mesendoderm marker brachyury with concurrent downregulation of the neuroectoderm marker Sox1. Brachyury expression indicated unwanted formation of mesodermal and endodermal cell types. To more effectively induce the NNE, we supplemented the BMP4 treatment with the small molecule SB431542 (SB) which inhibits the transforming growth factor beta (TGF-β) signaling pathway and mediates suppression of brachyury (Chambers et al. 2009). Day 5 analysis of aggregates treated with BMP4/SB demonstrated expression of the NNE marker Tfap2a mainly localized to the outer epithelium (E-cadherin^{+}) of the aggregates along with markedly reduced brachyury expression (Fig. 4.3e–f). Within the outer epithelium, we also identified a layer that was positive for Sox1 and N-cadherin (Ncad), consistent with neuroectoderm formation. The core of each aggregate was positive for Nanog, a factor necessary for maintaining pluripotency of ESCs (Misui et al. 2003). Additionally, qPCR analysis of day 5 aggregates treated with BMP4/SB demonstrated upregulation of the NNE marker Dlx3 alongside downregulation of the neuroectodermal marker Sox1 as compared to aggregates treated with vehicle controls. It was therefore determined that day 5 aggregates treated with BMP4/SB represent an outer epithelium of NNE surrounding a layer of mesendodermal and neuroectodermal components and an inner core of still-pluripotent stem cells. Morphologically, the outer epithelium of aggregates treated with BMP4/SB was noticeably thicker and relatively undisrupted when compared to aggregates treated with BMP4 or SB alone. As a confirmation of the NNE identity of this outer epithelium, long-term culture of these aggregates for 20 days in basal medium resulted in expression of Krt5 and p63, which is consistent with epidermis formation from NNE. The apical aspect of this epidermis is oriented toward the aggregate interior.

4.3.2 Formation of the Preplacodal Region (PPR)

The next step toward otic differentiation is formation of the PPR, which later gives rise to the cranial placodes, including the otic placode. Based on studies of in vivo PPR formation, it was determined that the BMP signaling necessary for induction of the NNE must be attenuated for PPR induction, which is also dependent on activation of the FGF signaling pathway. Based upon these observations from in vivo organogenesis, the aggregates were treated with a combination of the BMP inhibitor LDN-193189 (LDN) and human recombinant FGF2. We observed that BMP inhibition on day 4.5 did not counteract specification of the NNE given that aggregates treated with LDN at that time point continued to express Dlx3. Support for a morphology consistent with the PPR was found in the presence of thickened epithelial patches in aggregates treated with BMP4/SB/LDN which were not present in aggregates in the BMP4/SB treatment group. Consistent with in vivo data, this thickening is apparently contingent on internal FGF signaling; this was proven when application of the FGF inhibitor SU5402 prohibited thickened epithelial patches from forming, though Tfap2 expression was maintained. Exogenous application of FGF2 concurrent with LDN administration was shown to significantly increase the epithelial thickening compared to aggregates not treated with FGF2. Nearly all aggregates treated with BMP4/SB/LDN/FGF2 demonstrated a thickened epithelium positive for Gata3, Six1, and Tfap2a, which subsequently ruffled and formed ovoid vesicles from days 6 through 8. It is based on these findings that this outer epithelium was determined to represent the PPR.

4.3.3 Formation of the Otic Placode

Cranial sensory placodes can be distinguished by expression of various Pax transcription factors; the otic placode arises from the OEPD and is defined by co-expression of Pax2, Pax8, and Ecad. Quantitative polymerase chain reaction (qPCR) showed increased Pax2 and Pax8 expression in aggregates treated with BMP4/SB/LDN/FGF2 as compared to the other conditions; by day 6, these aggregates also demonstrated cells positive for Pax8 and Ecad throughout the outer epithelial layer consistent with formation of the OEPD. Ecad expression is important in recognition of derived OEPD given that Pax2 and Pax8 co-expression is also found in the optic stalk, mid-hindbrain, and kidney in addition to the otic vesicle. The extent of Pax8 expression was directly related to the FGF2 concentration. $Pax8^+$ $Ncad^+$ cells were also present within the core of each aggregate, indicating mid-hindbrain tissue formation in the aggregate interior. Development of βIII-$tubulin^+$ neurons within this region further confirmed the neural identity. Further development of the outer epithelium with increased expression of Pax8 and Ecad recapitulated otic placode induction between days 6 and 8 (Fig. 4.3g–h), indicating the importance of LDN/FGF2 administration for in vitro otic placode formation. Pax6, which identifies the anterior placodes, was not expressed in aggregates treated with BMP4/SB/LDN/FGF2. Pax3 co-expression with Pax8 was noted within the aggregate interior—Pax3 marks the trigeminal placode, which is closer toward the $Pax8^+$ PPR. The expression of more caudally oriented Pax proteins is potentially caused by insulin in Knockout Serum Replacement within the media, given that insulin has been utilized for mid-hindbrain induction in floating culture (Muguruma et al. 2010).

4.3.4 Self-Guided Differentiation to the Inner Ear Sensory Epithelium

In order to facilitate self-organization and further differentiation of the aggregates following otic placode induction, on day 8, aggregates are transferred to serum-free floating culture. By day 9, the interior cell mass migrates toward the outside of the aggregate causing the outer epithelium to become an inner epithelium lining the central core of the aggregate. Between days 9 and 12, continuous evagination of vesicles into the outer cell layer occurs (Fig. 4.3i). These vesicles, appearing as translucent cysts, comprise cells from the presumptive OEPD epithelium that are Ecad$^+$ Pax2$^+$ Pax8$^+$ Sox2$^+$, consistent with formation of the prosensory domain. Treatment of aggregates for 48 hours with a Wnt inhibitor, XAV939, suppresses the formation of prosensory and Pax2$^+$ vesicles, suggesting the importance of Wnt signaling in otic vesicle formation (Groves and Fekete 2012). The evagination of vesicles from the aggregate is inconsistent with the observed invagination of in vivo otic vesicles into the head mesenchyme. This may be attributed to the basal component of the epithelium being oriented toward the superficial aspect of the aggregate as determined by analysis of aggregates allowed to develop epidermal tissue and confirmed through staining for the polarity markers aPKC and laminin. Some of the large vesicles grew to a size (Fig. 4.3j–l) comparable to typical adult mouse utricles (~1000 μm) (Li et al. 2008).

4.3.5 Morphologic and Functional Analysis of Inner Ear Sensory Epithelia

The prosensory domain of the otic vesicle further develops into the sensory epithelium with Sox2$^+$ supporting cells and Atoh1/Myo7a$^+$ hair cells (Fig. 4.3m–o). By day 14 of the 3D culture system described, vesicles contained Sox2$^+$Jag1$^+$ cells (analogous to embryonic day 9.5 mouse otic vesicles), and, over the next 48 hours, aggregates contained an average of 15 vesicles harboring cells resembling sensory hair cells (Myo7a$^+$ Sox2$^+$ with basal nuclei and elongated apical ends) (Fig. 4.4a–b). These cells are arrayed alongside one another with apical ends facing toward the vesicle lumen. Further consistent with sensory epithelium formation was a layer of Sox2$^+$ cells representing supporting cells present at the basal aspect of the sensory hair cell-like cells (Fig. 4.4c). Tight cell-cell junctions, which stain positive for F-actin, were found at the luminal surface with F-actin$^+$ espin$^+$ stereocilia bundles (Fig. 4.4d–e), each with a single acetylated-α-tubulin$^+$ kinocilium and associated basal body. Transmission electron microscopy of the kinocilium revealed an internal configuration consisting of two central microtubules surrounded by nine microtubule doublets, characteristic of a kinocilium in a native sensory hair cell. The height of these stereocilia bundles increases between day 20 and 24 to reach that of bundles in the adult mouse utricle (Li et al. 2008). The hair cell identity of these cells was further confirmed by rapid uptake of FM1–43 dye, which is specifically taken up through mechanotransduction channels (Meyers et al. 2003). Moreover, these cells displayed membrane properties comparable to those in native hair cells in the inner ear with various voltage-dependent currents. All hair cells exhibited outwardly rectifying potassium currents, and some cells displayed a transient inward current consistent with calcium channel activity. Based on all these morphological, biochemical, and functional properties, it was concluded that Myo7a$^+$ cells in ESC-derived organoids are hair cells analogous to those found in the inner ear.

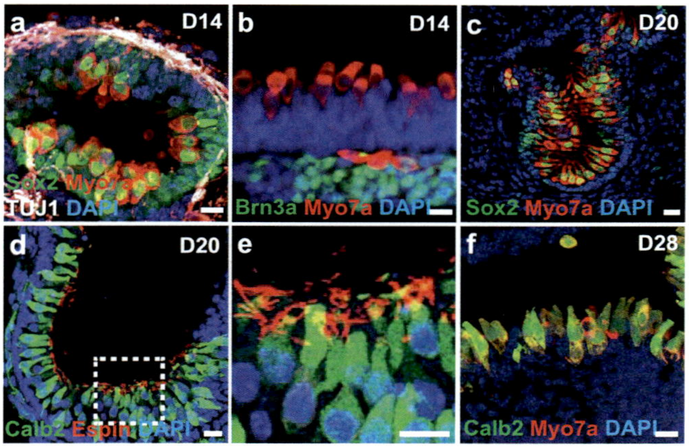

Fig. 4.4 Immunofluorescence analysis of day 14–28 inner ear organoids **(a–f)**. Myo7a; Myosin 7a, Calb2; Calbindin 2. Scale bars: 10 μm

The next step was to determine the type of hair cells arising in our culture. Co-expression of Pax2 and Sox2 has been shown to be specific to vestibular hair cells (Warchol and Richardson 2009). If these cells also express calbindin 2 (Calb2) (Fig. 4.4f), they are likely to be type II vestibular hair cells (Oesterle et al. 2008). Analysis of day 20 aggregates reveals that derived hair cells are positive for Sox2/Pax2/Calb2, indicating a homogenous population of type II vestibular hair cells. The overall morphology of the derived vesicles with a tightly packed bilayer of hair cells and supporting lining on the luminal surface resembles that of a vestibular end organ such as the utricle or saccule. Additionally, we detected $Brn3a^{+}Tuj1^{+}$ sensory neurons forming ribbon synapses with hair cells in the culture. Moreover, the number of ribbon synapses increased with time, mimicking in vivo maturation of inner ear hair cells (Lysakowski et al. 2011).

4.4 Potential Applications and Limitations

The inner ear organoid we have established faithfully recapitulates the formation and differentiation of the inner ear in vitro, thus serving as a potent model system to study both normal and pathological development of the inner ear. Utilizing mouse embryos for studying early inner ear development is extremely difficult due to the size and inaccessibility of the tissues, which cannot be maintained for extended periods of time ex vivo. The ability to generate a large amount of developmentally synchronized inner ear tissues by the organoid system also provides an unmet opportunity for various biochemical assays, such as chromatin immunoprecipitation-based assays.

The inner ear organoid can also serve as a potent model system to study pathophysiology of various forms of hereditary inner ear disorders. Patient-derived induced pluripotent stem cells (iPSs) may be used to generate inner ear organoids, which could be used to study the underlying molecular mechanisms that lead to the clinical phenotypes. Alternatively, quickly evolving genome-editing technology such as zinc finger nucleases, transcription activator-like effector nucleases (TALENs), or clustered regularly interspaced short palindromic repeats (CRISPR) could be employed to model human genetic disorders in vitro (Li et al. 2013; Ding et al. 2013). These in vitro model systems can also serve as a platform for testing potential therapeutic interventions to treat or potentially cure clinical symptoms.

The scalability of this system is particularly advantageous for high-throughput drug screening or toxicity testing (Sacheli et al. 2013). Certain types of commonly utilized drugs such as antibiotics and diuretics have ototoxic side effects and damage sensory hair cells in the inner ear. Studies on ototoxicity have traditionally been conducted using animal models with results that could have little or no correlation with the susceptibility of human patients to ototoxic agents (Forge and Schacht 2000; Wu et al. 2001; Kalinec 2005; Ding et al. 2013). Furthermore, the vast majority of animal models are not amenable to high-throughput screening techniques (Sacheli et al. 2013). The mechanisms underlying ototoxicity or otoprotection of certain compounds are currently poorly understood and may be better studied within the in vitro setting, circumventing challenges such as the limited availability and access to inner ear tissue from animal models (Kalinec et al. 2003; Kalinec 2005).

The feasibility of using stem cell-derived inner ear cells in cell-based therapies has been under intense investigation (Géléoc and Holt 2014; Park 2015; Müller and Barr-Gillespie 2015). Transplantation of inner ear progenitor cells derived from murine embryonic stem cells has been shown to successfully integrate at sites of injured sensory epithelium in vivo (Li et al. 2003). Another study describes in vitro generation of otic progenitor cells from human ESCs, which were successfully transplanted into gerbils with selective loss of spiral ganglion neurons. Transplanted animals showed improvements in hearing as determined by auditory evoked response testing (Chen et al. 2012). Despite these promising results, numerous challenges remain such as the varying electrolyte concentrations within inner ear compartments, integration of transplanted cells in the correct orientation and appropriate location, and precise connections between hair cells and sensory neurons (Park 2015). In order to circumvent these challenges, an alternative potential avenue for exploration is to transplant entire "sheets" of in vitro-derived sensory epithelium in the inner ear as has been done in rodent models of retinal degeneration (Seiler et al. 2010). Clearly technical hurdles will need to be overcome, yet the potential for successful autologous epithelial sheet transplantation is an exciting proposition.

Research into gene therapy targeting inner ear disease also continues to emerge, and an organoid model may be an efficient and effective system to investigate this therapeutic option (Sacheli et al. 2013; Géléoc and Holt 2014; Müller and Barr-Gillespie 2015). The 3D culture system described herein may be useful in testing transfection and transduction efficiency of various vectors, cellular toxicity, and the efficacy of proposed gene therapy agents. Specific gene therapy strategies, such as transdifferentiation of supporting cells, and the molecular mechanisms underlying them may also be investigated within this in vitro framework.

In the emerging era of precision medicine, organoids have already demonstrated an important role: one study demonstrated variable responses to drug administration in intestinal organoids derived from patients with cystic fibrosis. These findings have strong implications for prediction of specific therapeutic effects in individual patients with cystic fibrosis (Dekkers et al. 2013). Similar investigations aimed at tailoring interventions for inner ear disease would prove highly useful in investigating treatments for deafness or vestibular dysfunction.

One major limitation present within the current inner ear organoid protocol is the absence of cochlear cell types in the derived tissues. Further refinements to this protocol may expand the diversity of inner ear cell types within the organoid and more closely mimic the in vivo composition of the inner ear. This deficiency may be rectified through use of human ESCs; successful adaptation of a mouse ESC-derived protocol for optic cup formation into one using human ESCs resulted in human retinal organoids that are much larger in size and contain a greater number of layers as well as more diverse photoreceptors when compared to mouse organoids (Nakano et al. 2012). It will be interesting to assess if significant differences in overall morphology or types of derived cells are also observed between mouse and human inner ear organoids. Additionally, to accurately quantify the effects of experiments, a more robust endpoint analysis of organoids must be achieved. This may be accomplished by incorporating fluorescence-activated cell sorting or whole-mount immunolabeling techniques into organoid analysis.

Acknowledgments The authors would like to thank Atsushi Shimomura for the schematic drawing and Rachel DeJonge and Andrew Mikosz for some of the image data. This work was supported by a National Institutes of Health grant R01 DC013294 (to E.H.), an Indiana Clinical and Translational Research Institute predoctoral fellowship (to A.N.E.), and a Centralized Otolaryngology Research Effort (CORE) grant (to R.F.N.).

References

Ahrens K, Schlosser G (2005) Tissues and signals involved in the induction of placodal Six1 expression in *Xenopus laevis*. Dev Biol 288(1):40–59

Bailey AP, Streit A (2005) Sensory organs: making and breaking the pre-placodal region. In: Current topics in developmental biology, vol 72. Academic, San Diego/London, pp 167–204

Baker C, Bronner-Fraser M (2001) Vertebrate cranial placodes I. Embryonic induction. Dev Biol 232:1–61

Bhat N, Kwon H, Riley B (2013) A gene network that coordinates preplacodal competence and neural crest specification in zebrafish. Dev Biol 373(1):107–117

Bouchard M, Andersson E, Novitch B, Muhr J (2004) Tissue-specific expression of cre recombinase from the Pax8 locus. Genesis 38(3):105–109

Breneman K, Brownell W, Rabbitt R (2009) Hair cell bundles: flexoelectric motors of the inner ear. PLoS One 4(4):e5201

Brugmann S, Pandur P, Kenyon K, Pignoni F, Moody S (2004) Six1 promotes a placodal fate within the lateral neurogenic ectoderm by functioning as both a transcriptional activator and repressor. Development 131:5871–5881
Chambers S, Fasano C, Papapetrou E (2009) Highly efficient neural conversion of human ES and iPS cells by dual inhibition of SMAD signaling. Nat Nanotechnol 27:275–280
Chen W, Jongkamonwiwat N, Abbas L et al (2012) Restoration of auditory evodked responses by human ES-cell-derived otic progenitors. Nature 490:278–284
Chi N, Epstein JA (2002) Getting your Pax straight: PAX proteins in development and disease. Trends Genet 18(1):41–47
Christophorou N, Bailey A, Hanson S, Streit A (2009) Activation of Six1 target genes is required for sensory placode formation. Dev Biol 336:327–336
Christophorou N, Mende M, Lleras-Forero L, Grocott T, Streit A (2010) Pax2 coordinates epithelial morphogenesis and cell fate in the inner ear. Dev Biol 345:180–190
Dekkers J, Wiegerinck C, de Jonge H et al (2013) A functional CFTR assay using primary cystic fibrosis intestinal organoids. Nat Med 19(7):939–948
Ding Q, Regan S, Xia Y, Oostrom L, Cowan C, Musunuru K (2013) Enhanced efficiency of human pluripotent stem cell genome editing through replacing TALENs with CRISPRs. Cell Stem Cell 12(4):393–394
Eiraku M, Watanabe K, Matsuo-Takasaki M et al (2008) Self-Organized formation of polarized cortical tissues from ESCs and its active manipulation by extrinsic signals. Cell Stem Cell 3(5):519–532
Eiraku M, Takata N, Ishibashi H et al (2011) Self-organizing optic-cup morphogenesis in three-dimensional culture. Nature 472:51–56
Esterberg R, Fritz A (2009) dlx3b/4b are required for the formation of the preplacodal region and otic placode through local modulation of BMP activity. Dev Biol 325(1):189–199
Forge A, Schacht J (2000) Aminoglycoside antibiotics. Audiol Neurootol 5(1):3–22
Freter S, Muta Y, Mak S, Rinkwitz S, Ladher R (2008) Progressive restriction of otic fate: the role of FGF and Wnt in resolving inner ear potential. Development 135:3415–3424
Géléoc G, Holt J (2014) Sound strategies for hearing restoration. Science 344(6184):596–605
Glavic A, Honoré SM, Feijóo CG, Bastidas F, Allende ML, Mayor R (2004) Role of BMP signaling and the homeoprotein Iroquois in the specification of the cranial placodal field. Dev Biol 272(1):89–103
Groves A, Bronner-Fraser M (2000) Competence, specification and commitment in otic placode induction. Development 139:3489–3499
Groves A, Fekete D (2012) Shaping sound in space: the regulation of inner ear patterning. Development 139:1175–1187
Kalinec F (2005) High-throughput screening of ototoxic and otoprotective pharmacologic drugs. Volta Rev 105(3):383–406
Kalinec G, Webster P, Lim D, Kalinec F (2003) A cochlear cell line as an in vitro system for drug ototoxicity screening. Audiol Neurootol 8:177–189
Koehler K, Hashino E (2014) 3D mouse embryonic stem cell culture for generating inner ear organoids. Nat Protoc 9(6):1229–1244
Koehler K, Mikosz A, Molosh A, Patel D, Hashino E (2013) Generation of inner ear sensory epithelia from pluripotent stem cells in 3D culture. Nature 500:217–223
Kwon H, Bhat N, Sweet E, Cornell R, Riley B (2010) Identification of early requirements for preplacodal ectoderm and sensory organ development. PLoS Genet 6(9):e1001133
Ladher R, Anakwe K, Gurney A, Schoenwolf G, Francis-West P (2000) Identification of synergistic signals initiating inner ear development. Science 290:1965–1967
Ladher R, O'Neill P, Begbie J (2010) From shared lineage to distinct functions: the development of the inner ear and epibranchial placodes. Development 137:1777–1785
Lang D, Powell SK, Plummer RS, Young KP, Ruggeri BA (2007) PAX genes: Roles in development, pathophysiology, and cancer. Biochem Pharmacol 73(1):1–14

Leung A, Morest D, Li J (2013) Differential BMP signaling controls formation and differentiation of multipotent preplacodal ectoderm progenitors from human embryonic stem cells. Dev Biol 379(2):208–220

Li A, Xue J, Peterson E (2008) Architecture of the mouse utricle: macular organization and hair bundle heights. J Neurophysiol 99:718–733

Li H, Roblin G, Liu H, Heller S (2003) Generation of hair cells by stepwise differentiation of embryonic stem cells. PNAS 100(23):13,495–13,500

Li M, Suzuki K, Kim N, Liu G, Belmonte J (2013) A cut above the rest: targeted genome editing technologies in human pluripotent stem cells. J Biol Chem 289(8):4594–4599

Litsiou A, Hanson S, Streit A (2005) A balance of FGF, BMP and WNT signalling positions the future placode territory in the head. Development 132:4051–4062

Lysakowski A, Gaboyard-Niay S, Calin-Jageman I, Chatlani S, Price S, Eatock R (2011) molecular microdomains in a sensory terminal, the vestibular calyx ending. J Neurosci 31(27):10,101–10,114

Meyers J, MacDonald R, Duggan A et al (2003) Lighting up the senses: FM1–43 loading of sensory cells through nonselective ion channels. J Neurosci 23(10):4054–4065

Misui K, Tokuzawa Y, Itoh H, Segawa K et al (2003) The homeoprotein Nanog is required for maintenance of pluripotency in mouse epiblast and ES cells. Cell 30(113):631–642

Muguruma K, Nishiyama A, Ono Y et al (2010) Ontogeny-recapitulating generation and tissue integration of ES cell–derived Purkinje cells. Nat Neurosci 13:1171–1180

Müller U, Barr-Gillespie P (2015) New treatment options for hearing loss. Nat Rev Drug Discov 14:346–365

Nakano T, Ando S, Takata N (2012) Self-formation of optic cups and storable stratifed neural retina from humans ESCs. Cell Stem Cell 10:771–785

Oesterle E, Campbell S, Taylor R, Forge A, Hume C (2008) Sox2 and Jagged1 expression in normal and drug-damaged adult mouse inner ear. J Assoc Res Otolaryngol 9:65–89

Ohyama T, Mohamed O, Taketo M, Dufort D, Groves A (2006) Wnt signals mediate a fate decision between otic placode and epidermis. Development 133:865–875

Oshima K, Shin K, Diensthuber M, Peng A, Ricci A, Heller S (2010) Mechanosensitive hair cell-like cells from embryonic and induced pluripotent stem cells. Cell 141(4):704–716

Park Y-H (2015) Stem cell therapy for sensorineural hearing loss, still alive? J Audiol Otol 19(2):63–67

Pieper M, Ahrens K, Rink E, Peter A, Schlosser G (2012) Differential distribution of competence for panplacodal and neural crest induction to non-neural and neural ectoderm. Development 139(6):1175–1187

Reichert S, Randall R, Hill C (2013) A BMP regulatory network controls ectodermal cell fate decisions at the neural plate border. Development 140:4435–4444

Ruf R, Xu P, Silvius D, Otto E, Beekmann F, Muerb U et al (2004) SIX1 mutations cause branchio-oto-renal syndrome by disruption of EYA1–SIX1–DNA complexes. Proc Natl Acad Sci 101:8090–8095

Sacheli R, Delacroix L, Vandenackerveken P, Nguyen L, Malgrange B (2013) Gene transfer in inner ear cells: a challenging race. Gene Ther 20:237–247

Sai X, Ladher R (2015) Early steps in inner ear development: induction and morphogenesis of the otic placode. Front Pharmacol 6(19):1–8

Sai X, Yonemura S, Ladher R (2014) Junctionally restricted RhoA activity is necessary for apical constriction during phase 2 inner ear placode invagination. Dev Biol 394:206–216

Saint-Jeannet J, Moody S (2014) Establishing the pre-placodal region and breaking it into placodes with distinct identities. Dev Biol 389(1):13–27

Sasai Y, Eiraku M, Hidetaka S (2012) In vitro organogenesis in three dimensions: self-organising stem cells. Development 139(22):4111–4121

Sato S, Ikeda K, Shioi G, Ochi H, Ogino H, Yajima H et al (2010) Conserved expression of mouse Six1 in the pre-placodal region (PPR) and iden- tification of an enhancer for the rostral PPR. Dev Biol 344:158–171

Sato T, Vries R, Snippert H (2009) Single Lgr5 stem cells build crypt–villus structures in vitro without a mesenchymal niche. Nature 459:262–265

Schimmang T (2007) Expression and functions of FGF ligands during early otic development. Int J Dev Biol 51:473–481

Schlosser G (2006) Induction and specification of cranial placodes. Dev Biol 294:303–351

Schlosser G (2014) Early embryonic specification of vertebrate cranial placodes. Dev Biol 3:349–363

Schlosser G, Ahrens K (2004) Molecular anatomy of placode development in *Xenopus laevis*. Dev Biol 271(2):439–466

Seiler M, Aramant R, Thomas B, Peng Q, Sadda S, Keirstead H (2010) Visual restoration and transplant connectivity in degernerate rats implanted with retinal progenitor sheets. Eur J Neurosci 31:508–520

Streit A (2007) The preplacodal region: an ectodermal domain with multipotential progenitors that contribute to sense organs and cranial sensory ganglia. Int J Dev Biol 51(6–7):447–461. doi:10.1387/ijdb.072327as

Suga H (2011) Self-formation of functional adenohypophysis in three-dimensional culture. Nature 480:57–62

Urness L, Paxton C, Wang X, Schoenwolf G, Mansour S (2010) FGF signaling regulates otic placode induction and refinement by controlling both ectodermal target genes and hindbrain Wnt8a. Dev Biol 340:595–604

Vendrell V, Vázquez-Echeverría C, López-Hernández I, Alons B, Marinez S, Pujades C et al (2013) Roles of Wnt8a during formation and patterning of the mouse inner ear. Mech Dev 130:160–168

Warchol M, Richardson G (2009) Expression of the pax2 transcription factor is associated with vestibular phenotype in the avian inner ear. Dev Neurobiol 69:191–202

Wright T, Mansour S (2003) Fgf3 and Fgf10 are required for mouse otic placode induction. Development 130:3379–3390

Wu W, Sha S, McLaren J, Kawamoto K, Raphael Y, Schacht J (2001) Aminoglycoside ototoxicity in adult CBA, C57BL and BALB mice and the Sprague-Dawley rat. Hear Res 158(1–2):165–178

Xinaris C, Brizi V, Remuzzi G (2015) Organoid models and applications in biomedical research. Nephron 130:191–199

Chapter 5
Functional Tooth Regeneration

Masamitsu Oshima and Takashi Tsuji

Abstract Fundamental problems in the oral tissues caused by dental caries, periodontal disease and trauma affect the oral and general health issues. Current regenerative medicine has been developed by the accumulation of important findings in embryonic development, stem-cell biology and tissue engineering technology. Tooth regenerative therapy including both of tooth-tissue repair and whole-tooth replacement is expected as a novel therapeutic concept to enable a fully functional recovery of oral and maxillofacial region. Tooth tissue-derived stem cells and cell-activating cytokines are considered to be available approach for tooth-tissue regeneration based on the differentiation potential into several tooth tissues in vitro and in vivo. Whole-tooth replacement regenerative therapy is thought to be an attractive concept for next-generation regenerative therapy as a model of bioengineered organ replacement. To realise the whole-tooth regeneration, we have developed a novel three-dimensional cell-manipulation method designated the 'organ germ method'. This method involves compartmentalisation of epithelial and mesenchymal stem cells at a high cell density to replicate the epithelial-mesenchymal interactions in organogenesis. The bioengineered tooth germ generates a structurally correct tooth in vitro and erupted successfully with correct tooth structure when transplanted into the oral cavity. Bioengineered teeth were able to perform physiological tooth functions such as mastication, response to mechanical stress and a perceptive potential for noxious stimuli. In this chapter, we describe recent findings and technological development underpinning tooth regenerative therapy.

Keywords Tooth regeneration • Organ replacement regenerative therapy • Bioengineered tooth • Organ germ method • Epithelial-mesenchymal interaction • Tissue engineering

---

M. Oshima
Department of Oral Rehabilitation and Regenerative Medicine, Graduate School of Medicine, Dentistry and Pharmaceutical Sciences, Okayama University, Okayama 700-8525, Japan

RIKEN Center for Developmental Biology, Kobe, Hyogo 650-0047, Japan
e-mail: m-oshima@cc.okayama-u.ac.jp

T. Tsuji (✉)
Laboratory for Organ Regeneration, RIKEN Center for Developmental Biology, 2-2-3, Minatojima-mimamimschi, Chuo-ku, Kobe, Hyogo 650-0047, Japan
e-mail: t-tsuji@cdb.riken.jp

T. Tsuji (ed.), *Organ Regeneration Based on Developmental Biology*,
DOI 10.1007/978-981-10-3768-9_5

5.1 Introduction

The tooth is of ectodermal organ whose development is regulated by reciprocal epithelial-mesenchymal interactions (Jussila et al. 2013; Tucker and Sharpe 2004) and has characteristic hard tissues that include enamel, dentin and cementum. To maintain tooth homeostasis, a tooth also has soft connective tissues including pulp and a periodontal ligament (PDL) that contains nerve fibres and blood vessels (Avery 2002; Nanci 2012). Thus, tooth has a three-dimensional multicellular structure that establishes functional cooperation with the maxillofacial region (Avery 2002; Nanci 2012). Tooth loss or the onset of oral disease, such as dental caries, periodontal disease and traumatic injury, causes fundamental problems for oral function, e.g. enunciation, mastication, occlusion and associated general health issues (Peoffit et al. 2004). To restore the occlusal function after tooth loss, several dental therapies that replace the tooth with artificial materials, such as fixed dental bridges or removable dentures, have been widely performed as the conventional dental treatment (Rosenstiel et al. 2015; Ponkorny et al. 2008). Currently, osseointegrated dental implants that can stand alone in the jawbone without affecting the adjacent teeth have been used to treat tooth loss (Brenemark and Zarb 1985; Burns et al. 2003). Although these artificial therapies have been widely applied to rehabilitation of tooth loss, further technological improvements based on biological findings are expected to restore tooth physiological functions (Huang et al. 2008).

Recent advances in future regenerative therapies have been influenced by many previous research fields in embryonic development, stem-cell biology and tissue engineering technology (Brockes and Kumar 2005; Langer and Vacanti 1999; Atala 2005; Madeira et al. 2015). As an attractive regenerative concept, stem-cell transplantations using tissue-derived stem cells, embryonic stem (ES) cells or induced pluripotent stem (iPS) cells have been attempted for repairing damaged tissues due to structural and functional diseases (Trounson et al. 2013; Takebe et al. 2013; Addis and Epstein 2013; Kamao et al. 2014). Also, cytokine therapy is considered to have the potential to induce the activation and differentiation of stem/progenitor cells in various tissues (Gurtner et al. 2008). In the dental field, tooth tissue-derived stem cells and the cytokine network that regulates tooth development have been identified/characterised at the molecular level (Huang et al. 2009). These advances can be applied to the regenerative approach of dental pulp and periodontal tissues and contributed to the functional repair of partial tooth-tissue damage (Huang et al. 2009).

Ideally, the regenerative therapy is to develop fully functional bioengineered tissues/organs that can replace lost or damaged organs following disease, injury or ageing (Ikeda and Tsuji 2008). Organ replacement regenerative therapy, unlike stem-cell transplantation, has great potential for the replacement of dysfunctional organ by a fully functional bioengineered organ, which is reconstructed by in vitro three-dimensional cell manipulation using candidate stem cells (Atala 2005; Purnell 2008). In dentistry, tooth regenerative therapy would involve replacement with a bioengineered tooth built using stem cells that have the capacity to form a functional tooth unit containing periodontal tissue (Yen and Sharpe 2008; Volponi et al. 2010;

Sharpe and Young 2005). It is anticipated that whole-tooth replacement therapy will be established in the near future as a novel treatment that contributes to functional recovery by meeting aesthetic and physiological requirements (Volponi et al. 2010; Sharpe and Young 2005). Many approaches to replace missing teeth have been evaluated in the past three decades, including three-dimensionally bioengineered teeth and tooth germ generation using biodegradable materials and cell aggregation methods (Volponi et al. 2010; Sharpe and Young 2005; Duailibi et al. 2006; Liu et al. 2014). The first report of a fully functioning bioengineered tooth replacement with the correct tissue structure, masticatory function, responsiveness to mechanical stress and perceptive potential following transplantation into a tooth-loss region was published (Nakao et al. 2007; Ikeda et al. 2009; Oshima et al. 2011). In this chapter, we will describe the recent findings and technologies for partial tooth-tissue repair and whole-tooth regeneration in future that can provide functional recovery and ultimately replace the current dental treatments using artificial materials.

5.2 Tooth Organogenesis

Ectodermal organs, such as the teeth, hair and mammary glands, arise from their respective organ germs through reciprocal epithelial-mesenchymal interactions. This interaction is the principal mechanism that regulates almost all organogenesis via signalling molecules and transcription factors (Bei 2009; Nakatomi et al. 2010; Thesleff 2003). During early craniofacial development in mice, tooth-forming fields are specified at embryonic day (ED) 10.5–11.5 by the expression of homeobox genes, such as *Lhx8*, *Msx1*, *Msx2* and *Barx1*, and secretory molecules including bone morphogenetic proteins (BMPs) and fibroblast growth factors (FGFs), in the embryonic jaw (Nakatomi et al. 2010; Thesleff 2003; Jernvall and Thesleff 2012). At ED 11.5, the oral epithelium elongates into the mesenchymal tissue region, and then a tooth bud is formed by the condensation of neural crest-derived mesenchymal tissue (Nakatomi et al. 2010; Thesleff 2003; Jernvall and Thesleff 2012). At ED13.5–14.5, the first enamel knot, which acts as a transient signalling centre to orchestrate tooth developmental process by controlling the gene expression of various signalling molecules and transcription factors, appears in the dental epithelium. This signalling centre is thought to regulate individual cell fate and epithelial-mesenchymal interactions. The secondary enamel knots are formed; these regions play an important role in regulating the position and number of tooth cusps at ED16.5 (Nakatomi et al. 2010; Thesleff 2003; Jernvall and Thesleff 2012). After ED18.5, the epithelial and mesenchymal cells in the tooth germ terminally differentiate into the tooth tissue-forming cells such as ameloblasts and odontoblasts, respectively. These cells secrete a collagenous extracellular matrix to mineralise into the enamel or dentin matrix at the interface between epithelium and mesenchyme (Fukumoto and Yamada 2005). In the same period, the outer mesenchymal cells around tooth germ form the dental follicle tissue that can generate periodontal tissue including cementum, periodontal ligaments and alveolar bone (Saito et al. 2009).

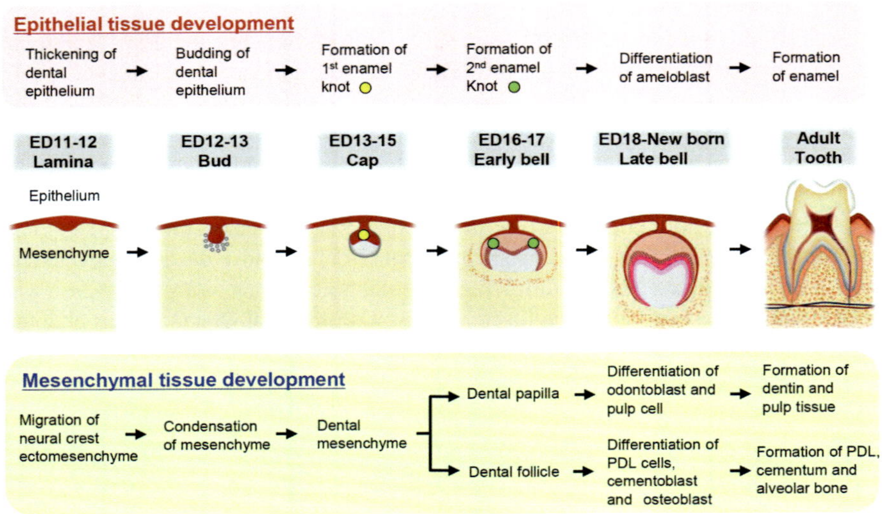

Fig. 5.1 Schematic diagram of tooth development. The oral epithelium elongates into the mesenchymal tissue region, and then a tooth bud is formed by the condensation of neural crest-derived mesenchymal tissue (ED11–13). The first enamel knot, which acts as a signalling centre to orchestrate tooth development by controlling the gene expression of various signalling molecules and transcription factors, appears in the dental epithelium (ED13–15). The secondary enamel knots are generated; these regions play an important role in regulating the position and number of tooth cusps (ED16–17). The epithelial and mesenchymal cells in the tooth germ terminally differentiate into the tooth tissue-forming cells such as ameloblasts and odontoblasts, respectively. These cells secrete a collagenous extracellular matrix to mineralise into the enamel or dentin matrix at the interface between epithelium and mesenchyme (ED18–newborn). In the same period, the outer mesenchymal cells around tooth germ form the dental follicle tissue that can generate periodontal tissue including cementum, periodontal ligaments and alveolar bone. Tooth root formation is initiated after tooth crown formation, and the mature teeth erupt into the oral cavity

Tooth root formation is initiated after tooth crown formation, and the mature teeth erupt into the oral cavity (Avery 2002; Nanci 2012). In mature tooth, after tooth development, immature cells seem to be maintained as adult tissue stem cells that are thought to act as a self-repair system for dental tissues and supply a wide variety of each dental cell type after the dental tissue injury (Egusa et al. 2012) (Fig. 5.1).

5.3 Tissue Repair and Engineering by Using Dental Tissue-Derived Stem Cells and Cytokines

Tissue-derived stem cells, including haematopoietic stem cells, neural stem cells, skin stem cells and mesenchymal stem cells, allow for self-renewal and differentiation to maintain homeostasis and to repair injured tissues. Recent studies of stem/progenitor cells have provided new insights that have furthered our understanding of tooth tissue-derived stem cells, which can differentiate along various dental-cell lineages, forming odontoblasts, pulp cells, PDL cells, cementoblasts and osteoblasts (Huang et al. 2009). These dental stem cells and the associated cytokine networks that regulate tooth development have been well identified at the molecular level (Shi et al. 2005; Bartold and Narayanan 2006). The stem-cell transplantation and cytokine approach based on stem-cell biology in dental field will be available for tooth-tissue repair that can achieve one of the promising concepts in the regenerative therapy.

5.3.1 Stem-Cell Therapy

5.3.1.1 Dentin-Pulp Tissue Repair

In the tooth development, dental pulp tissue develops from the dental papillae and comprises of odontoblasts, fibroblasts, blood vessels and peripheral nerves (Avery 2002; Nanci 2012). Dental pulp stem cells (DPSCs) and stem cells from human exfoliated deciduous teeth (SHED) have been identified from the dental pulp in human permanent third molars and exfoliated deciduous teeth, respectively (Gronthos et al. 2000; Miura et al. 2003). DPSCs and SHED represent CD146/STRO-1 expression as an undifferentiated marker and are thought to be dental stem cells with high proliferative capacities and multi-differentiation into odontoblasts, adipocytes and neural cells (Yang et al. 2015). These stem cells are available for stem cell-mediated dentin-pulp complex regeneration. Furthermore, DPSCs and pulp stem-cell subfractions, including $CD31^-/CD146^-$ side population (SP) cells and $CD105^+$ cells, which can generate pulp tissue, are candidate cell source for dentin repair and dental pulp regeneration (Nakashima et al. 2009). During tooth root development, the dental papilla is located apically to the developing pulp which is known as the apical papilla that is gelatinous soft tissue with few blood vessels and cellular components. That tissue contains the stem cells of the apical papilla (SCAP), which have a high proliferative potential based on their high levels of telomerase activity and ability to differentiate into odontoblasts or adipocytes (Sonoyama et al. 2006, 2008). SCAP is also able to generate typical dentin-pulp complex structures after transplantation in vivo and is thus considered a valuable stem-cell source for tissue repair and tissue engineering (Fig. 5.2a).

5.3.1.2 Periodontal Tissue Repair

Periodontal tissue, which is composed of cementum, PDL and alveolar bone, develops from the dental follicle tissue, and that establishes a biological connection through the insertion of the PDL fibre into the cementum and alveolar bone during tooth root formation (Saito et al. 2009; Shimono et al. 2003; Foster et al. 2007). PDL-derived stem cells (PDLSCs), which have been identified in adult human PDL

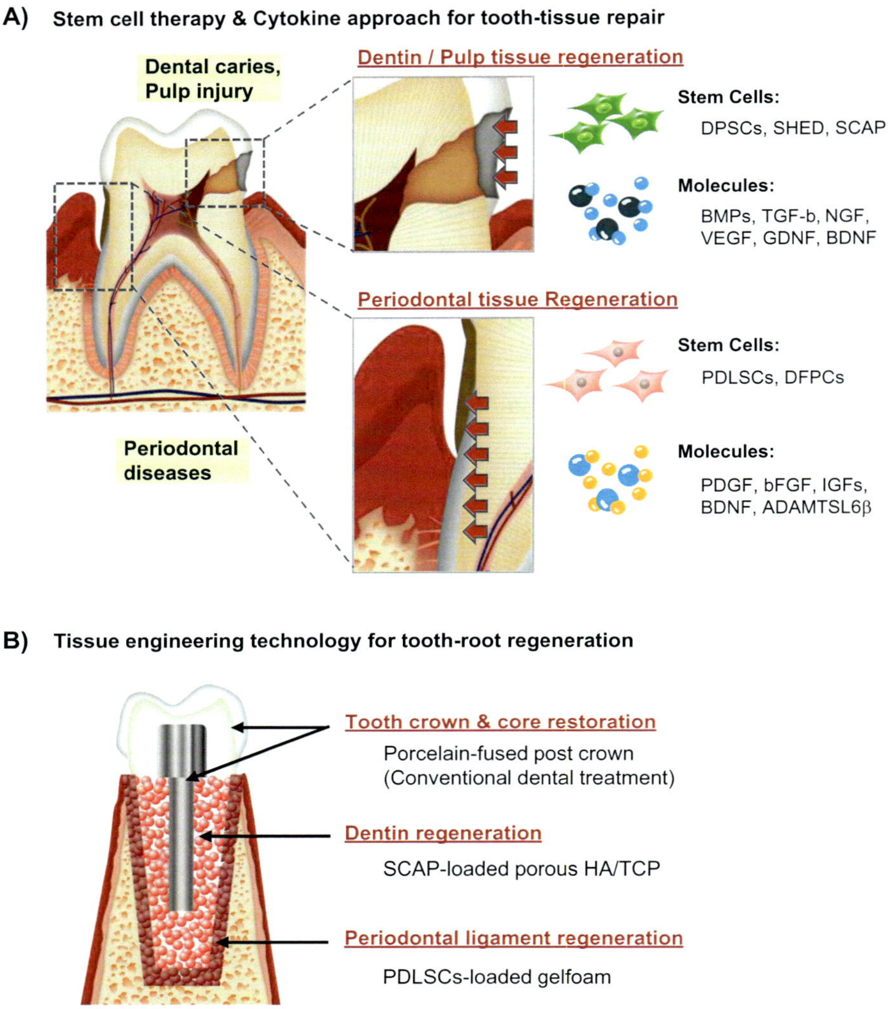

Fig. 5.2 Tissue repair and engineering by using dental tissue-derived stem cells and cytokines. (**a**) Tooth tissue-derived stem-cell transplantation and cytokine approach based on stem-cell biology in dental field will be available for tooth-tissue repair that can achieve one of the promising concepts in the regenerative therapy. As a regenerative approach for dental caries or pulp injury, DPSCs, SHED and SCAP are candidate cell source for dentin repair and dental pulp regeneration. Furthermore, the bioactive growth factors including TGF-β superfamily, IGFs, bFGF, BMPs and various angiogenic growth factors had a potential for stimulating odontoblast-like cell differentiation and dentine matrix formation. As a regenerative therapy for periodontal disease, PDLSCs and DFSCs are thought to be useful cell types for the regeneration of periodontal tissue injury. Several molecular approaches can be administered via the local application of human recombinant cytokines, such as PDGF, IGFs, BDNF, bFGF and ADAMTSL6β, to accelerate periodontal regeneration

tissue, are able to have a highly proliferative capacities and to differentiate into all of the periodontal cell types after transplantation in vivo (Seo et al. 2004). Upon in vivo transplantation into an immunocompromised animal, PDLSCs were able to generate cementum and PDL complex structure (Yang et al. 2009; Lei et al. 2014). Dental follicle stem cells (DFSCs) were first identified as mesenchymal stem/progenitor cells in the first molars of the neonatal rat, and they have been shown to be able to differentiate into osteoblasts, cementoblasts, adipocytes and neural cells (Saito et al. 2005; Morsczeck et al. 2005; Luan et al. 2006; Kémoun et al. 2007; Mehmet et al. 2009). These cells are thought to be useful cell types for the regeneration of periodontal tissue injury. In addition, the cell-sheet engineering using stem/progenitor cells have been developed for clinical use in periodontal tissue regeneration (Guo et al. 2013) (Fig. 5.2a).

5.3.2 Cytokine Approach

Deletions of enamel and dentin tissue are most commonly recognised as a tooth pathological disease due to dental caries or trauma, and the standard treatment involves the substitution of the natural/physiological dental tissue with artificial material. Following dental tissue damage, tertiary dentine is deposited with a biological stimulation in response to dental injury (Smith et al. 1995; Bjørndal 2008; Bjørndal and Darvann 1999). In particular, reparative dentine formation is a more complex process requiring initial progenitor/stem-cell recruitment and stimulating to odontoblast-like cell differentiation and dentine secretion (Sloan and Smith 2007; Smith and Lesot 2001). Critical molecules, which are likely important in the regulation of dentin tissue regeneration, have been identified based on epithelial-mesenchymal interactions in tooth developmental process. It was reported that bioactive growth factors including transforming growth factor-β (TGF-β) superfamily, insulin-like growth factors (IGFs), basic fibroblast growth factor (bFGF), bone morphogenic proteins (BMPs) and various angiogenic growth factors had a potential for stimulating odontoblast-like cell differentiation and dentine matrix formation

Fig. 5.2 (continued) * **Abbreviations:** *PDLSCs* periodontal ligament stem cells; *DFSCs* dental follicle stem cells; *PDGF* platelet-derived growth factor; *IGFs* insulin-like growth factors; *BDNF* brain-derived neurotrophic factor; *bFGF* basic fibroblast growth factor; *ADAMTSL6β* A disintegrin-like metalloprotease domain with thrombospondin type I motifs like 6β; *DPSCs* dental pulp stem cells; *SHED* stem cells from human exfoliated deciduous teeth; *SCAP* stem cells from apical papilla; *BMPs* bone morphogenetic proteins; *TGF-β* transforming growth factor-β; *NGF* nerve growth factor; *VEGF* vascular endothelial growth factor; *GDNF* glial cell line-derived neurotrophic factor.
(**b**) Bioengineered root regeneration using a root-shaped hydroxyapatite/tricalcium phosphate (HA/TCP) carrier, which was loaded with SCAP cells that were covered with gelfoam/PDLSCs, has been reported to form a root-like structure to which the porcelain crown and core were restored, resulting in normal tooth function

(Finkelman et al. 1990; Cassidy et al. 1997; Roberts-Clark and Smith 2000; Zhao et al. 2000; Sloan and Smith 1999; Dobie et al. 2002; Lovschall et al. 2001; Tziafas et al. 1998; Tziafas and Papadimitriou 1998; Hu et al. 1998; Rutherford et al. 1993; Jepsen et al. 1997; Cooper et al. 2010) (Fig. 5.2a).

Periodontal tissue often suffers irreversible inflammation caused by periodontal disease or excessive occlusal force; however, the effective treatments for periodontal tissue regeneration have not yet been established. Several molecular approaches can be administered via the local application of human recombinant cytokines, such as platelet-derived growth factor (PDGF), IGFs, brain-derived neurotrophic factor (BDNF) and bFGF, to accelerate periodontal regeneration (Marcopoulou et al. 2003; Raja et al. 2009; Dereka et al. 2006). Furthermore, periodontal tissue is composed of an extracellular matrix that contains collagen fibres and microfibrils, which play a critical role in periodontal tissue formation. Recently, it was reported that the local administration of a fibrillin-1-associated protein, ADAMTSL6β, accelerated wound healing in periodontal tissues by accumulating microfibrils (Tsutsui et al. 2010; Saito et al. 2011) (Fig. 5.2a).

5.3.3 Bioengineered Root Regeneration by Tissue Engineering Technology

Dental implants, which can be stand alone in the jawbone unlike conventional dental treatments with the invading adjacent teeth, have been widely applied for the rehabilitation of tooth loss (Burns et al. 2003; Petrie et al. 2009). However, because of the absence of the natural periodontal tissue, the currently available dental implant that is directly connected to the surrounding alveolar bone cannot exhibit biological tooth functions, such as the reduction of excessive occlusal force, orthodontic tooth movement via bone remodelling and the ability to perceive noxious stimuli (Brenemark and Zarb 1985; Burns et al. 2003). In order to regenerate the tooth root including periodontal tissues for maintaining physiological tooth functions, a stem cell-based tissue engineering application has been attempted. A unique approach for tooth root regeneration employing a root-shaped hydroxyapatite/tricalcium phosphate (HA/TCP) carrier loaded with gelfoam/PDLSC-covered SCAP cells has been reported (Sonoyama et al. 2006; Wei et al. 2013) (Fig. 5.2b). This tissue engineering technology generated a bioengineered tooth root using a combination of tooth root and periodontal tissues, and it may contribute to establish the next-generation dental regenerative technology that will integrate stem cell-mediated tissue regeneration, biomaterial engineering and current dental treatment. The root regeneration approach is expected to be applied in the clinic more early compared to the whole-tooth regeneration methods.

5.4 Technological Development for Whole-Tooth Regeneration Utilising a Novel Three-Dimensional Cell-Manipulation Method

Stem-cell transplantation and cytokine therapy are now considered an effective approach to restore partial organ functions at local damaged sites. However, these therapeutic technologies have not yet achieved the ideal goal of reconstructing complex organs which can recover from an extensive organ injury or a severe organ dysfunction. The ultimate goal of regenerative therapy in future is to develop organ replacement regenerative therapies that will restore lost or damaged tissues with a fully functioning bioengineered organ (Oshima and Tsuji 2014). Current bioengineering technology for regenerating three-dimensional organs has progressed to the replication of organogenesis through epithelial-mesenchymal interactions that occur in the developing embryo, thereby enabling the development of fully functional bioengineered organs using bioengineered organ germs that are generated from immature stem cells via three-dimensional cell manipulation in vitro (Ikeda and Tsuji 2008; Oshima and Tsuji 2014). In the dental field, an attractive concept involves transplanting a bioengineered tooth germ into the tooth-loss region that would develop into a functioning mature tooth (Nakao et al. 2007; Ikeda et al. 2009) (Fig. 5.3, upper panel). It is also expected that it will be possible to transplant a bioengineered tooth unit including a mature tooth, PDL and alveolar bone that can be physiologically integrated into the alveolar bone of the recipient's jaw (Nakao et al. 2007; Oshima et al. 2011) (Fig. 5.3, lower panel). For the realisation

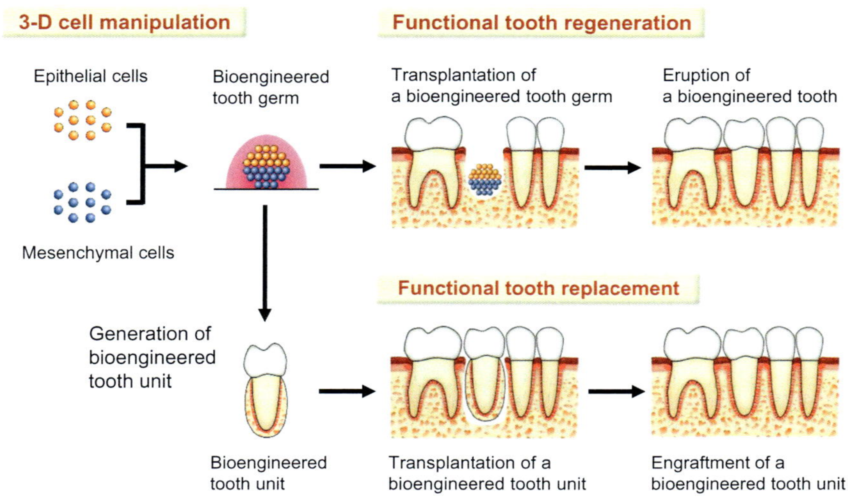

Fig. 5.3 Regenerative concept for whole-tooth replacement. Fully functioning teeth can be regenerated in vivo by transplanting bioengineered tooth germs reconstituted from epithelial and mesenchymal stem cells via the organ germ method or by transplanting bioengineered tooth units with a periodontal ligament and alveolar bone that developed from bioengineered tooth germs

of whole-tooth regeneration, the first major issue is considered to be developing a three-dimensional cell-manipulation technology using completely dissociated epithelial and mesenchymal stem cells in vitro. To date, two conventional approaches and a novel cell-manipulation method for generating bioengineered tooth germ or a mature tooth have been described below:

5.4.1 Biodegradable Scaffold Method

Conventional scaffold technology represented by three-dimensional tissue engineering has contributed to the large-scale tissue regeneration through seeding stem/progenitor cells on the degradable materials such as natural molecules and synthetic polymers. This method has been proven a high utility in three-dimensional tissue engineering technology, and these preparations have been used in clinical applications including bone and cartilage regenerative therapies (Quarto et al. 2001; Cao et al. 1997; Caplan and Bruder 2001). Previous studies using collagen/gelatine sponges or polyglycolic acid/poly-L-lactate-co-glycolide copolymers (PLA/PLGA) have reported the partial generation of tooth-tissue structure including enamel, dentin and dental pulp, through seeding epithelial and mesenchymal cells isolated from porcine tooth germ (Honda et al. 2003; Young et al. 2002; Iwatsuki et al. 2006; Duailibi et al. 2004; Yelick and Vacanti 2006; Sumita et al. 2006; Honda et al. 2007; Ohara et al. 2010). Although scaffold-based technology may be practical for controlling tooth shape and size, the fundamental problems regarding the regeneration of whole-tooth itself have not been resolved. The presence of residual scaffold material after in vivo transplantation is considered to be the cause of the low frequency of tooth formation and the irregularity of the resulting tooth-tissue structures, e.g. the enamel-dentin complex due to the orderly cell arrangement of ameloblast/odontoblast lineages (Honda et al. 2003; Young et al. 2002). Fully regeneration of proper tooth structure using scaffolds requires the formation of complex junctions between the enamel, dentin and cementum that result from accurate spatiotemporal cell gradients of ameloblasts, odontoblasts and cementoblasts as well as natural tooth development (Volponi et al. 2010; Nakao et al. 2007).

5.4.2 Cell Aggregation Method

The cell aggregation method is known as a typical bioengineering protocol employed for the reconstitution of a bioengineered organ germ to reproduce the epithelial-mesenchymal interactions that occur during organogenesis (Volponi et al. 2010). Previous studies have reported that transplanting bioengineered cell aggregates derived from the ectodermal organs such as hair follicle and mammary gland and demonstrated the regeneration of each reconstructed organ with the proper tissue structure and cellular arrangements (Zheng et al. 2005; Shackleton et al. 2006).

In the dental field, many researchers have isolated the dental epithelial and mesenchymal tissues from embryonic tooth germs of experimental animal and dissociated such tissues with surgical operation by using stereomicroscopic guidance and enzymatic treatments to obtain single stem cells. It has been reported that bioengineering cell aggregates reconstructing epithelial and mesenchymal tooth germ cells by using the cellular centrifugation have the potential for tooth formation after in vivo transplantation (Hu et al. 2006; Yamamoto et al. 2003). Even when bioengineered cell aggregate were mixed with epithelial and mesenchymal stem cells isolated from tooth germ, the correct tooth structure could be generated by the self-reorganisation through the cell rearrangement of epithelial and mesenchymal cells (Song et al. 2006). This technique is an innovative approach to replicate tooth organogenesis; however, the further improvements have been required in terms of the generating frequency of bioengineered tooth and the correct tissue formation.

5.4.3 Three-Dimensional Cell-Manipulation Method: The 'Organ Germ Method'

To achieve precise replication of the developmental processes in organogenesis, an in vitro three-dimensional novel cell-manipulation method designated as a bioengineered organ germ method has been recently established (Nakao et al. 2007). We demonstrated the possibility of developing a bioengineered tooth germ using completely dissociated single epithelial and mesenchymal cells from tooth germs at ED14.5 mice. The most important breakthrough using our method is the achievement of three-dimensional cell compartmentalisation of epithelial and mesenchymal cells at a high cell density in a collagen gel. This bioengineered tooth germ can achieve initial tooth development with the appropriate cell-to-cell compaction between epithelial and mesenchymal cells in vitro organ culture. Bioengineering tooth germ reproduced by our method successfully replicates the multicellular assembly, including ameloblasts, odontoblasts, pulp cells and dental follicle cells, underlying the epithelial-mesenchymal interactions as well as natural tooth development. The bioengineered tooth germ generates a structurally correct tooth after transplantation in an organ culture in vitro as well as following placement into a subrenal capsule in vivo (Nakao et al. 2007) (Fig. 5.4a). Tooth morphology is defined by not only the tooth length and crown size as macromorphological feature but also the cusp numbers/position as micromorphological feature. These morphological properties are regulated by specific gene expression at the boundary surface of immature oral epithelium and neural crest-derived mesenchyme in the embryonic jaw. Our bioengineered tooth germ could be reconstructed by adjusting various cell-to-cell contact length between the epithelial and mesenchymal cell layer, and thereby the crown width and cusp number of occurred bioengineered tooth were dependent on the contact length of epithelial and mesenchymal cell layer in bioengineered tooth germ (Ishida et al. 2011).

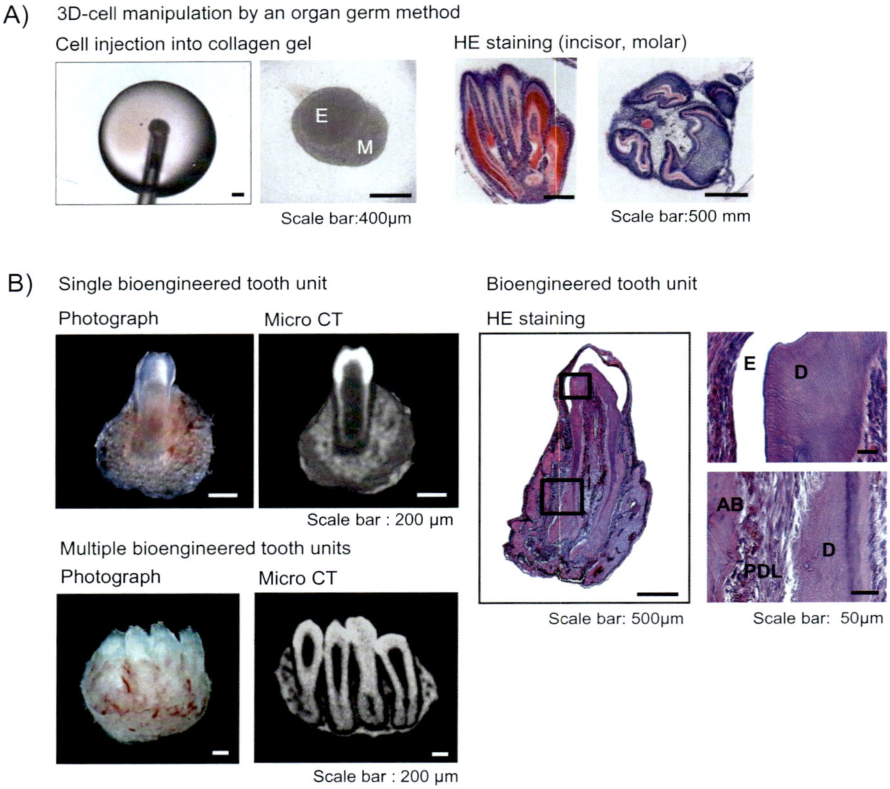

Fig. 5.4 The organ germ method: a novel three-dimensional cell-processing system. (**a**) Tooth germ-derived mesenchymal cells at a high density are injected into the centre of a collagen drop. Subsequently, tooth germ-derived epithelial cells are injected into the drop adjacent to the mesenchymal cell aggregate. Within 1 day of organ culture, the formation of bioengineered tooth germ with the appropriate compartmentalisation between the epithelial and mesenchymal cells and cell-to-cell compaction was observed (*left panels*). Bioengineered incisor and molar tooth germs, which were reconstituted using dissociated cells isolated from respective tooth germs, developed in organ culture over 14 days (*right panels*). (**b**) Transplantation of this bioengineered tooth germ into a subrenal capsule for 30 days (*left-upper panels*). Multiple bioengineered tooth units surrounded by alveolar bone can be generated by transplanting several tooth germs (*left-lower panels*). A bioengineered tooth unit comprising a mature tooth, periodontal ligament and alveolar bone with the correct tooth-tissue structure such as the enamel (*E*), the dentin (*D*), the periodontal ligament (*PDL*) and the alveolar bone (*right panels*)

Furthermore, an attractive concept for generating a bioengineered tooth unit composed of a mature tooth, PDL and alveolar bone through in vivo transplantation into the subrenal capsule has been developed (Fig. 5.4b). Multiple bioengineered tooth units surrounded by alveolar bone could also be generated through transplanting several tooth germs in a size-control device (Oshima et al. 2011). Each bioengineered tooth had the correct tooth structure including the pulp cavities and the partitioned periodontal tissue structure (Fig. 5.4b). In the case of edentulous jaw or following segmental mandibulotomy, it would be possible to achieve the multiple teeth restoration by using this regenerative transplantation method (Oshima et al. 2011). These technologies have the potential to be adapted for functional tooth regeneration and also represent a substantial advance in bioengineered organ replacement regenerative therapy.

5.5 Functional Whole-Tooth Replacement Using the Bioengineered Tooth

Teeth have important oral functions such as mastication, pronunciation and facial aesthetics, which have a critical influence on general health and quality of life (Dawson 2006). These tooth-related functions are achieved along with masticatory muscles and temporomandibular joint under the control of the central nervous system (Dawson 2006). To desire a successful tooth replacement therapy, a bioengineered tooth must be capable of being grafted into the tooth-loss region under adult oral environment and achieving full functionality, including sufficient occlusal performance, biological cooperation with the periodontal tissues and afferent responsiveness to noxious stimuli in the maxillofacial region.

5.5.1 Successful Transplantation of a Bioengineered Tooth Germ or a Bioengineered Mature Tooth Unit for Whole-Tooth Replacement Therapy

The critical issue regarding whole-tooth regenerative therapy via the transplantation of a bioengineered tooth germ is whether the germ can erupt into the oral cavity and occlude correctly with the opposing tooth under the adult oral environment. Natural tooth eruption is autonomously progressed by a regulatory mechanism that involves the tooth germ-cell components and the surrounding alveolar/jawbone (Wise et al. 2002; Wise and King 2008). Dental follicle cells contribute to overlying bone resorption by enzymatic degeneration during growth and the axial movement of the teeth during the process of tooth eruption (Wise and King 2008). It has been reported that a transplanted natural tooth germ can erupt in a murine toothless diastema region (Nakao et al. 2007; Ohazama et al. 2004). We have recently reported that a bioengineered tooth germ can develop with the correct tooth structure and completely erupt in an oral cavity at 49 days after transplantation (Nakao et al. 2007; Ikeda et al. 2009) (Fig. 5.5a). Additionally, the erupted bioengineered tooth not only reached the occlusal plane but also performed occlusal function with the opposing natural tooth (Ikeda et al. 2009) (Fig. 5.5a).

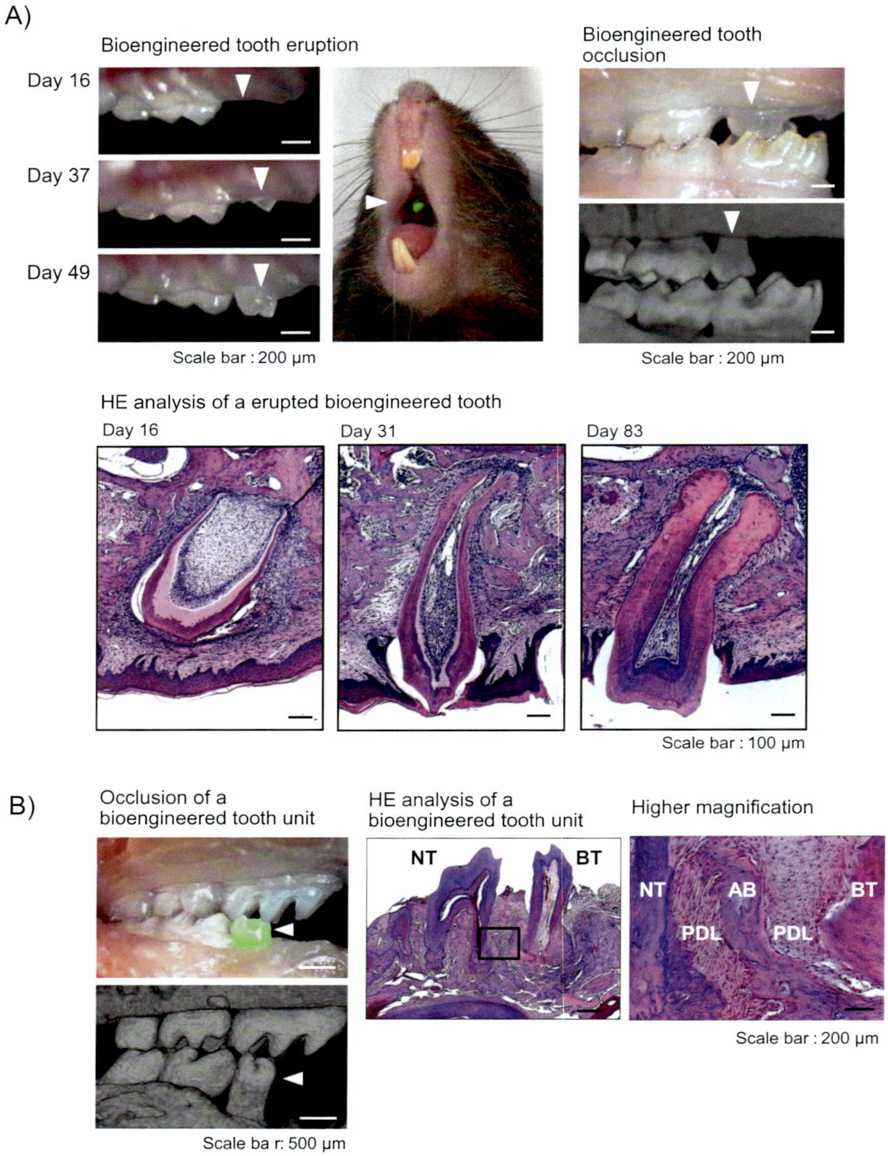

Fig. 5.5 Whole-tooth regeneration through the transplantation of a bioengineered tooth germ or a bioengineered tooth unit. (**a**) The transplanted bioengineered tooth germ erupted (*arrowheads, upper-left panels*) and reached the occlusal plane with the opposing lower first molar at 49 days after transplantation (*upper-right panels*). The bioengineered tooth also formed a correct tooth structure comprising enamel, dentin, dental pulp and periodontal tissue (*lower panels*). (**b**) The bioengineered tooth unit was engrafted through bone integration and reached the occlusal plane with the opposing upper first molar at 40 days after transplantation (*arrowheads, left panels*). The engrafted bioengineered tooth unit also had the correct tooth structure in the recipient jaw. NT, natural tooth; BT, bioengineered tooth; AB, alveolar bone; PDL, periodontal ligament (*centre and right panels*)

The success in the transplantation of bioengineered mature organs is whether they can exert the immediate organ performance of its full functions in vivo (Gridelli and Remuzzi 2000). In dental regenerative therapy, the most critical consideration is whether a bioengineered tooth unit composed of mature tooth, PDL and alveolar bone can be engrafted into the tooth-loss region, which involves natural bone remodelling in the recipient. We have previously demonstrated the successful engraftment of a bioengineered tooth unit with bone integration between the alveolar bone of the tooth unit and that of the recipient 40 days after transplantation, and the PDL that originated in the bioengineered tooth unit was sufficiently maintained during the bone integration and regeneration (Oshima et al. 2011) (Fig. 5.5b).

Furthermore, the enamel and dentin hardness of the bioengineered tooth components were equivalent to those of natural teeth based on the Knoop hardness test (Ikeda et al. 2009; Oshima et al. 2011). These tooth regenerative approaches demonstrated the potential for successfully restoring masticatory performance.

5.5.2 *Exertion of Physiological Tooth Functions in Bioengineered Tooth*

The structural properties of periodontal tissue contribute to the physiological tooth functions, including the absorption of occlusal loading, the maintenance of alveolar bone height and the orthodontic tooth movement involving bone remodelling (Avery 2002; Fukumoto and Yamada 2005). Autologous tooth transplantation studies have indicated that the remaining periodontal tissue near the tooth root could successfully restore physiological tooth functions including bone remodelling and the prevention of ankylosis (Tsukiboshi 1993). By contrast, the absence of a PDL in a dental implant is associated with deficits in essential tooth functions requiring the coordination of the teeth and the components of the maxillofacial region through the connection of PDL (Lindhe et al. 2008). Therefore, the functional cooperation between teeth and the maxillofacial region through the biological connection of periodontal ligament is essential for the exertion of oral functions (Avery 2002; Lindhe et al. 2008). We have demonstrated that bioengineered teeth are able to successfully undergo functional tooth movement equivalent to that of natural teeth underlying the proper localisation of osteoclasts and osteoblasts in response to mechanical stress (Ikeda et al. 2009; Oshima et al. 2011). These findings indicate that a bioengineered tooth can replicate periodontal ligament function and lead to the re-establishment of functional teeth within the maxillofacial region (Ikeda et al. 2009; Oshima et al. 2011).

Teeth are the peripheral organs of the sensory trigeminal and sympathetic nerves, and the afferent nervous system in the maxillofacial region plays important roles regarding the regulation of tooth physiological functions and the perception of noxious stimuli (Luukko et al. 2005). It is anticipated that neuronal regeneration, including the re-entry of nerve fibres after the transplantation of tooth germs or

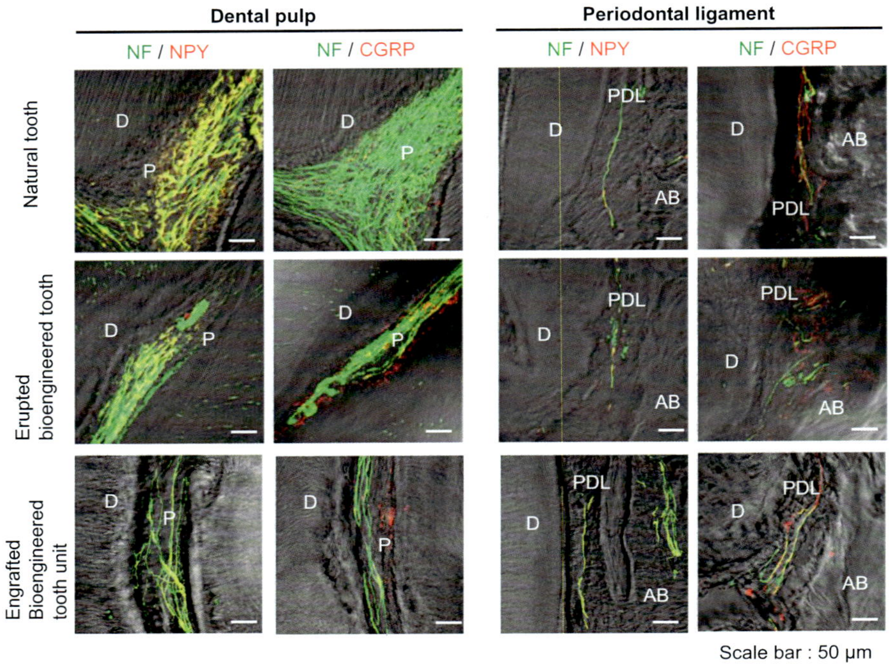

Fig. 5.6 Neural innervation in a bioengineered tooth. The peripheral nerve innervation of the dental pulp and periodontal ligament area in the natural and bioengineered teeth were represented by the immunostaining for neurofilament H. Neuropeptide Y (*NPY*), which is synthesised in sympathetic nerves, was detected in the pulp and the PDL neurons. Furthermore, calcitonin gene-related peptide (*CGRP*), which is synthesised in sensory nerves and is involved in sensing tooth pain, was also observed in both the pulp and the PDL neurons. *D* dentin; *P* pulp; *PDL* periodontal ligament; *AB* alveolar bone

tooth units, is required for the functional recovery of perceptive potential against noxious stimuli in future tooth regenerative therapy (Luukko et al. 2005). Our studies have demonstrated that sensory and sympathetic nerve fibres can innervate both the pulp and the PDL region of an engrafted bioengineered tooth (Ikeda et al. 2009; Oshima et al. 2011) (Fig. 5.6). Bioengineered teeth offer the potential for pain stimuli perception due to pulp injury and orthodontic movement and the potential to properly transduce these peripheral stimulations to the central nervous system through c-Fos immunoreactive neurons (Ikeda et al. 2009; Oshima et al. 2011). These findings indicate that bioengineered teeth can restore the proprioceptive responses to noxious stimuli within the maxillofacial region (Fig. 5.7).

Regenerated teeth developed from bioengineered germs or transplanted bioengineered mature tooth units have successfully demonstrated the physiological tooth functions in cooperation with the maxillofacial region such as sufficient masticatory performance, biological connections via periodontal tissues and afferent

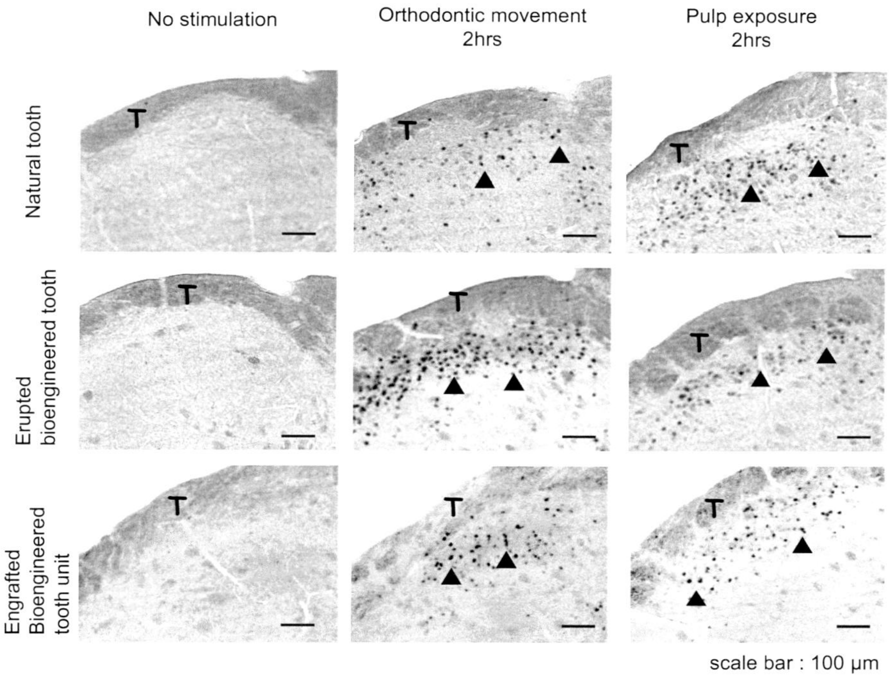

Fig. 5.7 Perceptive potential to mechanical stress in bioengineered teeth. Natural and bioengineered teeth were analysed by c-Fos immunoreactive neurons in the medullary dorsal horns of transplanted mice after 0 h (no stimulation, control; *left column*) and 2 h of stimulation via orthodontic movement (*centre column*) and pulp exposure (*right column*). c-Fos (arrowhead) was detectable after these stimulations were applied in natural (*upper*), erupted bioengineered tooth (*middle*) and engrafted bioengineered tooth units (*lower*)

responsiveness to noxious stimuli (Figs. 5.6 and 5.7). Our bioengineering technology for fully functional tooth regeneration could contribute to the realisation of whole-tooth replacement regenerative therapy in the future (Nakao et al. 2007; Ikeda et al. 2009; Oshima et al. 2011).

5.6 Future Perspectives for Tooth Regenerative Therapy

The progress made in regenerative technology is remarkable, and many patients expect the practical application of tooth-tissue repair and whole-tooth regenerative therapy. To address the future clinical application of tooth regenerative therapy, one of the major research hurdles remaining is the identification of appropriate cell sources (Ikeda and Tsuji 2008). These cell sources may be optimised by using the patient's own cells for regenerative therapy to avoid immunological rejection. In the dental field, recent studies of stem-cell biology have led to the identification of candidate cell sources for tooth-tissue repair and whole-tooth replacement therapy (Huang et al. 2009; Egusa et al. 2012). Tooth tissue-derived stem cells, including DPSCs, SHED, SCAP, PDLSCs and dental follicle stem cells, can differentiate into several dental-cell lineages and also contribute to the turnover and supply of various progenitor cells (Huang et al. 2009; Egusa et al. 2012). These cell linages would be useful for stem-cell transplantation therapy aimed towards tooth-tissue repair; however, the tooth-inducible cells, which are able to replicate an epithelial-mesenchymal interaction in tooth organogenesis, have not yet been identified (Egusa et al. 2012, 2013). In the future, the tooth-inducible cell sources for whole-tooth regeneration must be identified from the various tissue-derived stem-cell populations isolated from patients (Egusa et al. 2013). Pluripotent stem cells including ES cells and iPS cells are also candidate cell sources that can differentiate into endodermal, ectodermal and mesodermal cell linages (Yan et al. 2010). Recently, iPS cells have been established from several oral tissues such as pulp, PDL, gingiva and oral mucosa, and they have represented the ability to differentiate into dental epithelial or mesenchymal cells (Egusa et al. 2010; Arakaki et al. 2012; Otsu et al. 2012). Furthermore, the identification of critical regulated genes for reprogramming non-dental cells to differentiate into dental epithelial-mesenchymal stem cells is considered to be an important direction for future tooth regenerative therapy.

The different tooth types such as incisors, canines, premolars and molars have distinctive morphological features that are programmed at predetermined sites in the oral cavity during tooth development. Several studies have proposed molecular mechanisms for tooth morphology regulation (Cai et al. 2007). Not only tooth size/length but also crown/root shape are important aspects upon generating a bioengineered tooth with proper functional occlusion and aesthetics. Further studies are required to develop a bioengineering technology that can regulate tooth morphology, including tissue engineering using scaffolds and the identification of morphogenesis-related genes/cytokines to achieve the appropriate morphogenesis. Tooth regenerative therapy is now regarded as a viable model for studying future organ replacement regenerative therapies that can be applied to several complex organs, and it will contribute substantially to the knowledge and technology of organ regeneration (Volponi et al. 2010; Oshima and Tsuji 2014).

Acknowledgements This work was partially supported by the Health and Labour Sciences Research Grants from the Ministry of Health, Labour, and Welfare (no. 21040101) awarded to Akira Yamaguchi (Tokyo Medical and Dental University), a Grant-in-Aid for Scientific Research (A) (no. 20249078) awarded to T. Tsuji (2008–2010) and a Grant-in-Aid for Young Scientists (B) awarded to M. Oshima from the Ministry of Education, Culture, Sports and Technology, Japan. This work was also partially supported by the Organ Technologies, Inc.

Conflict of Interest M. Oshima and T. Tsuji have no competing interest.

References

Addis RC, Epstein JA (2013) Induced regeneration--the progress and promise of direct reprogramming for heart repair. Nat Med 19:829–836

Arakaki M, Ishikawa M, Nakamura T, Iwamoto T, Yamada A, Fukumoto E, Saito M, Otsu K, Harada H, Yamada Y, Fukumoto S (2012) Role of epithelial-stem cell interactions during dental cell differentiation. J Biol Chem 287:10590–10601

Atala A (2005) Tissue engineering, stem cells and cloning: current concepts and changing trends. Expert Opin Biol Ther 5:879–892

Avery JK (2002) Oral development and histology. Thieme Press, New York

Bartold P, Narayanan A (2006) Molecular and cell biology of healthy and diseased periodontal tissues. Periodontol 40:29–49

Bei M (2009) Molecular genetics of tooth development. Curr Opin Genet Dev 19:504–510

Bjørndal L (2008) The caries process and its effect on the pulp the science is changing and so is our understanding. J Endod 34:S2–S5

Bjørndal L, Darvann T (1999) A light microscopic study of odontoblastic and non-odontoblastic cells involved in tertiary dentinogenesis in well-defined cavitated carious lesions. Caries Res 33:50–60

Brenemark PI, Zarb GA (1985) Tissue-integrated prostheses. In: Albrektsson T (ed) Osseointegration in clinical dentistry. Quintessence Pub Co Press, Berlin, pp 211–232

Brockes JP, Kumar A (2005) Appendage regeneration in adult vertebrates and implications for regenerative medicine. Science 310:1919–1923

Burns DR, Beck DA, Nelson SK (2003) A review of selected dental literature on contemporary provisional fixed prosthodontic treatment: report of the Committee on Research in Fixed Prosthodontics of the Academy of Fixed Prosthodontics. J Prosthet Dent 90:474–497

Cai J, Cho SW, Kim JY et al (2007) Patterning the size and number of tooth and its cusps. Dev Biol 304:499–507

Cao Y, Vacanti JP, Paige KT, Upton J, Vacanti CA (1997) Transplantation of chondrocytes utilizing a polymer-cell construct to produce tissue-engineered cartilage in the shape of a human ear. Plast Reconstr Surg 100:297–304

Caplan AI, Bruder SP (2001) Mesenchymal stem cells: building blocks for molecular medicine in the 21st century. Trends Mol Med 7:259–264

Cassidy N, Fahey M, Prime SS, Smith AJ (1997) Comparative analysis of transforming growth factor-beta isoforms 1-3 in human and rabbit dentine matrices. Arch Oral Biol 42:219–223

Cooper PR, Takahashi Y, Graham LW, Simon S, Imazato S, Smith AJ (2010) Inflammation-regeneration interplay in the dentine-pulp complex. J Dent 38(9):687–697

Dawson PE (2006) Functional occlusion: from TMJ to smile design. Mosby, St. Louis

Dereka XE, Markopoulou CE, Vrotsos IA (2006) Role of growth factors on periodontal repair. Growth Factors 24(4):2607

Dobie K, Smith G, Sloan AJ, Smith AJ (2002) Effects of alginate hydrogels and TGF-beta 1 on human dental pulp repair in vitro. Connect Tissue Res 43:387–390

Duailibi SE, Duailibi MT, Vacanti JP, Yelick PC (2006) Prospects for tooth regeneration. Periodontol 41:177–187

Duailibi MT, Duailibi SE, Young CS, Bartlett JD, Vacanti JP, Yelick PC (2004) Bioengineered teeth from cultured rat tooth bud cells. J Dent Res 83:523–528

Egusa H, Okita K, Kayashima H, Yu G, Fukuyasu S, Saeki M, Matsumoto T, Yamanaka S, Yatani H (2010) Gingival fibroblasts as a promising source of induced pluripotent stem cells. PLoS One 5:e12743

Egusa H, Sonoyama W, Nishimura M, Atsuta I, Akiyama K (2012) Stem cells in dentistry–part I: stem cell sources. J Prosthodont Res. 56(3):151–165

Egusa H, Sonoyama W, Nishimura M, Atsuta I, Akiyama K (2013) Stem cells in dentistry–part II: clinical applications. J Prosthodont Res 56(4):229–248

Finkelman RD, Mohan S, Jennings JC, Taylor AK, Jepsen S, Baylink DJ (1990) Quantification of growth factors IGF-I, SGF/IGFII, and TGF-beta in human dentin. J Bone Miner Res 5:717–723
Foster BL, Popowics TE, Fong HK, Somerman MJ (2007) Advances in defining regulators of cementum development and periodontal regeneration. Curr Top Dev Biol 78:47–126
Fukumoto S, Yamada Y (2005) Review: extracellular matrix regulates tooth morphogenesis. Connect Tissue Res 46:220–226
Gridelli B, Remuzzi G (2000) Strategies for making more organs available for transplantation. N Engl J Med 343:404–410
Gronthos S, Mankani M, Brahim J, Robey PG, Shi S (2000) Postnatal human dental pulp stem cells (DPSCs) in vitro and in vivo. Proc Natl Acad Sci USA 97:13625–13630
Guo S, Guo W, Ding Y et al (2013) Comparative study of human dental follicle cell sheets and periodontal ligament cell sheets for periodontal tissue regeneration. Cell Transplant 22:1061–1073
Gurtner GC, Werner S, Barrandon Y, Longaker MT (2008) Wound repair and regeneration. Nature 453:314–321
Honda M, Morikawa N, Hata K, Yada T, Morita S, Ueda M, Kimata K (2003) Rat costochondral cell characteristics on poly (L-lactide-co-epsilon-caprolactone) scaffolds. Biomaterials 24:3511–3519
Honda MJ, Tsuchiya S, Sumita Y, Sagara H, Ueda M (2007) The sequential seeding of epithelial and mesenchymal cells for tissue-engineered tooth regeneration. Biomaterials 28:680–689
Hu CC, Zhang C, Qian Q, Tatum NB (1998) Reparative dentin formation in rat molars after direct pulp capping with growth factors. J Endod 24:744–751
Hu B, Nadiri A, Kuchler-Bopp S, Perrin-Schmitt F, Peters H, Lesot H (2006) Tissue engineering of tooth crown, root, and periodontium. Tissue Eng 12:2069–2075
Huang GT, Sonoyama W, Liu Y, Liu H, Wang S, Shi S (2008) The hidden treasure in apical papilla: the potential role in pulp/dentin regeneration and bioroot engineering. J Endod 34:645–651
Huang GT, Gronthos S, Shi S (2009) Mesenchymal stem cells derived from dental tissues vs. those from other sources: their biology and role in regenerative medicine. J Dent Res 88:792–806
Ikeda E, Tsuji T (2008) Growing bioengineered teeth from single cells: potential for dental regenerative medicine. Expert Opin Biol Ther 8:735–744
Ikeda E, Morita R, Nakao K et al (2009) Fully functional bioengineered tooth replacement as an organ replacement therapy. Proc Natl Acad Sci USA 106:13475–13480
Ishida K, Murofushi M, Nakao K, Morita R, Ogawa M, Tsuji T (2011) The regulation of tooth morphogenesis is associated with epithelial cell proliferation and the expression of Sonic hedgehog through epithelial-mesenchymal interactions. Biochem Biophys Res Commun 405:455–461
Iwatsuki S, Honda MJ, Harada H, Ueda M (2006) Cell proliferation in teeth reconstructed from dispersed cells of embryonic tooth germs in a three-dimensional scaffold. Eur J Oral Sci 114:310–317
Jepsen S, Albers HK, Fleiner B, Tucker M, Rueger D (1997) Recombinant human osteogenic protein-1 induces dentin formation: an experimental study in miniature swine. J Endod 23:378–382
Jernvall J, Thesleff I (2012) Tooth shape formation and tooth renewal: evolving with the same signals. Development 139:3487–3497
Jussila M, Juuri E, Thesleff I (2013) Tooth morphogenesis and renewal. Stem cells in craniofacial development and regeneration. Wiley-Blackwell, Hoboken, pp 109–134
Kamao H, Mandai M, Okamoto S et al (2014) Characterization of human induced pluripotent stem cell-derived retinal pigment epithelium cell sheets aiming for clinical application. Stem Cell Rep 2:205–218
Kémoun P, Laurencin-Dalicieux S, Rue J et al (2007) Human dental follicle cells acquire cementoblast features under stimulation by BMP-2/−7 and enamel matrix derivatives (EMD) in vitro. Cell Tissue Res 329:283–294

Langer RS, Vacanti JP (1999) Tissue engineering: the challenges ahead. Sci Am 280:86–89
Lei M, Li K, Li B, Gao L-N, Chen F-M, Jin Y (2014) Mesenchymal stem cell characteristics of dental pulp and periodontal ligament stem cells after in vivo transplantation. Biomaterials 35:6332–6343
Lindhe J, Lang NP, Karring T (2008) Clinical periodontology and implant dentistry, 5th edn. Blackwell, New Jersey
Liu P, Zhang Y, Chen S et al (2014) Application of iPS cells in dental bioengineering and beyond. Stem Cell Rev 10:663–670
Lovschall H, Fejerskov O, Flyvbjerg A (2001) Pulp-capping with recombinant human insulin-like growth factor I (rhIGF-I) in rat molars. Adv Dent Res 15:108–112
Luan X, Ito Y, Dangaria S, Diekwisch TG (2006) Dental follicle progenitor cell heterogeneity in the developing mouse periodontium. Stem Cells Dev 15:595–608
Luukko K, Kvinnsland IH, Kettunen P (2005) Tissue interactions in the regulation of axon pathfinding during tooth morphogenesis. Dev Dyn 234:482–488
Madeira C, Santhagunam A, Salgueiro JB, Cabral JMS (2015) Advanced cell therapies for articular cartilage regeneration. Trends Biotechnol 33:35–42
Marcopoulou CE, Vavouraki HN, Dereka XE, Vrotsos IA (2003) Proliferative effect of growth factors TGFbeta1, PDGFBB and rhBMP2 on human gingival fibroblasts and periodontal ligament cells. J Int Acad Periodontol 5(3):6370
Mehmet E, Ramazanoglu M, Gumru OZ et al (2009) Comparison and optimisation of transfection of human dental follicle cells, a novel source of stem cells, with different chemical methods and electro-poration. Neurochem Res 34:1272–1277
Miura M, Gronthos S, Zhao M, Lu B, Fisher LW, Robey PG, Shi S (2003) SHED: stem cells from human exfoliated deciduous teeth. Proc Natl Acad Sci U S A 100:5807–5812
Morsczeck C, Götz W, Schierholz J, Zeilhofer F, Kühn U, Möhl C, Sippel C, Hoffmann KH (2005) Isolation of precursor cells (PCs) from human dental follicle of wisdom teeth. Matrix Biol 24:155–165
Nakao K, Morita R, Saji Y et al (2007) The development of a bioengineered organ germ method. Nat Methods 4:227–230
Nakashima M, Iohara K, Sugiyama M (2009) Human dental pulp stem cells with highly angiogenic and neurogenic potential for possible use in pulp regeneration. Cytokine Growth Factor Rev 20:435–440
Nakatomi M, Wang XP, Key D, Lund JJ, Turbe-Doan A, Kist R, Aw A, Chen Y, Maas RL, Peters H (2010) Genetic interactions between Pax9 and Msx1 regulate lip development and several stages of tooth morphogenesis. Dev Biol 340:438–449
Nanci A (2012) Ten Cate's oral histology: development, structure, and function. Mosby Press, St. Louis
Ohara T, Itaya T, Usami K et al (2010) Evaluation of scaffold materials for tooth tissue engineering. J Biomed Mater Res – Part A 94:800–805
Ohazama A, Modino SA, Miletich I, Sharpe PT (2004) Stem-cell-based tissue engineering of murine teeth. J Dent Res 83:518–522
Oshima M, Tsuji T (2014) Functional tooth regenerative therapy: tooth tissue regeneration and whole-tooth replacement. Review. Odontology 102:123–136
Oshima M, Mizuno M, Imamura A et al (2011) Functional tooth regeneration using a bioengineered tooth unit as a mature organ replacement regenerative therapy. PLoS One 6:e21531
Otsu K, Kishigami R, Oikawa-Sasaki A, Fukumoto S, Yamada A, Fujiwara N, Ishizeki K, Harada H (2012) Differentiation of induced pluripotent stem cells into dental mesenchymal cells. Stem Cells Dev 21:1156–1164
Petrie TA, Reyes CD, Burns KL, García AJ (2009) Simple application of fibronectin-mimetic coating enhances osseointegration of titanium implants. J Cell Mol Med 13:2602–2612
Pokorny PH, Wiens JP, Litvak H (2008) Occlusion for fixed prosthodontics: a historical perspective of the gnathological influence. J Prosthet Dent 99:299–313

Proffit WR, Fields HW Jr, Sarver DM (2004) Contemporary orthodontics. Mosby Press, St. Louis, pp 78–83
Purnell B (2008) New release: the complete guide to organ repair. Introduction. Science 322:1489
Quarto R, Mastrogiacomo M, Cancedda R, Kutepov SM, Mukhachev V, Lavroukov A, Kon E, Marcacci M (2001) Repair of large bone defects with the use of autologous bone marrow stromal cells. N Engl J Med 344:385–386
Raja S, Byakod G, Pudakalkatti P (2009) Growth factors in periodontal regeneration. Int J Dent Hyg 7(2):82–89
Roberts-Clark DJ, Smith AJ (2000) Angiogenic growth factors in human dentine matrix. Arch Oral Biol 45:1013–1016
Rosenstiel SF, Land MF, Fujimoto J (2015) Contemporary fixed prosthodontics. Mosby Press, St. Louis, pp 209–366
Rutherford RB, Wahle J, Tucker M, Rueger D, Charaette M (1993) Induction of reparative dentin formation in monkeys by recombinant human osteogenic protein-1. Arch Oral Biol 38:571–576
Saito M, Handa K, Kiyono T, Hattori S, Yokoi T, Tsubakimoto T, Harada H, Noguchi T, Toyoda M, Sato S, Teranaka T (2005) Immortalization of cementoblast progenitor cells with Bmi-1 and TERT. J Bone Miner Res 20:50–57
Saito M, Nishida E, Sasaki T, Yoneda T, Shimizu N (2009) The KK-Periome database for transcripts of periodontal ligament development. J Exp Zool B Mol Dev Evol 312B:495–502
Saito M, Kurokawa M, Oda M, Oshima M, Tsutsui K, Kosaka K, Nakao K, Ogawa M, Manabe R, Suda N, Ganjargal G, Hada Y, Noguchi T, Teranaka T, Sekiguchi K, Yoneda T, Tsuji T (2011) ADAMTSL6β protein rescues fibrillin-1 microfibril disorder in a Marfan syndrome mouse model through the promotion of fibrillin-1 assembly. J Biol Chem 286:38602–38613
Seo BM, Miura M, Gronthos S, Bartold PM, Batouli S, Brahim J, Young M, Robey PG, Wang CY, Shi S (2004) Investigation of multipotent postnatal stem cells from human periodontal ligament. Lancet 364:149–155
Shackleton M, Vaillant F, Simpson KJ, Stingl J, Smyth GK, Asselin-Labat ML, Wu L, Lindeman GJ, Visvader JE (2006) Generation of a functional mammary gland from a single stem cell. Nature 439:84–88
Sharpe PT, Young CS (2005) Test-tube teeth. Sci Am 293:34–41
Shi S, Bartold PM, Miura M, Seo BM, Robey PG, Gronthos S (2005) The efficacy of mesenchymal stem cells to regenerate and repair dental structures. Orthod Craniofacial Res 8:191–199
Shimono M, Ishikawa T, Ishikawa H, Matsuzaki H, Hashimoto S, Muramatsu T, Shima K, Matsuzaka K, Inoue T (2003) Regulatory mechanisms of periodontal regeneration. Microsc Res Tech 60:491–502
Sloan AJ, Smith AJ (1999) Stimulation of the dentine–pulp complex of rat incisor teeth by transforming growth factor beta isoforms 1–3 in vitro. Arch Oral Biol 44:149–156
Sloan AJ, Smith AJ (2007) Stem cells and the dental pulp: potential roles in dentine regeneration and repair. Oral Dis 13:151–157
Smith AJ, Lesot H (2001) Induction and regulation of crown dentinogenesis: embryonic events as a template for dental tissue repair? Crit Rev Oral Biol Med 12:425–437
Smith AJ, Cassidy N, Perry H, Be'gue-Kirn C, Ruch JV, Lesot H (1995) Reactionary dentinogenesis. Int J Dev Biol 39:273–280
Song Y, Zhang Z, Yu X, Yan M, Zhang X, Gu S, Stuart T, Liu C, Reiser J, Zhang Y, Chen Y (2006) Application of lentivirus-mediated RNAi in studying gene function in mammalian tooth development. Dev Dyn 235:1334–1344
Sonoyama W, Liu Y, Fang D, Yamaza T, Seo BM, Zhang C, Liu H, Gronthos S, Wang CY, Wang S, Shi S (2006) Mesenchymal stem cell-mediated functional tooth regeneration in swine. PLoS One 1:e79
Sonoyama W, Liu Y, Yamaza T, Tuan RS, Wang S, Shi S, Huang GT (2008) Characterization of the apical papilla and its residing stem cells from human immature permanent teeth: a pilot study. J Endod 34:166–171

Sumita Y, Honda MJ, Ohara T, Tsuchiya S, Sagara H, Kagami H, Ueda M (2006) Performance of collagen sponge as a 3-D scaffold for tooth-tissue engineering. Biomaterials 27:3238–3248
Takebe T, Sekine K, Enomura M et al (2013) Vascularized and functional human liver from an iPSC-derived organ bud transplant. Nature 499:481–484
Thesleff I (2003) Epithelial-mesenchymal signalling regulating tooth morphogenesis. J Cell Sci 116:1647–1648
Trounson A, Daley GQ, Pasque V, Plath K (2013) A new route to human embryonic stem cells. Nat Med 19:820–821
Tsukiboshi M (1993) Autogenous tooth transplantation: a reevaluation. Int J Periodontics Restorative Dent 13:120–149
Tsutsui K, Manabe R, Yamada T, Nakano I, Oguri Y, Keene DR, Sengle G, Sakai LY, Sekiguchi K (2010) ADAMTSL-6 is a novel extracellular matrix protein that binds to fibrillin-1 and promotes fibrillin-1 fibril formation. J Biol Chem 285:4870–4882
Tucker A, Sharpe P (2004) The cutting-edge of mammalian development; how the embryo makes teeth. Nat Rev Genet 5:499–508
Tziafas D, Papadimitriou S (1998) Role of exogenous TGF-beta in induction of reparative dentinogenesis in vivo. Eur J Oral Sci 106:192–196
Tziafas D, Alvanou A, Papadimitriou S, Gasic J, Komnenou A (1998) Effects of recombinant basic fibroblast growth factor, insulin-like growth factor-II and transforming growth factor-beta 1 on dog dental pulp cells in vivo. Arch Oral Biol 43:431–444
Volponi AA, Pang Y, Sharpe PT (2010) Stem cell-based biological tooth repair and regeneration. Trends Cell Biol 20:715–722
Wei F, Song T, Ding G et al (2013) Functional tooth restoration by allogeneic mesenchymal stem cell-based bio-root regeneration in swine. Stem Cells Dev 22:1752–1762
Wise GE, King GJ (2008) Mechanisms of tooth eruption and orthodontic tooth movement. J Dent Res 87:414–434
Wise GE, Frazier-Bowers S, D'Souza RN (2002) Cellular, molecular, and genetic determinants of tooth eruption. Crit Rev Oral Biol Med 13:323–334
Yamamoto H, Kim EJ, Cho SW, Jung HS (2003) Analysis of tooth formation by reaggregated dental mesenchyme from mouse embryo. J Electron Microsc 52:559–566
Yan X, Qin H, Qu C, Tuan RS, Shi S, Huang GT (2010) iPS cells reprogrammed from human mesenchymal-like stem/progenitor cells of dental tissue origin. Stem Cells Dev 19(4):469–480
Yang Z, Jin F, Zhang X, Ma D, Han C, Huo N, Wang Y, Zhang Y, Lin Z, Jin Y (2009) Tissue engineering of cementum/periodontal-ligament complex using a novel three-dimensional pellet cultivation system for human periodontal ligament stem cells. Tissue Eng Part C Methods 15:571–581
Yang J, Zhang Y, Wan C et al (2015) Autophagy in SDF-1α-mediated DPSC migration and pulp regeneration. Biomaterials 44:11–23
Yelick PC, Vacanti JP (2006) Bioengineered teeth from tooth bud cells. Dent Clin N Am 50:191–203
Yen AH, Sharpe PT (2008) Stem cells and tooth tissue engineering. Cell Tissue Res 331:359–372
Young CS, Terada S, Vacanti JP, Honda M, Bartlett JD, Yelick PC (2002) Tissue engineering of complex tooth structures on biodegradable polymer scaffolds. J Dent Res 81:695–700
Zhao S, Sloan AJ, Murray PE, Lumley PJ, Smith AJ (2000) Ultrastructural localisation of TGF-beta exposure in dentine by chemical treatment. Histochem J 32:489–494
Zheng Y, Du X, Wang W, Boucher M, Parimoo S, Stenn K (2005) Organogenesis from dissociated cells: generation of mature cycling hair follicles from skin-derived cells. J Invest Dermatol 124:867–876

Chapter 6
Functional Hair Follicle Regeneration

Koh-ei Toyoshima and Takashi Tsuji

Abstract The hair organ plays biologically important roles in thermoregulation, physical insulation, waterproofing, tactile sensation, protection, camouflage, and social communication. Only the hair follicle, which is induced by reciprocal epithelial and mesenchymal interactions in the skin field, has various organ-inductive potential stem cells and their niches and undergoes repeated organogenesis after birth through interactions between epithelial stem cells in the bulge region and dermal papillae. Hair loss disorders such as androgenetic alopecia are psychologically distressing and have negative effects on the quality of life. Therefore, the development of hair follicle regeneration therapy is expected to be a next-generation organ replacement regenerative therapy. Previously, many studies have reported technological approaches to reproduce de novo folliculogenesis. Recently, we successfully developed fully functional hair regeneration via intracutaneous transplantation of a bioengineered hair follicle germ, which was reconstituted from adult hair follicle-derived stem cells using our developed organ germ method. Here, we review hair follicle regeneration studies and discuss the potential of our functional hair follicle regeneration for the realization of future hair follicle organ regenerative therapy.

Keywords Hair follicle • Organ replacement regenerative therapy • Bioengineered hair follicle • Organ germ method

---

K.-e. Toyoshima
Department Regenerative Medicine, Plastic and Reconstructive Surgery, Kitasato University School of Medicine, Sagamihara, Kanagawa 252-0374, Japan

RIKEN Center for Developmental Biology, Kobe, Hyogo 650-0047, Japan

Organ Technologies Inc., Tokyo 101-0048, Japan
e-mail: k-toyoshima@cdb.riken.jp

T. Tsuji (✉)
Laboratory for Organ Regeneration, RIKEN Center for Developmental Biology, 2-2-3, Minatojima-mimamimschi, Chuo-ku, Kobe, Hyogo 650-0047, Japan
e-mail: t-tsuji@cdb.riken.jp

T. Tsuji (ed.), *Organ Regeneration Based on Developmental Biology*,
DOI 10.1007/978-981-10-3768-9_6

6.1 Introduction

The hair organ plays biologically important roles in thermoregulation, physical insulation from UV radiation, waterproofing, tactile sensation, protection against noxious stimuli, camouflage, and social communication (Hardy 1992; Schneider et al. 2009). The initiation of hair follicle organogenesis is induced in the hair germ through reciprocal epithelial-mesenchymal interactions and, ultimately, development into the hair follicle (Hardy 1992; Stenn and Paus 2001) (Fig. 6.1a). The hair follicle is composed of an upper permanent region and a lower variable region that

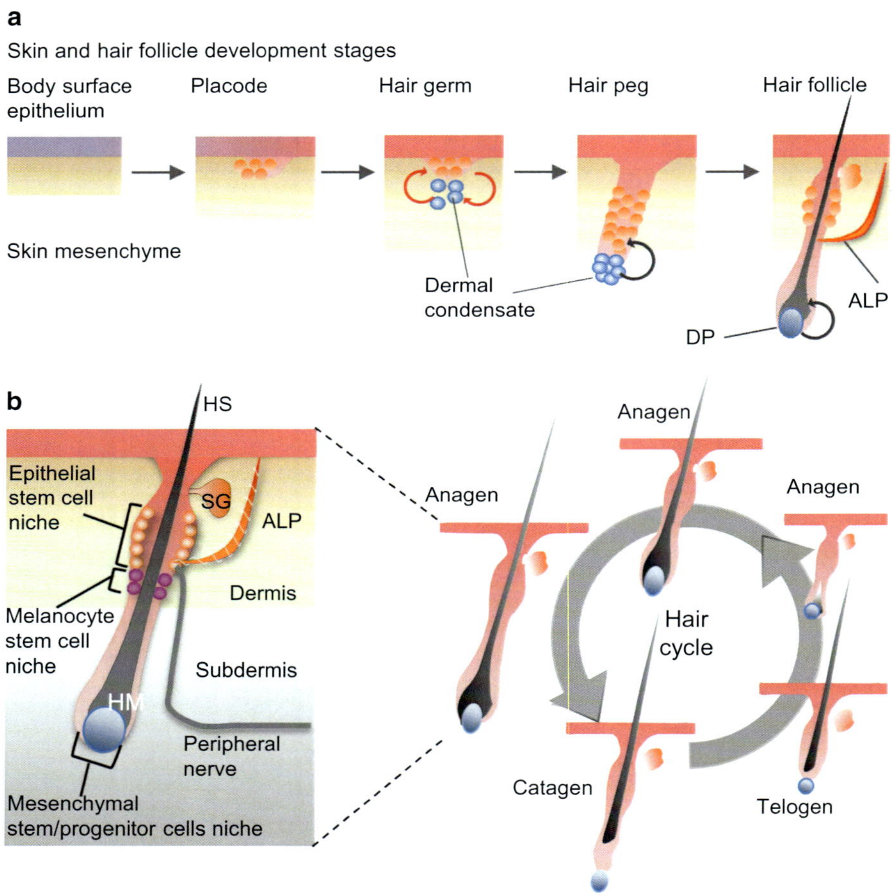

Fig. 6.1 Organogenesis of the hair follicle during embryogenesis and the hair cycle of the adult hair follicle. (**a**) Hair follicle development during embryonic skin development. *Red round arrows* and *black round arrows* represent epithelial-mesenchymal interactions and hair signaling, respectively. (**b**) Various stem cells and their niches in the adult hair follicle and hair cycle. *DP* dermal papilla, *HM* hair matrix, *SG* sebaceous gland, *ALP* arrector pili muscle, *HS* hair shaft

includes the hair bulb, which is in fact a hair shaft factory (Schneider et al. 2009; Stenn and Paus 2001). After morphogenesis, various stem cell types are maintained in particular regions: follicle epithelial cells in the follicle stem cell niche of the bulge region (Oshima et al. 2001; Hsu et al. 2014), multipotent mesenchymal precursors among DP and the dermal sheath cup (Jahoda and Reynolds 2001; Jahoda et al. 2003; Rahmani et al. 2015), neural crest-derived melanocyte progenitors in the bulge and sub-bulge region (Nishimura et al. 2002, 2005), and follicle epithelial stem cells in the bulge region connected to the arrector pili muscle (Hardy 1992; Schneider et al. 2009; Fujiwara et al. 2011) (Fig. 6.1b). These follicle stem cells contribute to repetitive regeneration of the variable region, in which hair follicle morphogenesis and the hair growth phase in postnatal hair follicles have many similar features, and both processes are characterized by the activation of cell differentiation programs that lead to the construction of hair shaft-producing epithelial hair bulbs (Schneider et al. 2009; Stenn and Paus 2001).

Hair loss disorders, such as alopecia areata and androgenetic alopecia, are psychologically distressing and have negative effects on the quality of life in both sexes (Mounsey and Reed 2009). Current pharmacological treatments do not achieve ideal control of hair loss, even in common conditions such as androgenetic alopecia or alopecia areata (Mounsey and Reed 2009). Early studies have shown that tissues or cultured cells derived from rodent dermal papilla could induce a new growth phase leading to a functionally and structurally proper hair bulb when experimentally implanted into the permanent region of the hair follicle or afollicular ear skin (Jahoda et al. 1984). Furthermore, Raynolds et al. (1999) raise the possibility that de novo hair follicle induction can be achieved through allogenic transplantation of hair follicle-inducible mesenchymal tissues into the skin. It has been hoped that the development of bioengineering technologies will enable future regenerative therapy for hair loss (Chuong et al. 2007).

To achieve the realization of hair follicle regenerative therapy for hair loss, it is most important to develop a highly efficient hair follicle germ regeneration technique that can provide a structurally proper and fully functional hair follicle and apply this technique for clinical usage (Chuong et al. 2007; Toyoshima et al. 2012). Over 30 years, many studies have described technologies to reconstitute the variable lower region of the hair follicle (Toyoshima et al. 2012), to reproduce de novo folliculogenesis via replacement with hair follicle-inductive dermal cells (Stenn et al. 2007) (Fig. 6.2a), and to direct the self-assembly of skin-derived epithelial and mesenchymal cells (Weinberg et al. 1993; Zheng et al. 2005; Stenn et al. 2007; Lichti et al. 2008) (Fig. 6.2b). These technologies provide the basic potential to reconstruct a regenerated hair follicle for hair regeneration in vivo (Chuong et al. 2007). However, several technical issues, including precise cell processing methods in a three-dimensional stem cell culture, eruption of a bioengineered hair by intracutaneous transplantation in vivo, restoration of the correct connection to surrounding tissues such as arrector pili muscle and nerve fibers, and an enduring hair cycle over a lifetime, must be resolved (Chuong et al. 2007; Lee and Chuong 2009; Toyoshima et al. 2012). Recently, a novel bioengineering method, designated the organ germ method, was developed to generate a bioengineered organ germ by multicellular

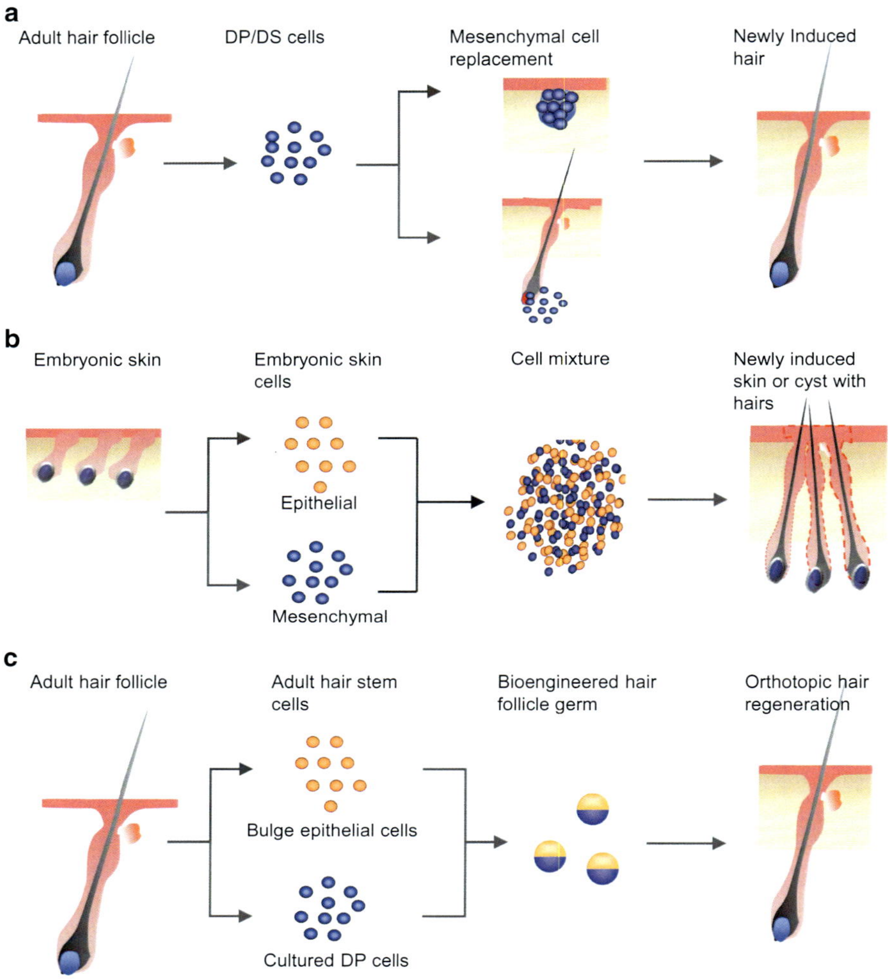

Fig. 6.2 The method used to reproduce the bioengineered hair follicle germ relied on reproducing the epithelial-mesenchymal interactions of the natural hair follicle. (**a**) Replacement of hair follicle-inductive mesenchymal cell in afollicular skin or hair follicle via the implantation of hair follicle-derived mesenchymal cells. (**b**) De novo skin regeneration method via the engraftment of cell aggregates. (**c**) Orthotopic hair follicle regeneration by intracutaneous transplantation of the bioengineered hair follicle germ reconstituted using the organ germ method

organization of epithelial and mesenchymal cells in a 3D stem cell culture (Nakao et al. 2007; Ikeda et al. 2009) (Fig. 6.2c). The bioengineered organ germs reconstituted using the organ germ method could regenerate the fully functional bioengineered hair follicles in vivo. Hair regeneration has been successfully demonstrated by intracutaneous transplantation of a bioengineered pelage and vibrissa hair follicle germ or ectopically regenerated mature hair follicle, which was regenerated with embryonic skin-derived cells and adult vibrissa follicle-derived stem cells, respectively (Toyoshima et al. 2012; Asakawa et al. 2012).

In this chapter, we describe the features of the hair follicle as a main target of organ regenerative therapy for the medical cure of alopecia and the changing in technical developments. We also provide several results using bioengineered hair follicle germs for hair regeneration and a perspective for the realization of organ regenerative therapy of the hair follicle.

6.2 Organogenesis of the Hair Follicle During Embryogenesis

Hair follicle organogenesis, in principle, takes place in the developing skin (Hardy 1992; Stenn and Paus 2001; Fuchs 2007; Schneider et al. 2009) (Fig. 6.1). When the embryonic epidermis and dermis are both intact, complex signaling between them is initiated that leads to fate changes in both tissue layers, ultimately resulting in epidermal patterning and the development of a hair follicle (Paus 2007; Benitah and Frye 2012). To better understand this complex biological multicellular system and the technical and clinical subjects, both hair follicle organogenesis and regeneration processes should be reconsidered in light of the notion of self-organization, which is found universally in nature and defined as the spontaneous formation of ordered patterns and structures from a population of elements with no or minimal patterns (Sasai 2013a). Sasai has advocated that the principle process of biological self-organization can be defined and classified into the self-assembly stage, self-patterning stage, and self-driven morphogenesis (Sasai 2013a).

In the context of the notion of biological self-organization, primitive skin develops as a hair follicle organogenetic field at the preceding outset of hair follicle organogenesis (Fuchs 2007; Lim and Nusse 2013). Post-gastrulation, the embryonic surface cells emerge as an ectodermal cell monolayer that is regionally specified and differentiates into the neurogenic and body surface ectoderm, which will ultimately develop into the central nervous and integumentary system (i.e., skin organ system) (Fuchs 2007; Lim and Nusse 2013). Mesoderm-derived mesenchymal cells underlie the body surface ectoderm (Fuchs 2007). Hair morphogenesis initiates the specialization of the epidermis at regularly spaced intervals, leading to the formation of epidermal placodes (Lim and Nusse 2013) (Fig. 6.1a). A large number of studies examining hair follicle patterning have focused on a reaction-diffusion model to understand the underlying mechanisms. Canonical Wnt/β-catenin signaling in the ectodermal epithelium and its inhibitors behave as a reaction-diffusion model resulting in an activation status or a lack of β-catenin expression, thus determining the primary hair follicle spacing and distribution (Lim and Nusse 2013). Subsequently, almost immediately after placode formation, the mesenchymal cells underlying the placode give rise to a cluster and condense into a dermal condensate (Hardy 1992; Fuchs 2007; Schneider et al. 2009). The signals from the dermal condensates induce epidermal placode cells to rapidly divide in a downward direction and invade the dermis, enwrapping the dermal condensate, which becomes the dermal papilla (Carlsen 1974; Hardy 1992; Schneider et al. 2009; Wang et al. 2012).

Wnt signaling is essential for inducing the dermal condensate (Atit et al. 2006; Ohtola et al. 2008; Lim and Nusse 2013). The expression profile of the Wnt family in the homogeneous primitive ectodermal epithelium has been evaluated by in situ hybridization, indicating that Wnt gene expression patterns in the embryonic skin epithelium during the placode promotion stage are classified as uniformly expressing Wnt3, 4, and 10a, localizing Wnt10b in the placode (Andl et al. 2002). Wnt10b signaling regulates the differentiation from neural crest-derived mesenchyme into DP and the surrounding fat tissue (Fu and Hsu 2013; Ouji et al. 2013; Ross et al. 2000). Wnt10b is a critical epithelial signal for fate decisions in the hair follicle mesenchymal cell population (Lim and Nusse 2013). The dermal condensate induces the placodal epithelium to proliferate, develop into a multilayer, and invade the dermis, resulting in the formation of a hair germ (Hardy 1992; Stenn and Paus 2001; Schneider et al. 2009; Wang et al. 2012).

The canonical process of hair follicle development after the hair germ stage is thought to be a self-driven morphogenesis process because isolated embryonic skin fragments, including the hair follicle germ, can progressively develop into a hair follicle in an in vitro organ culture without intact hormonal and neural circuits (Hardy 1949; Schneider et al. 2009). Elongation of the hair germ and augmentation of cell proliferation at the proximal end of the hair peg toward the surrounding dermal condensate, which will become dermal papilla, are evoked by the reciprocal dermal signaling from the hair peg to the bulbous hair peg stage (Hardy 1992; Stenn and Paus 2001; Schneider et al. 2009). In the latest stages of folliculogenesis, the structurally mature hair follicle has two distinct components, a variable region, which comprises the portion growing downward that contains the hair bulb referred to as the actual hair factory, and a constant region (Schneider et al. 2009). The constant region of the hair follicle connects to the peripheral nerve and arrector pili muscle and functionally matures in the postnatal skin (Paus et al. 1999) (Fig. 6.1b). Human hair follicle development is first recognizable in the fetal craniofacial region of the skin by 9–10 weeks gestational age (GA), and it globally progresses throughout organogenesis of hair follicle by 20 weeks GA (Carlsen 1974; Akiyama et al. 2000). Interestingly, similar to rodent vibrissae, the upper lip portion of the human fetus precedes that of any other region of the body in terms of hair follicle organization (Akiyama et al. 2000).

6.3 Hair Cycle of the Adult Hair Follicle

After morphogenesis in the postnatal stage, the hair follicle undergoes the hair cycle of the regression (catagen), resting (telogen), and reproduction of hair follicle organogenesis and hair growth (anagen) phases throughout the lifetime of the organism (Fig. 6.1b) (1–3). A cyclical recapitulation of organ regeneration similar to that of the hair follicle is observed only in the case of deer antlers (Randall et al. 1993; Thornton et al. 1996; Polykandriotis et al. 2010). The germinative epithelial cells in the proximal region of the hair matrix are described as transit-amplifying cells because they only survive through anagen (Reynolds and Jahoda 1991 and 1996). At the end of anagen, cell division in the germinative hair matrix supplies the hair differentiation of the hair shaft and the slowly developing IRS, and consequently the hair follicle enters catagen (Stenn and Paus 2001; Millar 2002). During catagen, the entire variable region is reduced to an epithelial column to form a secondary hair germ, mainly by apoptosis of matrix, IRS and ORS epithelial cells, while bulge hair follicle stem cells evade apoptosis (Millar 2002). Following catagen, the hair follicles enter a telogen phase, which is acknowledged as a relatively quiescent and resting phase (Stenn and Paus 2001; Millar 2002; Schneider et al. 2009).

The initiation of a new anagen phase in the adult hair cycle and embryonic folliculogenesis are appreciated for sharing many features that are considered, at least in part, to be different phenomena (Stenn and Paus 2001; Botchkarevn and Kishimoto 2003; Schneider et al. 2009). The processes are distinguished as follows: one is the organogenesis of a whole hair organ, and the other is the partial reproduction of a variable region of adult hair follicle (Botchkarevn and Kishimoto 2003; Stenn et al. 2007). Both processes are characterized by the induction and control of cell proliferation and differentiation to construct the hair-producing system and maintain hair growth through epithelial-mesenchymal interactions, which are mediated by similar signaling molecules (Botchkarevn and Kishimoto 2003). During early anagen, the secondary hair germ forms a hair bulb, which grows progressively downward and away from the bulge region (Botchkarevn and Kishimoto 2003). Although the ORS cells of the bulge region readopt a quiescent, undifferentiated epithelial cell phenotype in the lower portion of the hair matrix, the proliferative potential is maintained through active Wnt signaling from dermal papilla cells and consequent stabilization of β-catenin in matrix epithelial cells throughout anagen (Stenn and Paus 2001; Botchkarevn and Kishimoto 2003; Schneider et al. 2009). In the upper part of the hair matrix, proliferation is arrested and terminal differentiation is initiated (Botchkarevn and Kishimoto 2003). Many pioneering investigations have revealed multiple molecules that regulate hair lineages, such as Wnt/b-catenin/Lef1, Notch1/Jagged/Delta, BMPs/BMPR1a, Msx and Foxn1 for hair shaft development and IGF1 and HGF production, and VEGF for the maintenance of anagen (Botchkarevn and Kishimoto 2003; Schneider et al. 2009; Lim and Nusse 2013).

The anagen-catagen transition, which is associated with a remarkable decrease in the exchange of epithelial-mesenchymal interaction signals in anagen, determines the hair length of the corresponding hair types (Randall et al. 1993; Hibberts et al. 1998). Several studies suggest that this transition timing is controlled by an internal biological clock because ectopically transplanted hair follicles can retain their own inherent hair cycling (39). During catagen, the lower part of the hair follicle rapidly regresses via apoptosis of epithelial cells as a consequence of the increased expression of apoptosis stimulation factors, such as FGF5, TGFβ, TNFα, and neurotrophin (Stenn and Paus 2001; Botchkarevn and Kishimoto 2003; Wang et al. 2012). Upon completion of catagen, the hair follicle enters telogen, which can be separated into two stages: refractory to the induction of anagen and a competent stage wherein secondary hair germ cells become highly sensitive to anagen-inducing factors (Plikus et al. 2008). The BMP2 and 4 signals derived from dermal papilla cells inhibit activation of the Wnt/β-catenin pathway in the epithelial cells of the secondary hair germ (Plikus et al. 2008). Conclusive studies have demonstrated that extra-follicular BMP signals, which are provided by undifferentiated progenitors of adipocytes in the subcutaneous adipose, also override the entry of competent telogen into anagen and are negatively regulated by epithelial adipogenetic factors (Plikus et al. 2008; Festa et al. 2011; Rivera-Gonzalez et al. 2014; Hsu et al. 2014). Additionally, several observations have suggested that the skin peripheral nerve connected to the hair follicle controls enlargement of the hair follicle during anagen (Peters et al. 2006; Hsu et al. 2014).

6.4 Epithelial and Mesenchymal Stem Cells in the Adult Hair Follicle

To achieve iteration of the hair cycle, maintenance of the various adult stem cells and their niches in the mature hair follicle is considered to be essential throughout the lifetime of the organism (Cotsarelis et al. 1990; Greco et al. 2009). The most essential hair follicle stem cells (i.e., the minimal cell elements for the reconstruction of the entire hair follicle) are epithelial and mesenchymal stem cells (Botchkarevn and Kishimoto 2003). When the lower half of the hair follicle is surgically ablated, the remaining upper portion cannot regenerate a hair bulb (Jahoda et al. 1984; Horne et al. 1986). Surgical destruction of the area between a sebaceous gland and the arrector pili muscle connecting the portion of the human hair follicle, which is referred to as the bulge region, causes irreparable damage to the hair follicle and is applied clinically for hair removal operations (Unger et al. 2010). Moreover, reconstruction experiments of the bulge epithelium and dermal papilla or dermal sheath cup demonstrated that a structurally and functionally complete hair bulb can be regenerated by reproducing the epithelial-mesenchymal interactions, similarly to anagen of the natural hair follicle (Jahoda et al. 1984; Horne et al. 1986). These observations directly indicate that the hair follicle epithelial stem cells and mesenchymal stem cells are localized in stem cell niches in the bulge region and lower variable region, respectively (Lavker et al. 2003; Schneider et al. 2009).

The major biological characteristics of stem cells in vivo are quiescence, a high proliferative capacity, and multipotency to produce multiple lineage cells (Claudinot et al. 2005a, b). Slow cycling epithelial cells of the adult hair follicle are detected as label-retaining cells in the bulge region, and lineage analysis has shown that all follicular epithelial cells originate from bulge ORS epithelial cells (Oshima et al. 2001; Claudinot et al. 2005a, b; Ohyama et al. 2006; Waters et al. 2007). Several researchers have reported that certain markers, such as cytokeratin 15 (CK15), CD34, and

CD49f, are preferentially expressed in bulge ORS cells of the murine pelage hair follicle (Claudinot et al. 2005a, b; Wang et al. 2012). The combination of antibodies recognizing the cell surface molecules of bulge cells and transgenic mice that express fluorescent reporter proteins under the control of these markers provides a means for the characterization and isolation of living bulge cells (Ohyama et al. 2006). Living bulge cells that are enzymatically dissociated into single cells have enabled stem cell biologists to quantitatively evaluate in vitro colony-forming ability and to analyze the clonogenic cell dynamics of hair epithelial stem cells (Ohyama et al. 2006). Subsequent studies have indicated that hair follicle epithelial stem cells can differentiate into not only hair follicles but also the interfollicular epidermis (Oshima et al. 2001). The multiple subpopulations of epithelial stem cells and progenitor cells, which are distinguishable by their expression of specific molecules such as Sox9, LGR5, LGR6, LRIG1, and Gli1, have been shown to be structurally and functionally organized as a cellular hierarchical system in the murine pelage bulge region (Morris et al. 2004; Vidal et al. 2005; Jensen et al. 2009).

In the variable region of the hair follicle, two histologically distinct mesenchymal cell populations have been shown: the dermal papilla and the dermal sheath (Hardy 1992; Jahoda and Reynolds 2001; Rahmani et al. 2015). The dermal papilla is surrounded by the hair matrix epithelium and continues histologically into the dermal sheath, which consists of loose connective tissue overlying the outermost hair follicle (Hardy 1992; Rahmani et al. 2015). Isolated dermal papilla cells can be propagated in an in vitro culture system (Jahoda et al. 1984; Osada et al. 2007). Previous studies have indicated that freshly isolated and cultured adult dermal papilla cells provide unique and critical signals for the induction of anagen (Stenn and Paus 2001; Botchkarevn and Kishimoto 2003). Mesenchymal stem/progenitor cell populations, which can differentiate into dermal sheath, adipose, cartilage, and dermal fibroblasts, are found among the dermal papilla (Jahoda et al. 2003). However, analysis of the cell dynamics of the dermal papilla and dermal sheath cells suggested that dermal sheath cells proliferate and migrate to provide the dermal papilla cells (Rahmani et al. 2015). In contrast to follicular epithelial stem cells, hair mesenchymal cells have been shown to express versican, alkaline phosphatase (ALP), and α-smooth muscle actin (αSMA) as relative specific markers for the dermal papilla and dermal sheath cells (Kishimoto et al. 1999). However, an in vivo fate mapping study of adult hair follicle dermal sheath cells using αSMA-fluorescent transgenic mice indicated that dermal sheath cup cells possess critical stem cell properties (Rahmani et al. 2015). ALP activity is strongly detected in the dermal papilla in early anagen, although differences in ALP activity between these components increase during all other hair cycle phases (Iida et al. 2007). It has been suggested that the hair follicle-inducing ability of dermal papilla cells is closely related to ALP expression (Iida et al. 2007). These findings indicate that the cellular heterogeneity of dermal papilla cells, which consists of various stages of cell commitment, may be altered and governed according to the hair cycle phases (Iida et al. 2007).

6.5 The Human Hair Follicle as a Clinical Target of Organ Regenerative Therapy

Human hairs are commonly classified into a terminal hair and a vellus hair, according to the diameter, length, and internal structure (Hojiro 1972; Driskell et al. 2009). The temporal parameters of the hair cycle are dependent on the types of hairs and on the means of the maximum length and density of the hair shaft (Toyoshima et al. 2012). On the typical normal human scalp, there are 100,000 terminal hairs, approximately 90% of which are normally in anagen and 10% in telogen (Ramos and Miot 2015). During the late fetal stage to birth, fetal hair follicles are converted from a lanugo type to mainly terminal hair and vellus hair types, or they involute into vestigial organs in the labial gingiva (Akiyama et al. 2000). In the context of hair type conversion, terminal hairs grow at secondary sexual sites before puberty (Randall et al. 1993). It is thought that a single hair follicle retains the inherent hair type and the biological program of hair type conversion determined by fate determination during hair follicle organogenesis, although the underlying mechanism remains unexplained (Randall 1992; Thornton et al. 1996).

Throughout the human lifetime, although up to 50% of the population is affected by hair loss, which is a common symptom of various hair-related disorders, traumatic injury, psychiatric disorder, and age-related physiological changes, 95% of clinical hair loss in men and women is caused by androgenetic alopecia (Mounsey and Reed 2009). It is possible that hair loss can occur anywhere in human skin, although most patients commonly complain of terminal hair trouble, especially scalp hair loss (Mounsey and Reed 2009). Based on the focal pathosis of alopecia, it is clinically valuable to divide hair loss into an acute or a chronic scarring alopecia and a non-scarring alopecia (Mounsey and Reed 2009; Qi 2015; Knopp 2015). Acute scarring alopecia is characterized by extrinsic causes, such as active inflammation, irreversible destruction of the hair follicle, and fibrous tissue replacement of hair follicles, and it requires a swift effective medical cure during early disease stages. In contrast, non-scarring alopecia is characterized by significant alterations in follicular size, abnormalities in the hair cycle, and reversible anatomical changes in the follicles (Sperling 2001; Mounsey and Reed 2009). The stable region of the hair follicle is commonly irreversibly degenerated or damaged in scarring alopecia (Sperling 2001).

Androgenetic alopecia in humans is pathologically characterized by a gradual reduction in size and shortening of the growth phase of focal hair follicles, which are intrinsically fate determined and genetically programmed to alter vellus hair types mediated by the increasing expression of 5α-reductase and the conversion of testosterone into dihydrotestosterone in the dermal papilla cells of susceptible hair follicles (Randall et al. 1993). Thus, the surgical approach to androgenetic alopecia, which is based on follicular unit transplantation (FUT) and these minimal surrounding tissues dissected from the normal scalp to focal areas, can provide highly clinically effective outcomes for male and female patients (Unger et al. 2010). Drugs targeting 5α-reductase, which can inhibit a trigger of the intrinsic program of androgenetic alopecia, are also useful for the clinical treatment of male pattern baldness, but they have marginal effects in female patients (Mounsey and Reed 2009; Unger et al. 2010; Qi 2015; Knopp 2015). These observations suggest that androgenetic alopecia arises from alterations of the dermal papilla cells and may reflect a primary aberration in either the hair follicle compartment or its surrounding tissues (Rivera-Gonzalez 2016).

Several lines of evidence from animal experiments indicate that the hair loss pattern is attributed to the differential embryological origins of the frontal and occipital scalp mesenchyme (Qi and Garza 2014). The neural crest-derived mesenchymal cells exhibit extensive diversity in molecular markers that depend on the embryonic development stages (Qi and Garza 2014). Recently, neural crest-derived dermal stem cells, which have been defined as multipotent stem cells that are enriched in the dermal sheath, can differentiate into the entire hair follicular mesenchyme and the subcutaneous adipose (Qi and Garza 2014). The undifferentiated adipogenetic progenitors in the subcutaneous tissue turn off the termination of hair follicle telogen phase, resulting in expansion of the telogen phase (Qi and Garza 2014). Conversely, during the transition from telogen to anagen, hair follicle epithelial cells induce progenitor differentiation into adipocytes, which can trigger the initiation of anagen (Qi and Garza 2014). These findings suggest that a capability of the interaction between the hair follicle and adipose tissue is a modulation of the duration of telogen, a synchronous enlargement of mature adipose tissue with hair cycling, and a differential reinforcement of dermal fibrosis, corresponding to the pathology of chronic scarring alopecia (Qi and Garza 2014).

6.6 Methodology for the Reconstruction of a Fully Functional Bioengineered Hair Follicle

The reproduction method for the bioengineered hair follicle germ relies on the ability to reproduce the epithelial-mesenchymal interactions that occur during embryonic organogenesis, and adult cyclic regeneration of the hair follicle is essential for fundamental hair follicle organ regeneration strategies (Chuong et al. 2007). To establish de novo regeneration of functional hair follicle, several animal models of hair formation are available and can be applied to dissect intact or partial hair follicle tissues, to culture follicular cells and tissues, and then to orthotopically or ectopically transplant them into an immunodeficient or syngeneic mouse (Chuong et al. 2007; Toyoshima et al. 2012). The orthotopic transplantation model can be used to visually estimate hair growth and continuous hair cycling of the bioengineered hair follicle in living animals (Chuong et al. 2007; Toyoshima et al. 2012). Oliver and colleagues demonstrated that a secondary hair follicle germ could be artificially reconstructed with the constant region of the host hair follicle and isolated dermal papilla cells and newly induce hair bulb and hair shaft growth (Oliver 1966; Jahoda et al. 1984; Horne et al. 1986) (Fig. 6.2a). The replacement of hair-inductive mesenchymal cells in an amputated hair follicle has served as an analogy for the reproduction of adult anagen (Jahoda et al. 1984). Many researchers have reported that the replacement of dermal cells in afollicular skin using either fresh or in vitro-expanded follicular mesenchymal cells, which are collected from adult hair bulbs in an anagen hair follicle, can newly induce hair follicle formation (Oliver 1966; Jahoda et al. 1984). In the context of mesenchymal cell replacement, Reynolds and colleagues provided the most dramatic finings for human hair follicle regeneration, which was achieved by allogeneic transplantation of connective tissue sheath tissues dissected from the male scalp hair follicle into the proximal epithelium of female forearm skin (Reynolds et al. 1999) (Fig. 6.2a). Although newly induced small hair follicles, including male donor-derived dermal papilla cells and thin hair growth, were found at the implanted site, the origin and hair-forming ability of the intrafollicular epithelial cells was not clear (Reynolds et al. 1999).

Based on the three categories of biological self-organization processes as described above, the three-dimensional cellular arrangement of the hair follicle organ germ is essential for inducing a functional hair follicle and achieved using a self-assembly process or artificial manipulation of follicular epithelial cells and mesenchymal cells (Sasai 2013a). The de novo hair regeneration methods in vitro and in vivo that rely on the reproduction of epithelial-mesenchymal interactions during the embryonic folliculogenesis and growth phases of the adult hair cycle have been attempted in several previous studies using isolated epithelial and mesenchymal cells (Weinberg et al. 1993; Zheng et al. 2005; Lichti et al. 2008). It has also been reported that dissociated follicular epithelial and mesenchymal cells are aggregated to form the hair follicle germ through a self-assembly process and that this model can utilize dermal and epidermal candidate cells (Sasai 2013b) (Fig. 6.2b). The most widely used de novo hair growth model is the silicone chamber assay (Weinberg et al. 1993; Lichti et al. 2008). In this model, the slurry of the target cell mixture is injected inside grafting chambers that have been surgically implanted into an excision area on the back of the mouse. As a positive control, 10^7 epidermal and dermal cells each derived from newborn mice induce the de novo development of hairy skin in each chamber at approximately 3 weeks after grafting. This model is useful for evaluating the follicular formation ability, and either cell component can be replaced with candidate cells such as multipotent keratinocytes isolated from adult mice (Mannik et al. 2010), cultured dermal papilla cells (Lichti et al. 2008), and human keratinocytes (Ehama et al. 2007).

The silicone chamber assay is limited by the requirement for the surgical implantation of a special apparatus (Lee and Chuong 2009). Surgical difficulties and hair regeneration efficiency were improved by the development of the patch assay, in which a similar high-density mixture of $>10^5$ dissociated epithelial and mesenchymal cells was injected at certain ratios into the hypodermis of host mice (Zheng et al. 2005) (Fig. 6.2b). The multipotency of CK15-positive epidermal cells isolated from transgenic mouse skin, including pelage hair follicles, was assessed by combining them with neonatal dermal cells in the patch assay (Zheng et al. 2005; Stenn et al. 2007). Furthermore, the hair-inducing abilities of follicular cells from various mammalian species and transgenic animals and under different culture conditions can be evaluated using this assay model (Stenn et al. 2007; Ehama et al. 2007; Mannik et al. 2010). Using the rotation-culture method, aggregated cells from the neonatal rodent skin epithelium and mesenchyme can induce immature hair follicles in vitro and after transplantation into a receptive animal and also readily regenerate mature hair follicles using the silicon chamber and patch assays (Ihara et al. 1991; Takeda et al. 1998). Limitations of the patch assay and in vitro rotary culture assay are that the de novo hairs grow into the inside of cystic structures that are not associated with the natural skin, making it difficult to observe the hair growth without surgical dissection; the control of the orientation and direction of the bioengineered hair follicles, which are generally random; and the lack of restoration of the extra-follicular environment (Lee and Chuong 2009; Toyoshima et al. 2012). Although these bioengineering techniques provide an easy practicable model for examining hair follicle formation, the phenomena cannot be directly translated clinically because the self-assembly of epithelial and mesenchymal cells into the bioengineered hair follicle germ is inefficient and requires a large number of cells and an uncontrollable hair density and arrangement to achieve the clinical application of organ regenerative therapy for the hair follicle (Toyoshima et al. 2012).

6.7 Functional Hair Follicle Regeneration

As discussed above, before attaining clinically effective hair regeneration in human skin, technological breakthroughs for the major obstacles must be achieved by unravelling practical high-throughput methods to reproduce the bioengineered hair follicle germ and developing an in vivo evaluation model that is not only widely appropriate for various functions of the bioengineered hair follicle but also readily available for clinical application (Lee and Chuong 2009). To achieve useful hair follicle regeneration for a clinical cure, the bioengineered hair organ should have a structurally correct architecture and result in a fully grown hair shaft with a histologically proper arrangement and connection in the skin (Chuong et al. 2007; Lee and Chuong 2009); have a repeated enduring hair cycle, which is considered essential for the regeneration of the various stem cells and their niches; and exhibit the cooperative functions of the hair organ in the natural skin environment (Toyoshima et al. 2012). The organ germ method has facilitated these goals, particularly the practical high-throughput assay (Chuong et al. 2007; Lee and Chuong 2009; Toyoshima et al. 2012). The bioengineered hair follicle germ is thought to spontaneously drive organ development via an intrinsic folliculogenesis process that is reproduced from hair-inductive epithelial and mesenchymal stem cells by cell manipulation techniques to generate a highly efficient and useful hair follicle regeneration model for clinical applications (Toyoshima et al. 2012). Nakao et al. have previously developed an organ germ method and demonstrated that a bioengineered hair follicle germ, which was reconstituted with dissociated epithelial and mesenchymal cells (approximately 50,000–100,000 cells each) derived from the embryonic hair germ of murine vibrissae, can ectopically regenerate a bioengineered hair follicle producing a hair shaft under the kidney capsule (Nakao et al. 2007; Asakawa et al. 2012) (Fig. 6.2c).

6.7.1 *Hair Follicle Regeneration by Intracutaneous Transplantation of the Bioengineered Hair Follicle Germ*

It is essential to evaluate whether the bioengineered hair follicle germ can develop into a structurally proper bioengineered hair follicle in the adult intracutaneous environment to provide fully functional hair organ regeneration, including hair eruption growth, hair cycles, and connections with surrounding tissues, based on the functional regeneration of stem cell niches (Toyoshima et al. 2012). Toyoshima et al. successfully demonstrated that the bioengineered hair follicle germ was reconstituted with considerably small number of adult murine vibrissa hair follicle-derived epithelial stem cells (10^4 cells) and cultured dermal papilla cells (3×10^3 cells) using an organ germ method and regenerated the hair follicle in the murine skin environment through the novel development of the intracutaneous transplantation method, which was based on the FUT surgical operation procedure for ready availability for clinical application (Toyoshima et al. 2012; Tezuka et al. 2016) (Fig. 6.2c and 6.3).

Surgical specialists in FUT therapy generally dissect normal hair follicles that maintain their structural integrity in an epithelium of the hair opening and hair shaft and then translocationally engraft them into the bold region by maintaining the external direction so that it protrudes from the skin surface to achieve a high engraftment efficiency and no significant adverse events, such as serious infectious disease and postoperative cyst formation (Toyoshima et al. 2012; Tezuka et al. 2016). To prevent posttransplant cyst formation by the rapid closing between the host skin epithelium and the bioengineered hair follicle germ, as a guide for the direction of the infundibulum, an inter-epithelial tissue-connecting nylon thread is highly effective for the accurate arrangement of the newly formed hair pore of the bioengineered hair follicle to the surrounding host skin epithelium (Toyoshima et al. 2012; Tezuka et al. 2016) (Fig. 6.3). In cases of bioengineered vibrissa follicles, the bioengineered hair shafts erupt at a resulting frequency of 74% at 21 days after engraftment and are almost unpigmented (Fig. 6.3). These results raise the possibility not only of hair organ regeneration in the adult skin environment but also reestablishment of the connections between the bioengineered hair follicle and the recipient skin, similar to the results obtained using FUT.

It is preferable for a clinically applicable hair regeneration therapy to utilize autologous FUT therapy, which is the most popular and effective surgical cure for various types of hair loss. The density, area, and direction of grafting of follicular units can be practically controlled in ranges to achieve therapeutic goals (Toyoshima et al. 2012). In the preparation of bioengineered follicle germs, it has been indicated

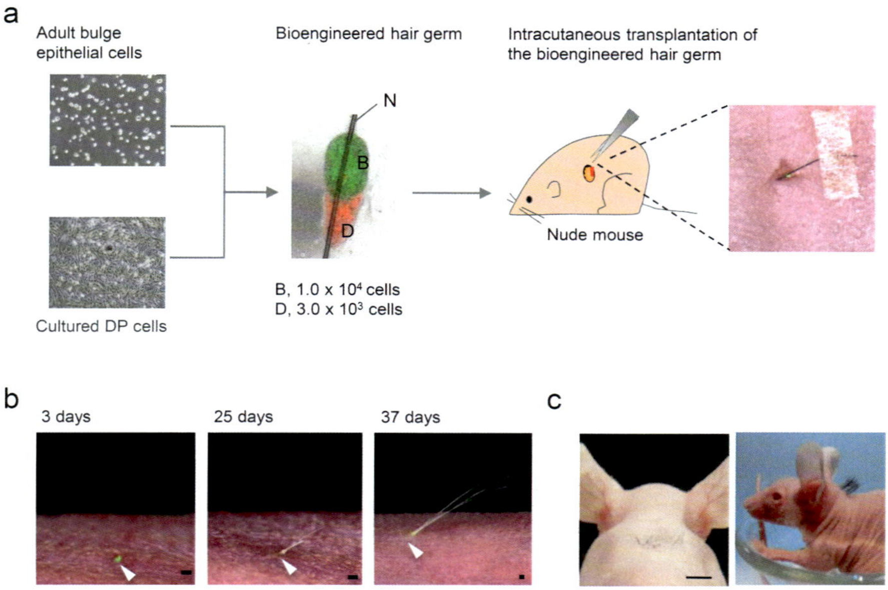

Fig. 6.3 Orthotopic hair follicle regeneration through intracutaneous transplantation of the bioengineered hair follicle germ. (**a**) Schematic representation of a preparation of the bioengineered hair follicle germ reconstructed with adult vibrissa and intracutaneous transplantation after insertion of a nylon thread as a guide for the direction of the infundibulum via insertion into the bioengineered germ. (**b**) Macro-morphological observations of the eruption and growth of bioengineered vibrissa hair shafts at day 3 after transplantation; eruption and growth of the hair shaft at days 25 and 37. (**c**) Controllable bioengineered hair density through high-density intracutaneous transplantation of the bioengineered follicle germs. A total of 28 independent bioengineered pelage hair follicle germs were transplanted into the neck skin of nude mice, and they displayed high-density hair growth at 21 days after transplantation (*left*) and a few weeks after hair eruption (*right*). Scale bar, 1 cm

that the number of bioengineered hair follicles correlates with the cell number in the bioengineered hair follicle germ and the contact area between the epithelial and mesenchymal cells. These results suggest that the number of bioengineered vibrissa follicles depends on the quantity of follicle-forming cells in bulge-derived epithelial and primary-cultured DP cells, and they show that this procedure can provide a precise transplantation technology for future hair regeneration therapy. To achieve hair follicle regeneration at hair densities of 120 hairs/cm^2 in normal scalp or 60–100 hairs/cm^2 using FUT treatment, the 28 bioengineered hair germs were successfully transplanted into an approximately 1 cm^2 occipital skin region and erupted 21 days after grafting at a high density of 124.0 ± 17.3 hair shaft cm^2 (Fig. 6.3). These results indicate that, similar to FUT therapy, the transplantation of bioengineered hair follicle germs can be applied as a treatment for androgenic alopecia (Toyoshima et al. 2012; Tezuka et al. 2016).

6.7.2 Reproduction of the Hair Follicle Architecture and Hair Cycle

Toyoshima et al. also analyzed the hair cycles of the bioengineered pelage and vibrissa shafts that had erupted from bioengineered follicles in immunodeficient murine skin over 80 days. The bioengineered pelage and vibrissa follicles repeated the hair cycle at least 3 times during the 80-day period, and no significant differences in the hair cycle periods were found between the natural and bioengineered follicles. These results indicated that the bioengineered hair follicle could undergo proper hair cycles according to the cell types of origin. It has been suggested that the bioengineered pelage and vibrissa follicles can reproduce these hair cycles, which are maintained by stem cells and provide a stem cell niche. The distinct hair follicle types, which are classified into awl/auchene, guard, zigzag, and vibrissa hair shafts in murine skin based on properties such as length, thickness, kinks, and hardness, are thought to be specified by dermal papilla cell types through communication between DP cells and overlying epithelial cells. The bioengineered pelage follicle germs are found to produce all types of pelage hairs, as confirmed by the hair type-specific structural properties observed by light microscopy, in accordance with the follicle fate determined during embryonic development (Toyoshima et al. 2012; Asakawa et al. 2012). The bioengineered vibrissa follicle germ regenerated a vibrissa-type hair shaft with the appropriate structural properties at the light microscopic level (Toyoshima et al. 2012). Ultrastructural observation of the bioengineered hair shafts revealed the correct reproduction of the morphological and histological structures, including the hair medulla, cortex, and cuticle.

The bioengineered hair follicle formed the correct structures, comprising an infundibulum and sebaceous gland in the proximal region as well as a hair matrix, hair shaft, IRS, ORS, and dermal papilla (Fig. 6.4). The bioengineered vibrissa follicle germs can regenerate not only the variable region but also the infundibulum and sebaceous gland in the permanent region. By contrast, the vibrissa follicle-derived cells were not distribute among the surrounding cutaneous tissues (Fig. 6.4). Each natural and bioengineered vibrissa follicle contained 500~1000 DP cells. Thus, these findings provide new insights regarding the regulation of hair properties and strongly suggest that these characteristics can be properly restored by cell processing for organ regeneration and by transplantation of the bioengineered hair follicle germ.

It is well known that hair follicle organ-inductive epithelial and mesenchymal stem cells provide a source of differentiated hair follicle cells that enable hair cycling to occur over the lifetime of a mammal. It is also essential to rearrange these various stem cells and their niches in the bioengineered follicle to reproduce enduring hair cycles. The bioengineered follicles that were reproduced using an organ germ method and intracutaneous transplantation could reconstruct various niches, such as CD34- and CD49f-positive epithelial stem cells and SOX2-positive vibrissa

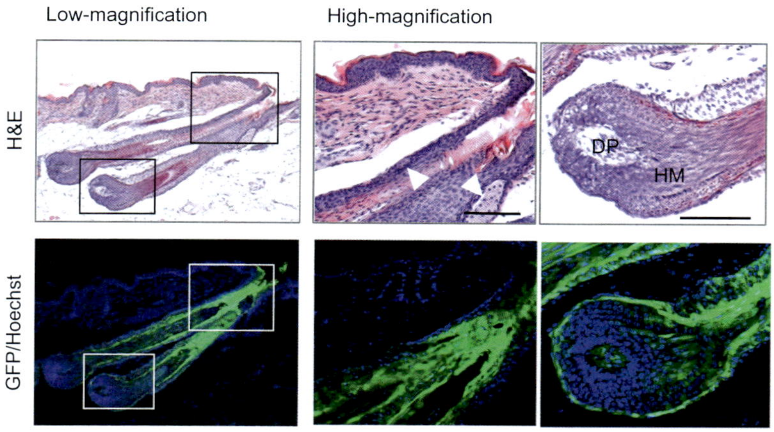

Fig. 6.4 Structural features of the bioengineered hair follicle. The GFP-labeled bioengineered vibrissa follicles were stained with H&E (*upper panels*) and Hoechst (*lower panels*). The *boxed areas* in the low-magnification H&E and fluorescent panel are shown at a higher magnification in the *center* and *right panels*. The *arrowhead* indicates a sebaceous gland. Scale bars, 100 μm

mesenchymal stem cells and progenitors. To evaluate the functional restoration of the hair follicle stem cells and their niches in the bioengineered hair follicle, the bioengineered hair cycles were analyzed for 80 days. The bioengineered pelage and vibrissa follicles repeated the hair cycle at least 3 times during the 80-day period, and there were no significant differences in the hair cycle periods between the natural and bioengineered follicles. These results indicated that the bioengineered hair follicle could undergo proper hair cycles according to the cell types of origin. The bioengineered pelage and vibrissa follicles have been suggested to reproduce these hair cycles, which are maintained by stem cells and provide a stem cell niche.

Hair shaft pigmentation is provided by melanocyte in the hair matrix, and melanocyte stem cells are maintained in the sub-bulge region of vibrissa hair follicles during anagen (Nishimura et al. 2002, 2005) (Fig. 6.5). Melanocyte progenitor cells, which are widely distributed between the ORS of the variable region and the proximal region of the hair matrix, can proliferate to provide melanogenic cells (Nishimura et al. 2002, 2005). The unpigmented bioengineered vibrissa hair follicle lacks melanocyte stem/progenitor cells, resulting in white hair shafts (Fig. 6.3c and 6.5). The addition of the proximal end of hair matrix cells, which contain melanocyte progenitor or stem cells, resulted in pigmented bioengineered vibrissae (Fig. 6.5). These results indicated that various follicle stem cells and their niches were successfully rearranged in the bioengineered hair follicles using appropriate epithelial and mesenchymal cell populations.

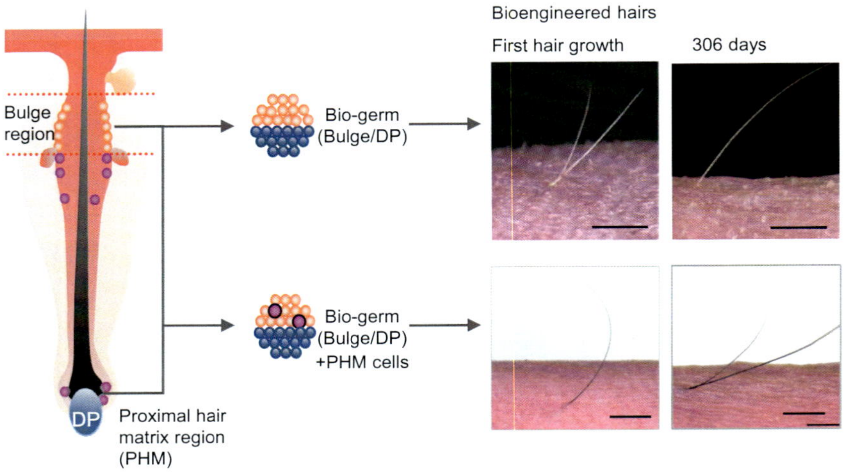

Fig. 6.5 Hair pigmentation of the bioengineered vibrissae by cell combinations with bioengineered vibrissa follicle cell components. The bioengineered vibrissa follicle germ, which was reconstituted between the bulge epithelial stem cells and the primary cultured DP cells (*upper*), was combined with the proximal region of the hair matrix (+PHM; *lower*). Unpigmented and pigmented bioengineered hairs at the first growth phase and at 306 days after transplantation are shown in *right* photographs, respectively. Bars, 1 mm

6.7.3 Reproduction of the Cooperative Function of the Bioengineered Hair Follicle

The peripheral nervous system has essential roles in organ function and the perception of noxious stimuli, such as pain and mechanical stress (Grant et al. 2009; Peters et al. 2006; Jahoda and Christiano 2011). Restoration of the nervous system is thus a critical issue that must be addressed by organ replacement regenerative therapy (Toyoshima et al. 2012; Asakawa et al. 2012; Tezuka et al. 2016). In the hair follicle, the follicles, pelage, and vibrissae achieve piloerection using the surrounding arrector pili muscle through activation of the sympathetic nerves and also function as a sensory organ (Peters et al. 2001, 2006; Sato et al. 2012). Toyoshima et al. also demonstrated that the nerve fibers and muscles can connect autonomously to the pelage and vibrissa follicle and that the bioengineered follicles exhibit piloerection. These results suggest that the transplantation of a bioengineered hair follicle germ can restore natural hair function and reestablish the cooperation between the follicle and the surrounding recipient muscles and nerves (Toyoshima et al. 2012; Asakawa et al. 2012; Tezuka et al. 2016) (Fig. 6.6). Thus, transplantation of the bioengineered hair follicle germ may be applicable for the future surgical treatment of alopecia (Toyoshima et al. 2012).

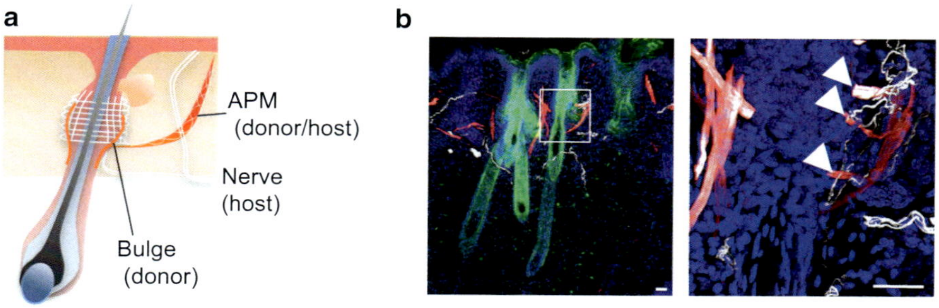

Fig. 6.6 Analyses of connections to surrounding tissues and the piloerection capability of bioengineered hair follicles. (**a**) Schematic representation of the connections of the bioengineered hair follicles with surrounding tissues in adult skin. (**b**) The connections of the follicles of a EGFP-labeled bioengineered pelage to arrector pili muscles (*red signal*) and nerve fibers (*white signal*) were analyzed by immunohistochemical staining. The boxed areas in the *left panel* are shown at a higher magnification on the *right*. The *arrowheads* indicate the muscle and nerve fibers connected to the pelage follicles. Scale bars, 100 μm at low magnification and 50 μm at high magnification

6.7.4 Human Bioengineered Hair Follicle

One of the ultimate goals of research investigating hair follicle regeneration is the clinical translation for the treatment of hair loss and various hair-associated disorders. A number of researchers have reported the hair follicle-forming potential of dissected tissues and cultured cells derived from human biopsies; however, no one has provided conclusive evidence for the regeneration of a fully functional human hair follicle. The human bioengineered hair follicle germ, which consists of dissociated bulge region-derived epithelial cells and scalp hair follicle-derived intact DPs in an androgenetic alopecia patient, could regenerate the human hair follicle and support the growth of a pigmented hair shaft in the intracutaneous transplantation area of nude mice. This bioengineered hair follicle was composed of correct structures, consisting minimally of an infundibulum and sebaceous gland, hair shaft, inner root sheath, ORS, hair matrix, and DP in the hair bulb structure, which were confirmed to be of human origin (Toyoshima et al. 2012). This result indicates that the bioengineered hair follicle germ method is applicable to human-derived cells and contributes to future developments in hair follicle regenerative medicine (Toyoshima et al. 2012).

6.8 Future Perspectives for the Hair Follicle

To achieve hair follicle regeneration using stem cells in the adult hair follicle, defining the regeneration of the various stem cells and their niches is considered essential. Based on the preceding studies, we can utilize the extensive knowledge of various types of follicular stem cells in the hair follicle (Chuong et al. 2007).

The bioengineered hair follicle germ can develop into a fully functional hair follicle through the rearrangement of various stem cells and their niches (Toyoshima et al. 2012). The dissected bioengineered hair follicles can also restore fully physiological hair functions in adult skin via ectopic transplantation of the bioengineered hair follicle germs (Toyoshima et al. 2012). Future studies of in vitro culture systems that can reproduce bioengineered hair follicles from bioengineered hair follicle germs permit the promotion of therapeutic systems such as FUT in the clinic.

To successfully provide clinically useful and effective hair follicle regenerative therapy, several issues persist, such as the optimization of human hair follicle-derived stem cell sources and in vitro expansion for clinical applications (Ohyama 2007). There are significant variations in follicular stem cell marker expression among species, hair types, skin surface regions, and individuals, although the clinical definition of useful stem cell markers is maintained (Ohyama 2007). There is no way to satisfactorily in vitro propagate human follicular stem cells using clinically available culture systems. The organ germ method and functional hair regeneration assay assist the development of a high-throughput system, which would facilitate techniques leading to the clinical application of hair follicle regenerative therapy.

Recent biological and technical innovations have achieved substantial progress in the development of a novel therapeutic model for hair follicle regenerative therapy for alopecia and organ replacement regenerative therapy. Further studies focused on the optimization of human hair follicle-derived stem cell sources for clinical applications of stem cell niches will contribute to the development of hair regenerative therapy as a prominent class of organ replacement regenerative therapy in the future.

Acknowledgments This work was partially supported by Organ Technologies, Inc.

Conflict of Interest K. Toyoshima and T. Tsuji have no competing interests.

References

Akiyama M, Smith LT, Shimizu HJ (2000) Changing patterns of localization of putative stem cells in developing human hair follicles. Invest Dermatol 114:321–327

Andl T, Reddy ST, Gaddapara T, Millar SE (2002) WNT Signals Are Required for the Initiation of Hair Follicle Development. Dev Cell 2:643–653

Asakawa K, Toyoshima K-E, Ishibashi N, Tobe H, Iwadate A, Kanayama T, Hasegawa T, Nakao K, Toki H, Noguchi S, Ogawa M, Sato A, Tsuji T (2012) Hair organ regeneration via the bioengineered hair follicular unit transplantation. Sci Rep 2:424. doi:10.1038/srep00424

Atit R, Sgaier SK, Mohamed OA et al (2006) β-Catenin activation is necessary and sufficient to specify the dorsal dermal fate in the mouse. Dev Biol 296:164–176

Benitah SA, Frye M (2012) Stem cells in ectodermal development. J Mol Med 90:783–790

Botchkarev VA, Kishimoto J (2003) Molecular Control of Epithelial-Mesenchymal Interactions During Hair Follicle Cycling. J Investig Dermatol Symp Proc 8:46–55

Carlsen RA (1974) Human fetal hair follicles: the mesenchymal component. J Invest Dermatol 63:206–211
Chuong CM, Cotsarelis G, Stenn K (2007) Defining hair follicles in the age of stem cell bioengineering. J Invest Dermatol 127:2098–2100
Claudinot S, Nicolas M, Oshima H, Rochat A, Barrandon Y (2005a) Long-term renewal of hair follicles from clonogenic multipotent stem cells. Proc Natl Acad Sci USA 102:14677–14682
Claudinot S, Nicolas M, Oshima H, Rochat A, Barrandon Y (2005b) Long-term renewal of hair follicles from clonogenic multipotent stem cells. Proc Natl Acad Sci U S A 102:14677–14682
Cotsarelis G, Sun TT, Lavker RM (1990) Label-retaining cells reside in the bulge area of pilosebaceous unit: implications for follicular stem cells, hair cycle, and skin carcinogenesis. Cell 61:1329–1337
Driskell RR, Giangreco A, Jensen KB, Mulder KW, Watt FM (2009) Sox2- positive dermal papilla cells specify hair follicle type in mammalian epidermis. Development 136:2815–2823
Ehama R, Ishimatsu-Tsuji Y, Iriyama S, Ideta R, Soma T, Yano K, Kawasaki C, Suzuki S, Shirakata Y, Hashimoto K, Kishimoto J (2007) Hair follicle regeneration using grafted rodent and human cells. J Invest Dermatol 127:2106–2115
Festa E, Fretz J, Berry R, Schmidt B, Rodeheffer M, Horowitz M, Horsley V (2011) Adipocyte lineage cells contribute to the skin stem cell niche to drive hair cycling. Cell 146:761–771
Fu J, Hsu W (2013) Epidermal Wnt controls hair follicle induction by orchestrating dynamic signaling crosstalk between the epidermis and dermis. J Invest Dermatol 133:890–898
Fuchs E (2007) Scratching the surface of skin development. Nature 445:834–842
Fujiwara H et al (2011) The basement membrane of hair follicle stem cells is a muscle cell niche. Cell 144:577–589
Grant RA, Mitchinson B, Fox CW, Prescott TJ (2009) Active touch sensing in the rat: anticipatory and regulatory control of whisker movements during surface exploration. J Neurophysiol 101:862–874
Greco V et al (2009) A two-step mechanism for stem cell activation during hair regeneration. Cell Stem Cell 4:155–169
Hardy MH (1949) The development of mouse hair in vitro with some observations on pigmentation. J Anat 83(4):364–384
Hardy MH (1992) The secret life of the hair follicle. Trends Genet 8:55–61
Hibberts NA, Howell AE, Randall VA (1998) Balding hair follicle dermal papilla cells contain higher levels of androgen receptors than those from non-balding scalp. Journal of Endocrinology 6:559–565
Hojiro O (1972) Fine structure of the mouse hair follicle. J Electron Microsc 21:127–138
Horne KA, Jahoda CA, Oliver RF (1986) Whisker growth induced by implantation of cultured vibrissa dermal papilla cells in the adult rat. J Embryol Exp Morphol 97:111–124
Hsu YC, Li L, Fuchs E (2014) Emerging interactions between skin stem cells and their niches. Nat Med 20(8):847–856
Ihara S, Watanabe M, Nagao E, Shioya N (1991) Formation of hair follicles from a single-cell suspension of embryonic rat skin by a two-step procedure in vitro. Cell Tissue Res 266:65–73
Iida M, Ihara S, Matsuzaki T (2007) Hair cycle-dependent changes of alkaline phosphatase activity in the mesenchyme and epithelium in mouse vibrissal follicles. Dev Growth Differ 49:185–195
Ikeda E et al (2009) Fully functional bioengineered tooth replacement as an organ replacement therapy. Proc Natl Acad Sci USA 106:13475–13480
Jahoda CA, Christiano AM (2011) Niche Crosstalk: Intercellular Signals at the Hair Follicle. Cell 146:678–681
Jahoda CA, Horne KA, Oliver RF (1984) Induction of hair growth by implantation of cultured dermal papilla cells. Nature 311:560–562
Jahoda CA, Reynolds AJ (2001) Hair follicle dermal sheath cells: unsung participants in wound healing. Lancet 358:1445–1448

Jahoda CA, Whitehouse J, Reynolds AJ, Hole N (2003) Hair follicle dermal cells differentiate into adipogenic and osteogenic lineages. Exp Dermatol 12:849–859
Waters JM, Richardson GD, Jahoda CAB (2007) Hair follicle stem cells. Seminars in Cell & Developmental Biology 18:245–254
Jensen KB et al (2009) Lrig1 expression de nes a distinct multipotent stem cell population in mammalian epidermis. Cell Stem Cell 4:427–439
Kishimoto J et al (1999) Selective activation of the versican promoter by epithelial-mesenchymal interactions during hair follicle development. Proc. Natl Acad Sci USA 96:7336–7341
Lavker RM, Sun TT, Oshima H, Barrandon Y, Akiyama M, Ferraris C, Chevalier G, Favier B, Jahoda CA, Dhouailly D, Panteleyev AA, Christiano AM (2003) Hair follicle stem cells. J Investig DermatolSymp Proc 8:28–38
Lee LF, Chuong CM (2009) Building complex tissues: high-throughput screening for molecules required in hair engineering. J Invest Dermatol 129:815–817
Lichti U, Anders J, Yuspa SH (2008) Isolation and short-term culture of primary keratinocytes, hair follicle populations and dermal cells from newborn mice and keratinocytes from adult mice for in vitro analysis and for graying to immunodeficient mice. Nat Protoc 3:799–810
Lim X, Nusse R (2013) Wnt Signaling in Skin Development, Homeostasis, and Disease. Cold Spring Harb Perspect Biol 5:a08029
Mannik J, Alzayady K, Ghazizadeh S (2010) Regeneration of multilineage skin epithelia by differentiated keratinocytes. J Invest Dermatol 130:388–397
Millar SE (2002) Molecular mechanisms regulating hair follicle development. J Invest Dermatol 116:216–8225
Morris RJ et al (2004) Capturing and profiling adult hair follicle stem cells. Nat Biotechnol 22:411–417
Mounsey AL, Reed SW (2009) Diagnosing and treating hair loss. Am Fam Physician 80:356–362
Nakao K et al (2007) The development of a bioengineered organ germ method. Nat Methods 4:227–230
Nishimura EK et al (2002) Dominant role of the niche in melanocyte stem-cell fate determination. Nature 416:854–860
Nishimura EK, Granter SR, Fisher DE (2005) Mechanisms of hair graying: incomplete melanocyte stem cell maintenance in the niche. Science 4:720–724
Ohtola J, Myers J, Akhtar-Zaidi B et al (2008) β-catenin has sequential roles in the survival and specification of ventral dermis. Development 135:2321–2329
Ohyama M (2007) Hair follicle bulge: a fascinating reservoir of epithelial stem cells. Journal of Dermatological Science 46:81–89
Ohyama M, Terunuma A, Tock CL, Radonovich MF, Pise-Masison CA, Hopping SB, Brady JN, Udey MC, Vogel JC (2006) Characterization and isolation of stem cell–enriched human hair follicle bulge cells. J Clin Invest 116(1):249–60
Oliver RF (1966) Whisker growth after removal of the dermal papilla and lengths of follicle in the hooded rat. J Embryol Exp Morphol 15:331–347
Osada A, Iwabuchi T, Kishimoto J, Hamazaki TS, Okochi H (2007) Long-term culture of mouse vibrissal dermal papilla cells and de novo hair follicle induction. Tissue Eng 13:975–982
Oshima H, Rochat A, Kedzia C, Kobayashi K, Barrandon Y (2001) Morphogenesis and renewal of hair follicles from adult multipotent stem cells. Cell 104:233–245
Ouji Y, Nakamura-Uchiyama F, Yoshikawa M (2013) Canonical Wnts, specifically Wnt-10b, show ability to maintain dermal papilla cells. Biochem Biophys Res Commun 438:493–499
Paus R, Muller-Rover S, Veen C et al (1999) A comprehensive guide for the recognition and classification of distinct stages of hair follicle morphogenesis. J Invest Dermatol 113:523–532
Peters EM, Botchkarev VA, Botchkareva NV, Tobin DJ, Paus R (2001) Hair-cycle-associated remodeling of the peptidergic innervation of murine skin, and hair growth modulation by neuropeptides. J Invest Dermatol 116:236–245

Peters EMJ, Arck PC, Paus R (2006) Hair growth inhibition by psychoemotional stress: a mouse model for neural mechanisms in hair growth control. Exp Dermatol 15:1–13
Plikus MV, Mayer JA, de la Cruz D, Baker RE, Maini PK, Maxson R, Chuong CM (2008) Cyclic dermal BMP signalling regulates stem cell activation during hair regeneration. Nature 451:340–344
Polykandriotis E, Popescu LM, Horch RE (2010) Regenerative medicine: then and now an update of recent history into future possibilities. J Cell Mol Med 14:2350–2358
Qi J, Garza LA (2014) An overview of Alopecias. Cold Spring Harb Perspect Med 4:a013615
Rahmani W, Abbasi S, Hagner A et al (2015) Hair follicle dermal stem cells regenerate the dermal sheath, repopulate the dermal papilla, and modulate hair type. Dev Cell 31:543–558
Ramos PM, Miot HA (2015) Female pattern hair loss: a clinical and pathophysiological review. An Bras Dermatol 90:529–543
Randall VA, Thornton MJ, Messenger AG et al (1993) Brinklow BR: hormones and hair growth: variations in androgen receptor content of dermal papilla cells cultured from human and red deer (Cervus elaphus) hair follicles. J Invest Dermatol 101:114S–120S
Reynolds AJ, Jahoda CA (1991) Hair follicle stem cells? A distinct germinative epidermal cell population is activated in vitro by the presence of hair dermal papilla cells. Journal of Cell Science 99:373–385
Reynolds AJ, Jahoda CA (1996) Hair matrix germinative epidermal cells confer follicle-inducing capabilities on dermal sheath and high passage papilla cells. Development 122:3085–3094
Reynolds AJ, Lawrence C, Cserhalmi-Friedman PB, Christiano AM, Jahoda CA (1999) Trans-gender induction of hair follicles. Nature 4 402(6757):33–34
Rivera-Gonzalez G, Shook B, Horsley V (2014) Adipocytes in skin health and disease. Cold Spring Harb Perspect Med 4:a015271
Ross SE, Hemati N, Longo KA et al (2000) Inhibition of adipogenesis by Wnt signaling. Science 289:950–953
Sasai Y (2013a) Cytosystems dynamics in self-organization of tissue architecture. Nature 493:318–326
Sasai Y (2013b) Next-generation regenerative medicine: organogenesis from stem cells in 3D culture. Cell Stem Cell 12:520–530
Sato A et al (2012) Single follicular unit transplantation reconstructs arrector pili muscle- and nerve-connections and restores functional hair follicle piloerection in preparing. J Dermatol 39:1–6
Schneider MR, Schmidt-Ullrich R, Paus R (2009) The hair follicle as a dynamic miniorgan. Curr Biol 19(3):R132–R142
Sperling LC (2001) Scarring alopecia and the dermatopathologist. J Cutan Pathol 28:333–342
Stenn K et al (2007) Bioengineering the hair follicle. Organogenesis 3:6–13
Stenn KS, Paus R (2001) Controls of hair follicle cycling. Physiol Rev 81:449–494
Takeda A, Matsuhashi S, Shioya N, Ihara S (1998) Histodifferentiation of hair follicles in grafting of cell aggregates obtained by rotation culture of embryonic rat skin. Scand J Plast Reconstr Surg Hand Surg 32:359–364
Tezuka K, Toyoshima KE, Tsuji T (2016) Hair follicle regeneration by transplantation of a bioengineered hair follicle germ. Methods Mol Biol 1453:71–84
Thornton MJ, Kato S, Hibberts NA, Brinklow BR, Loudon AS, Randall VA (1996) Ability to culture dermal papilla cells from red deer (Cervus elaphus) hair follicles with differing hormonal responses in vivo offers a new model for studying the control of hair follicle biology. J Exp Zool 275:452–458
Toyoshima KE, Asakawa K, Ishibashi N et al (2012) Fully functional hair follicle regeneration through the rearrangement of stem cells and their niches. Nat Commun 3:784
Unger W, Shapiro R, Unger MA, Unger U (2010) Hair transplantation, 5th edn. Informa Healthcare UK, New York

Vidal VP, Chaboissier MC, Lützkendorf S, Cotsarelis G, Mill P, Hui CC, Ortonne N, Ortonne JP, Schedl A (2005) Sox9 is essential for outer root sheath differentiation and the formation of the hair stem cell compartment. Curr Biol 15:1340–1351
Wang X, Tredget EE, Wu Y (2012) Dynamic signals for hair follicle development and regeneration. Stem Cells Dev 21(1):7–18
Weinberg WC et al (1993) Reconstitution of hair follicle development in vivo: determination of follicle formation, hair growth, and hair quality by dermal cells. J Invest Dermatol 100:229–236
Zheng Y et al (2005) Organogenesis from dissociated cells: generation of mature cycling hair follicles from skin-derived cells. J Invest Dermatol 124:867–876

Chapter 7
Functional Salivary Gland Regeneration

Miho Ogawa and Takashi Tsuji

Abstract Oral health and homeostasis are maintained by the functional interactions of many organs, including the salivary glands, teeth, and tongue. Salivary gland dysfunction leads to dry mouth diseases, such as dental caries, bacterial infection, swallowing dysfunction, and reduced quality of life. The current clinical therapies for dry mouth are temporary, and they cannot repair salivary gland dysfunction. Salivary gland regenerative therapy with tissue repair and whole salivary gland replacement is a novel organ regenerative therapy. To achieve the recovery of the salivary gland function, adult tissue stem cells may be used as a cell source for salivary gland tissue repair therapies. To attain the entire salivary gland replacement therapy, which represents the next-generation regenerative therapy, we developed a novel cell manipulation method that can regenerate the ectodermal organ germ. The bioengineered salivary gland germs successfully engrafted grew in the transplantation site, generating the correct structure. The bioengineered salivary glands were able to secrete saliva into the oral cavity and improve dry mouth symptoms. In this chapter, we describe the recent progress and developmental methods for salivary gland regeneration therapy.

Keywords Salivary gland regeneration • Salivary gland replacement regenerative therapy • Saliva • Bioengineered salivary gland • Organ germ method • Transplantation

M. Ogawa
Organ Technologies Inc., Chiyoda-ku, Tokyo 101-0048, Japan

RIKEN Center for Developmental Biology, Kobe, Hyogo 650-0047, Japan
e-mail: miho_ogawa@organ-technol.co.jp

T. Tsuji (✉)
Laboratory for Organ Regeneration, RIKEN Center for Developmental Biology,
2-2-3, Minatojima-mimamimschi, Chuo-ku, Kobe, Hyogo 650-0047, Japan
e-mail: t-tsuji@cdb.riken.jp

T. Tsuji (ed.), *Organ Regeneration Based on Developmental Biology*,
DOI 10.1007/978-981-10-3768-9_7

7.1 Introduction

The salivary gland is an exocrine organ that synthesizes and secretes saliva. There are three major pairs of glands: the parotids (PG), submandibular glands (SMG), and sublingual glands (SLG) (Fig. 7.1a). Additionally, there are many minor salivary glands. The SMG and PG secrete serous saliva, which mainly contains amylase proteins. SLG secrete mucous saliva, which contains glycoproteins, such as mucin proteins (Edgar et al. 2004; Tucker and Miletich 2010; Avery 2002). Saliva plays various roles, including food digestion, taste, swallowing, protection from dryness, and oral health maintenance and homeostasis. Thus, salivary gland dysfunction induces various clinical problems in oral health. Salivary gland dysfunction is attributed to acinar cell atrophy, which is caused by radiation therapy for patients with head and neck cancer, aging, and autoimmune diseases (such as Sjögren's syndrome), and can be a side effect of various medications. Acinar cell atrophy results in xerostomia (dry mouth syndrome) (Saleh et al. 2015; Vissink et al. 2010; Ship et al. 2002; Fox 2004).

Xerostomia causes various clinical oral problems, such as serious dental decay, oral bacterial infection, taste disorder, voice disorder, and swallowing disorder, which

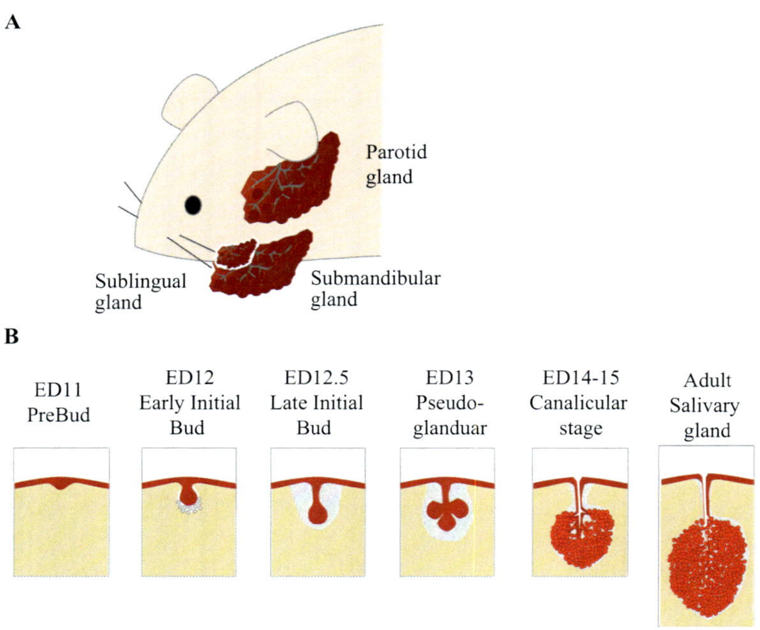

Fig. 7.1 Schematic representation of the salivary glands. (**a**) The three major salivary glands include the submandibular glands, sublingual glands, and parotid glands. (**b**) Development of submandibular glands which are produced from organ germ induced by the interaction of reciprocal epithelial and mesenchymal tissue (EDs 11–12). The epithelial tissue invaginates into the mesenchymal tissue and forms the epithelial stalk and terminal bulb (EDs 12–13), which form the duct and acinar cells (ED 14). The acinar cells mature and begin to synthesize and secrete secretory proteins (adult)

result in a general reduction in the quality of life (Atkinson et al. 2005). Current therapies for xerostomia include symptomatic treatments, the use of artificial saliva substitutes, and the administration of salivary gland stimulants and sialogogues, which enhance moisture retention in the oral cavity (Fox 2004; Nakamura et al. 2004). Parasympathetic stimulation drugs, such as pilocarpine and cevimeline, promote saliva secretion via the stimulation of residual acinar cells (Fox 2004). However, the effects of these therapies are temporary, and they cannot reproduce salivary gland dysfunction. Therefore, the development of alternative treatments that provide enduring effects or recover salivary gland function is expected (Kagami et al. 2008).

Recent regenerative therapy to restore organ function has been developed in many research fields, such as developmental biology, stem cell biology, and tissue engineering (Brockes and Kumar 2005; Langer and Vacanti 1999; Atala 2005; Madeira et al. 2015). Notably, transplantation therapy with tissue stem cells or cell sheet has been attempted for the repair of damaged tissues and organs in divergent diseases many years ago (Copelan 2006; Segers and Lee 2008). In salivary gland regeneration therapy, many research groups have reported various strategies including stem cell transplantation, gene modification, and tissue engineering to reproduce the damaged acinar tissue and restore saliva secretion (Yoo et al. 2014; Feng et al. 2009; O'Connell et al. 1999). Recently, ectodermal organ regeneration has been reported using bioengineered organ germ transplantation methods (see Chaps. 5, 6, and 8). In this chapter, we will discuss the recent findings and technologies for partial salivary gland tissue repair and whole salivary gland regeneration as a next-generation regenerative therapy that can recover function and prevent xerostomia.

7.2 Salivary Gland Development During Embryogenesis

The salivary gland is an exocrine organ arising from the salivary gland germ, which is generated by reciprocal interactions between the oral ectodermal epithelium and the neural crest-derived mesenchyme during embryogenesis (Tucker and Miletich 2010; Knosp et al. 2012; Patel et al. 2006; Knox and Hoffman 2008) (Fig. 7.1b). On embryonic day (ED) 11, the mesenchymal cells provide signals and induce oral epithelial thickening and invagination (Knosp et al .2012; Jaskoll and Melnick 2004). The expression of *Fgf10*, *Fgfr2b*, *Pitx1*, and *p63* is essential for initial salivary gland development. The epithelial bud grows and forms terminal bulbs and a stalk (initial bud), and then branching morphogenesis occurs, including cell proliferation, cleft formation, migration, and apoptosis, which proceed during EDs 12.5–14.5 (pseudoglandular) (Sakai 2009; Hsu and Yamada 2010; Harunaga et al. 2011). After ED 15.0, the salivary gland germ begins functional differentiation. The epithelial stalk differentiates into duct cells, including the excretory, striated, and intercalated ducts, and the terminal bulbs differentiate into acinar cells and mature (Denny and Denny 1999). There are three types of acinar cells: the serous, mucous, and seromucous cells. The seromucous cells secrete both serous and mucous saliva. In the excretory duct, adult tissue stem cells are maintained and supplied to the acinar and duct cells after the salivary gland tissue is injured (Man et al. 2011; Ihrler et al. 2002; Lombaert and Hoffman 2013).

7.3 Salivary Gland Tissue Repair Using Tissue-Derived Stem Cells

Adult tissue-derived stem cells have a general capacity for self-renewal and differentiation to repair injured tissue (Fig. 7.2). Salivary gland-derived stem cells have been isolated and characterized from the exocrine ducts of PG and SMG (Rotter et al. 2008; Lombaert et al. 2008; Jeong et al. 2013; Kawakami et al. 2013). The salivary gland-derived stem cells isolated from PG express mesenchymal stem cell (MSC) markers (CD44, CD49f, CD90, and CD105). These cells have the ability to differentiate into adipocytes, osteocytes, and chondrocytes and have the capacity to recover their function in radiation-damaged salivary glands (Rotter et al. 2008). The salivary gland-derived stem cells isolated from SMG express stem cell markers (c-kit and scal-1), and these cells can induce acinar and duct cells. Furthermore, these stem cells have the potential to differentiate into liver or pancreas tissues and to form salispheres in in vitro cultures. The salisphere can repair radiation-induced atrophied acinar cells by stem cell transplantation, restoring saliva flow (Lombaert et al. 2008). Additionally, bone marrow MSCs or extracts called "soups" have the potential to repair damaged tissues, increase the tissue regeneration ability of the surviving salivary gland tissue stem cells, and promote the regeneration of damaged acinar cells after radiation (Sumita et al. 2011; Tran et al. 2013). Tissue repair by adult tissue-derived stem cell transplantation therefore has the therapeutic potential to regenerate salivary glands.

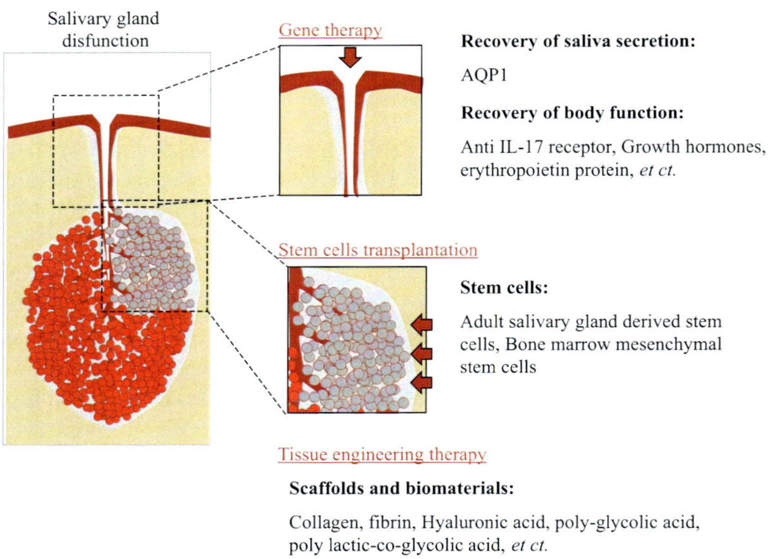

Fig. 7.2 Tissue repair by using stem cell transplantation and gene therapy. As a regenerative approach for salivary gland dysfunction or injury, stem cell transplantation, gene therapy, and tissue engineering therapy have been reported

7.4 Salivary Gland Tissue Repair Using Gene Therapy

Gene therapy is also a general technique for salivary gland regeneration (Rotter et al. 2008; Denny and Denny 1999; Horie et al. 1996; Sugito et al. 2004; Bücheler et al. 2002; Tran et al. 2005; Sun et al. 2006; Kishi et al. 2006) (Fig. 7.2). Because the salivary glands have ducts that open into the oral cavity and are close to the surface, it is very convenient to directly inject into the ductal epithelium with an adenovirus or adeno-associated virus that expresses a particular gene. Gene transfer of water channel aquaporin-1 (AQP1), which is important for transcellular water transport, can significantly restore saliva secretion in irradiated salivary glands (Delporte et al .1997). Additionally, interleukin-17 (IL-17) receptor antibodies, growth hormones, and erythropoietin protein have been altered using gene transfer in salivary gland tissue. Because the salivary glands have both exocrine and endocrine functions, substances are secreted into the bloodstream and can perform systemic functions (Kagami et al. 1996; He et al. 1998; Voutetakis et al. 2005). Gene therapy has progressed to phase I and is also expected to be a new strategy for the regeneration of salivary glands and other organs.

7.5 Salivary Gland Tissue Repair Using Tissue Engineering Therapy

The important aspects involved in the tissue engineering of salivary glands are cell-cell adhesion, cell-extracellular matrix (ECM) protein adhesion, and the biocompatible and biodegradable 3D scaffold used, which can maintain the adhesion (Aframian and Palmon 2008) (Fig. 7.2). ECM proteins, such as laminin and glycosaminoglycans, are important for salivary gland epithelial cell polarity and proliferation. A combination of scaffolds, including collagen gel, Matrigel, hyaluronic acid (HA), and polyglycolic acid, causes physical changes and promotes cell migration, polarity, and cell adhesion (Peters et al. 2014; Pradhan and Farach-Carson 2010; Lombaert et al. 2016). Therefore, it is important to select the suitable cells, ECM, and scaffold and to take into consideration the molecules involved in salivary gland development and branch morphogenesis.

7.6 Whole Salivary Gland Regeneration Using Organ Germ Methods

The ultimate goal of regenerative organs is the replacement of injured and dysfunctional organs with fully functional bioengineered organs. One concept for functional organ regeneration is to mimic the developmental process of organogenesis. Organ germ reconstruction using a cell aggregation method is a typical technique to

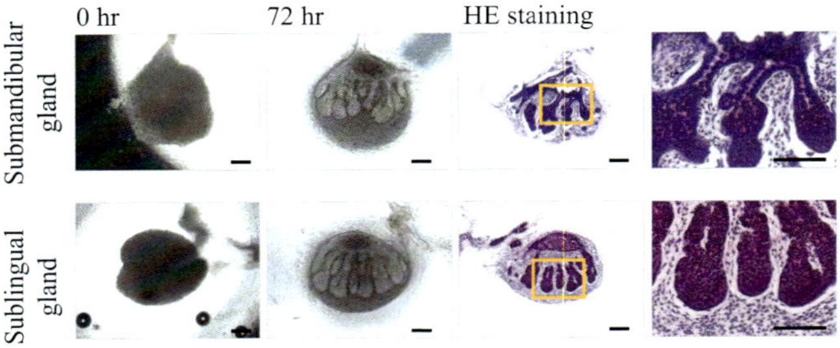

Fig. 7.3 Regeneration of salivary gland germ using organ germ methods. Phase-contrast images of the bioengineered submandibular and sublingual gland germ on 0 and 72 h of organ culture. The bioengineered salivary gland germ developed a blanching morphogenesis followed by stalk elongation and cleft formation within 72 h

reproduce self-organization and organ development. The salivary gland epithelial and mesenchymal cell mixture aggregates promote organ development and branching morphogenesis (Wei et al. 2007). Additionally, we demonstrated that a bioengineered organ germ, using organ germ methods, could regenerate ectodermal organs, such as teeth, hair follicles, and lacrimal glands (see Chaps. 5, 6, and 8). This method could be applied to achieve functional salivary gland regeneration (Ogawa et al. 2013). A bioengineered salivary gland germ was reconstructed using single epithelial and mesenchymal cells isolated from submandibular gland germs of ED 13.5 mice. The bioengineered submandibular gland germ successfully initiated salivary gland development, with branching morphogenesis followed by stalk and cleft formation in organ culture (Fig. 7.3). Bioengineered sublingual and parotid gland germs were also developed using organ germ methods, and they grew similarly to the submandibular gland.

7.6.1 Transplantation of the Bioengineered Salivary Gland Germs

Correct duct formation to connect the oral cavity and the bioengineered salivary gland germ is essential for correct acinar formation and saliva secretion. To achieve duct formation, bioengineered salivary gland germs were transplanted into the parotid gland ducts using an inter-epithelial tissue-connecting plastic method in a mouse model of salivary gland defects (Ogawa et al. 2013). Thirty days following transplantation, the bioengineered salivary gland and the host parotid duct were connected with nylon thread (Fig. 7.4a). The bioengineered submandibular gland regenerated serous acinar cells, and the sublingual gland regenerated mucous acinar cells. These bioengineered salivary glands had the correct organ structure, including localization of the water channel aquaporin-5 (AQP5), myoepithelial cells, and nerve fibers, which were similar to natural fibers (Fig. 7.4b).

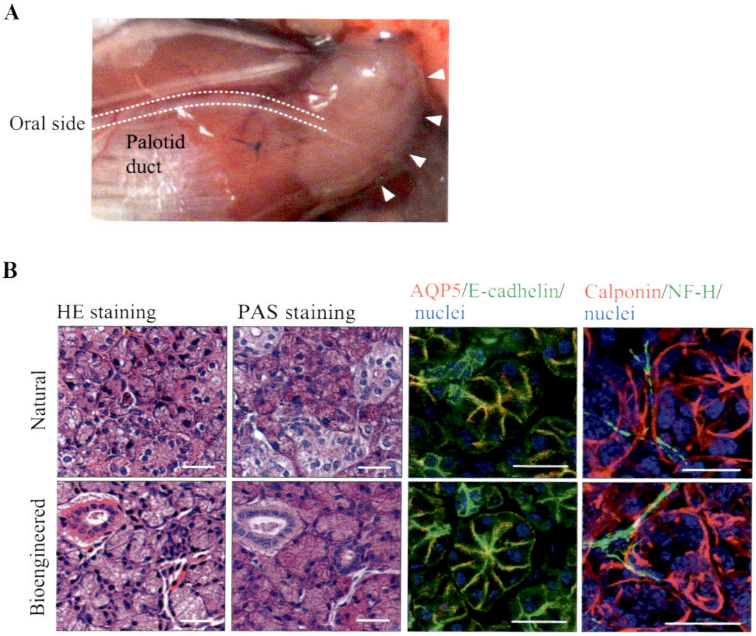

Fig. 7.4 Transplantation of bioengineered salivary gland germ. (**a**) Photographs of the bioengineered submandibular gland after 30-day transplantation. The bioengineered submandibular gland and parotid gland duct have established a correct connection. *Arrowhead* bioengineered submandibular gland. (**b**) Histological analysis of the natural (*upper columns*) and bioengineered (*lower columns*) submandibular gland. Images of HE staining (*left*) and periodic acid and Schiff (PAS) staining (*second from the left*). Immunohistochemical images of AQP5 (*red*) and E-cadherin (green; third from the *left*) and calponin (*red*) and NF-H (*green*; *right*) are shown. The bioengineered submandibular gland had a correct organ structure and regenerated serous acinar cells

7.6.2 *Saliva Secretion from Bioengineered Salivary Glands*

Restoration of the central nervous system is an important issue in organ regenerative therapy. Food, heat, and pain stimulation to the oral cavity induce saliva secretion via afferent and efferent nervous stimulation (Proctor and Carpenter 2014) (Fig. 7.5a). Additionally, saliva is essential for tasting; therefore, a saliva secretion was analyzed using gustatory tests, including sour (citrate), bitter (quinine hydrochloride), salty (NaCl), sweet (sucrose), and umami (glutamate) tastes (Matsuo 2000; Froehlich et al. 1987; Sasano et al. 2010; Ogawa et al. 2014). Citrate stimulation induced similar quantities of saliva secretion from the bioengineered salivary glands as natural salivary glands (Fig. 7.5b). All gustatory stimulation induced significant quantities of saliva secretion compared with non-stimulation, and the amount of saliva was dependent on the type of stimulus in the order of sour > bitter > umami > salty = sweet (Fig. 7.5c). These results indicate that saliva secretion from

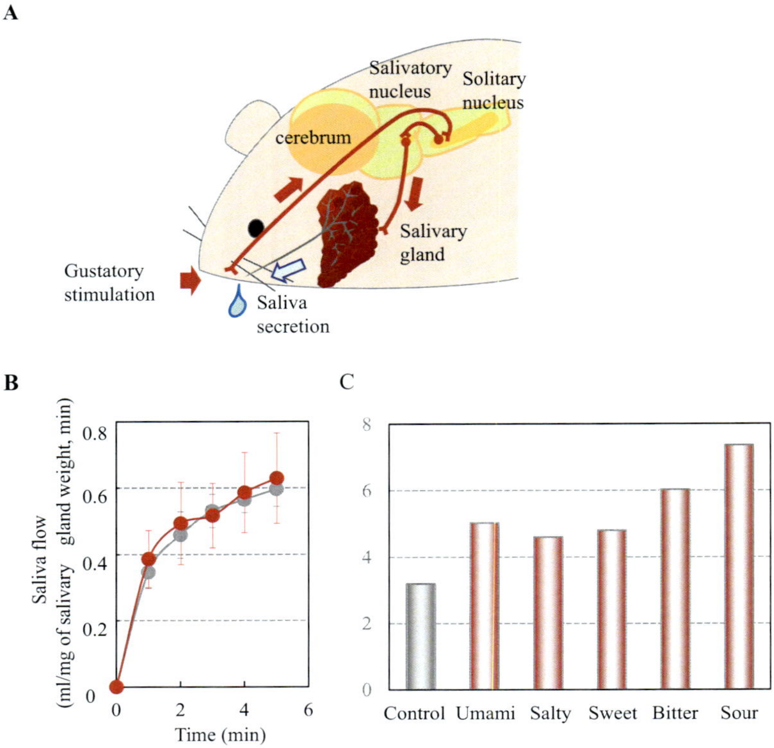

Fig. 7.5 Assessment of saliva secretion. (**a**) Schematic representation of saliva secretion via the central nervous system using gustatory stimulation. (**b**) The time course of the amount of saliva secretion associated with normal mice (*gray dots*) and bioengineered submandibular gland-engrafted mice (*red dots*) after the gustatory stimulation by citrate. The amount of secreted saliva was not significantly different. (**c**) The amount of saliva secretion after 5 min of stimulation was associated with water stimulation (*gray bar*) and gustatory stimulation, including umami (glutamic acid), salty (NaCl), sweet (sucrose), bitter (quinine hydrochloride), and sour (citrate) (*red bars*)

the bioengineered salivary glands occurred via proper nerve innervations and neurotransmission.

7.6.3 Protection from Bacterial Infection and Dry Mouth

Saliva contains numerous proteins and cytokines that are essential for the maintenance of oral health and homeostasis, including amylase, lysozyme, IgA, lactoferrin, myeloperoxidase, NGF, EGF, and parotin (Lamy et al. 2010, Cohen 1962). Saliva reduction induces various clinical problems, such as bacterial infection, dental caries, sleep disorders, and swallowing dysfunction. In mouse models of salivary

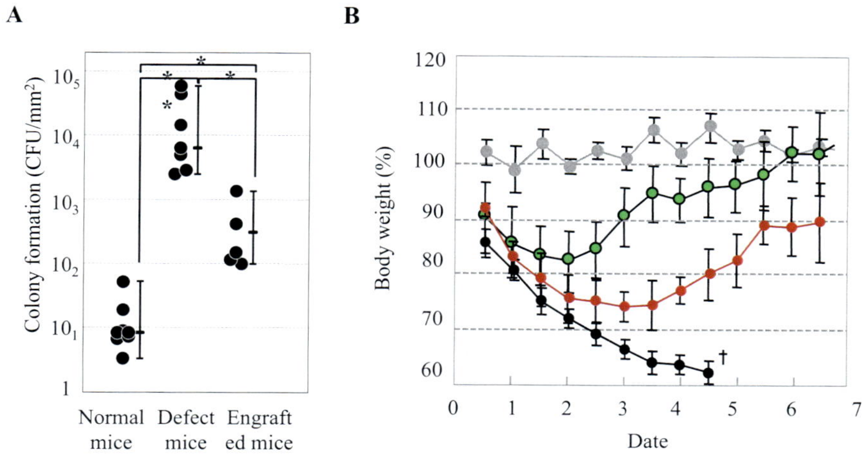

Fig. 7.6 Improvement of xerostomia by bioengineered salivary gland. (**a**) Assessment of bacterial propagation in the buccal mucosa of normal mice, salivary gland defect model mice, and bioengineered salivary gland-engrafted mice. $*P < 0.05$, $**P < 0.001$ by Student's t-test. (**b**) Measurement of body weight every 0.5 days after transplantation in normal mice (*gray dots*), salivary gland defect mice (*black dots*), salivary gland-engrafted mice (*red dots*), and salivary gland defect mice that were given high-viscosity water (*green dots*). All salivary gland defect mice died within 5 days (†) after the removal of all of the major salivary glands

gland defects, the volume of oral bacteria increased compared with normal mice. In contrast, it was significantly reduced in the bioengineered salivary gland-engrafted mice compared with salivary gland defect mice (Fig. 7.6a) (Ogawa et al. 2013). These results indicate that bioengineered saliva has a cleansing function that prevents bacterial growth and dryness of the oral cavity.

7.6.4 *Swallowing Function Recovery*

Among the salivary gland functions, swallowing is important for the absorption of nutrition and reduces the risk of aspiration, which can cause chronic lung disease (Sreebny and Schwartz 1997). Saliva promotes the formation of a bolus of food and results in swallowing reflex. In salivary gland defect mice, body weight decreased abnormally, and all of the mice died within 5 days, despite free access to food and water. However, high-viscosity water prevented the decrease in body weight and improved the survival rate in the salivary gland defect mice (Fig. 7.6b) (Ogawa et al. 2013).

High-viscosity water is usually used to support swallowing in dry mouth patients and geriatric nursing. Thus, the salivary gland defect mice may represent a useful animal model for studying difficulties in swallowing. In the bioengineered salivary gland-engrafted mice, their body weights increased 4 days after transplantation, and they survived (Fig. 7.6b). These findings indicate that saliva secretion from bioengineered salivary glands can improve the swallowing function associated with oral health maintenance.

7.7 Future Perspectives for Salivary Gland Regenerative Therapy

Organ regenerative technology has advanced significantly, and many patients can expect to be treated with salivary gland regenerative therapy. To address the future clinical applications of salivary gland replacement therapy, it is essential to identify suitable cell sources. One candidate cell source is the patient's own cells because there is no immunological rejection. Recent stem cell studies have revealed the presence of adult tissue stem cells in the salivary gland. These adult tissue-derived stem cells, which express stem cell markers or MSC markers, can repair injured acinar cells by stem cell transplantation. However, the possibility of utilizing these stem cells has not been studied with regard to inducing similar salivary glands to those induced by epithelial-mesenchymal interactions. In contrast, pluripotent stem cells, such as embryonic stem (ES) cells and induced pluripotent stem (iPS) cells, also have the capacity to become cell sources because these cells can differentiate into endodermal, ectodermal, and mesodermal cells (Wu and Hochedlinger 2011; Cohen and Melton 2011; Yan et al. 2010). The regeneration of some organs, such as the optic cup and pituitary gland, has been reported using ES cells or iPS cells. In the future, it is likely that methods for salivary gland regeneration using these pluripotent stem cells will be established.

In autoimmune diseases, atrophy of acinar cells and cell damage is caused by autoantigens. Because the transplanted regenerated acinar cells may also be affected by the autoimmune response, a genetic modification that decreases the expression of autoantigens against patient-derived stem cells must be performed to achieve future clinical applications of salivary gland replacement therapy in autoimmune disease. Current whole-organ regenerative therapy has the potential to become a future therapeutic technology for several diseases. Salivary gland replacement and regenerative therapy is expected to be realized by promoting fundamental technology development and the clinical application of regeneration research.

Acknowledgments This work was partially supported by a Grant-in-Aid for Kiban (A) from the Ministry of Education, Culture, Sports, Science and Technology (no. 25242041). `by Organ Technologies Inc.

Conflict of Interest M. Ogawa and T. Tsuji have no competing interests.

References

Aframian DJ, Palmon A (2008) Current status of the development of an artificial salivary gland. Tissue Eng Part B 14:187–198

Atala A (2005) Tissue engineering, stem cells and cloning: current concepts and changing trends. Expert Opin Biol Ther 5(7):879–892

Atkinson JC, Grisius M, Massey W (2005) Salivary hypofunction and xerostomia: diagnosis and treatment. Dent Clin N Am 49:309–326

Avery JK (2002) Oral development and histology. Thieme Press, New York, pp 292–330

Brockes JP, Kumar A (2005) Appendage regeneration in adult vertebrates and implications for regenerative medicine. Science 310(5756):1919–1923

Bücheler M, Wirz C, Schütz A, Bootz F (2002) Tissue engineering of human salivary gland organoids. Acta Otolaryngol 122(5):541–545

Cohen S (1962) Isolation of a mouse submaxillary gland protein accelerating incisor eruption and eyelid opening in the new-born animal. J Biol Chem 237:1555–1562

Cohen DE, Melton D (2011) Turning straw into gold: directing cell fate for regenerative medicine. Nat Rev Genet 12(4):243–252

Copelan EA (2006) Hematopoietic stem-cell transplantation. N Engl J Med 354:1813–1826

Delporte C, O'Connell BC, He X, Lancaster HE, O'Connell AC, Agre P, Baum BJ (1997) Increased fluid secretion after adenoviral-mediated transfer of the aquaporin-1 cDNA to irradiated rat salivary glands. Proc Natl Acad Sci U S A 94(7):3268–3273

Denny PC, Denny PA (1999) Dynamics of parenchymal cell division, differentiation, and apoptosis in the young adult female mouse submandibular gland. Anat Rec 254:408–417

Edgar M, Dawes C, Mullane OD (2004) Saliva and Oral Health, 3rd edn. British Dental Association, UK

Feng J, Van der Zwaag M, Stokman MA, Van Os R, Coppes RP (2009) Isolation and characterization of human salivary gland cells for stem cell transplantation to reduce radiation-induced hyposalivation. Radiother Oncol 92:466–471

Fox PC (2004) Salivary enhancement therapies. Caries Res 38:241–246

Froehlich DA, Pangborn RM, Whitaker JR (1987) The effect of oral stimulation on human parotid salivary flow rate and alpha-amylase secretion. Physiol Behav 41(3):209–217

Harunaga J, Hsu JC, Yamada KM (2011) Dynamics of salivary gland morphogenesis. J Dent Res 90(9):1070–1077

He X, Goldsmith CM, Marmary Y, Wellner RB, Parlow AF, Nieman LK, Baum BJ (1998) Systemic action of human growth hormone following adenovirus-mediated gene transfer to rat submandibular glands. Gene Ther 5(4):537–541

Horie K, Kagami H, Hiramatsu Y, Hata K, Shigetomi T, Ueda M (1996) Selected salivary-gland cell culture and the effects of isoproterenol, vasoactive intestinal polypeptide and substance P. Arch Oral Biol 41(3):243–252

Hsu JC, Yamada KM (2010) Salivary gland branching morphogenesis -- Recent progress and future opportunities. Int J Oral Sci 2(3):117–126

Ihrler S, Zietz C, Sendelhofert A, Lang S, Blasenbreu-Vogt S, Löhrs U (2002) A morphogenetic concept of salivary duct regeneration and metaplasia. Virchows Arch 440(5):519–526

Jaskoll T, Melnick M (2004) Embryonic salivary gland branching morphogenesis. Madame Curie:13–14

Jeong J, Baek H, Kim YJ, Choi Y, Lee H, Lee E, Kim ES, Hah JH, Kwon TK, Choi IJ, Kwon H (2013) Human salivary gland stem cells ameliorate hyposalivation of radiation-damaged rat salivary glands. Exp Mol Med 45:e58

Kagami H, O'Connell BC, Baum BJ (1996) Evidence for the systemic delivery of a transgene product from salivary glands. Hum Gene Ther 7(17):2177–2184

Kagami H, Wang S, Hai B (2008) Restoring the function of salivary glands. Oral Dis 14:15–24

Kawakami M, Ishikawa H, Tachibana T, Tanaka A, Mataga I (2013) Functional transplantation of salivary gland cells differentiated from mouse early ES cells in vitro. Hum Cell 26:80–90

Kishi T, Takao T, Fujita K, Taniguchi H (2006) Clonal proliferation of multipotent stem/progenitor cells in the neonatal and adult salivary glands. Biochem Biophys Res Commun 340(2):544–552

Knosp WM, Knox SM, Hoffman MP (2012) Salivary gland organogenesis. Wiley Interdiscip Rev Dev Biol 1(1):69–82

Knox S, Hoffman MP (2008) Salivary gland development. Blackwell Publications, Ames, IA

Lamy E, Graca G, Costa GD, Franco C, Silva FC, Baptista ES, Coelho AV (2010) Changes in mouse whole saliva soluble proteome induced by tannin-enriched diet. Proteome Sci 8:65

Langer RS, Vacanti JP (1999) Tissue engineering: the challenges ahead. Sci Am 280(4):86–89

Lombaert IM, Hoffman MP. (2013) Stem Cells in Salivary Gland Development and Regeneration. Stem Cells in Craniofacial Development and Regeneration. Hoboken, New Jersey, USA; Wiley-Blackwell. pp.271-284.

Lombaert IM, Brunsting JF, Wierenga PK, Faber H, Stokman MA, Kok T, Visser WH, Kampinga HH, de Haan G, Coppes RP (2008) Rescue of salivary gland function after stem cell transplantation in irradiated glands. PLoS One 3:e2063

Lombaert I, Movahednia MM, Adine C (2016) Ferreira JN. Therapeutic Approaches from Stem Cells to Tissue Organoids. Stem Cells, Salivary Gland Regeneration

Madeira C, Santhagunam A, Salgueiro JB, Cabral JM (2015) Advanced cell therapies for articular cartilage regeneration. Trends Biotechnol 33(1):35–42

Man YG, Ball WD, Marchetti L, Hand AR (2011) Contributions of intercalated duct cells to the normal parenchyma of submandibular glands of adult rats. Anat Rec 263(2):202–214

Matsuo R (2000) Role of saliva in the maintenance of taste sensitivity. Crit Rev Oral Biol Med 11:216–229

Nakamura T, Matsui M, Uchida K, Futatsugi A, Kusakawa S, Matsumoto N, Nakamura K, Manabe T, Taketo MM, Mikoshiba K (2004) M3 muscarinic acetylcholine receptor plays a critical role in parasympathetic control of salivation in mice. J Physiol 558:561–575

O'Connell AC, Baccaglini L, Fox PC, O'Connell BC, Kenshalo D, Oweisy H, Hoque AT, Sun D, Herscher LL, Braddon VR, Delporte C, Baum BJ (1999) Safety and efficacy of adenovirus-mediated transfer of the human aquaporin-1 cDNA to irradiated parotid glands of non-human primates. Cancer Gene Ther 6(6):505–513

Ogawa M, Oshima M, Imamura A, Sekine Y, Ishida K, Yamashita K, Nakajima K, Hirayama M, Tachikawa T, Tsuji T (2013) Functional salivary gland regeneration by transplantation of a bioengineered organ germ. Nat Commun 4:2498

Ogawa M, Yamashita K, Niikura M, Nakajima K, Toyoshima KE, Oshima M, Tsuji T (2014) Saliva secretion in engrafted mouse bioengineered salivary glands using taste stimulation. J Prosthodont Res 58(1):17–25

Patel VN, Rebustini IT, Hoffman MP (2006) Salivary gland branching morphogenesis. Differentiation 74(7):349–364

Peters SB, Naim N, Nelson DA, Mosier AP, Cady NC, Larsen M (2014) Biocompatible tissue scaffold compliance promotes salivary gland morphogenesis and differentiation. Tissue Eng Part A 20:1632–1642

Pradhan S, Farach-Carson MC (2010) Mining the extracellular matrix for tissue engineering applications. Regen Med 5:961–970

Proctor GB, Carpenter GH (2014) Salivary secretion: mechanism and neural regulation. Monogr Oral Sci 24:14–29

Rotter N, Oder J, Schlenke P, Lindner U, Bohrnsen F, Kramer J, Rohwedel J, Huss R, Brandau S, Wollenberg B, Lang S (2008) Isolation and characterization of adult stem cells from human salivary glands. Stem Cells Dev 17:509–518

Sakai T (2009) Epithelial branching morphogenesis of salivary gland: exploration of new functional regulators. J Med Investig 56(Suppl):234–238

Saleh J, Figueiredo MA, Cherubini K, Salum FG (2015) Salivary hypofunction: an update on aetiology, diagnosis and therapeutics. Arch Oral Biol 60(2):242–255

Sasano T, Satoh-Kuriwada S, Shoji N, Sekine-Hayakawa Y, Kawai M, Uneyama H (2010) Application of umami taste stimulation to remedy hypogeusia based on reflex salivation. Biol Pharm Bull 33(11):1791–1795

Segers VFM, Lee RT (2008) Stem-cell therapy for cardiac disease. Nature 451:937–942

Ship JA, Pillemer SR, Baum BJ (2002) (2002) Xerostomia and the geriatric patient. J Am Geriatr Soc 50:535–543

Sreebny LM, Schwartz SS (1997) A reference guide to drugs and dry mouth--2nd edition. Gerodontology 14(1):33–47

Sugito T, Kagami H, Hata K, Nishiguchi H, Ueda M (2004) Transplantation of cultured salivary gland cells into an atrophic salivary gland. Cell Transplant 13(6):691–699

Sumita Y, Liu Y, Khalili S, Maria OM, Xia D, Key S, Cotrim AP, Mezey E, Tran SD (2011) Bone marrow-derived cells rescue salivary gland function in mice with head and neck irradiation. Int J Biochem Cell Biol 43:80–87

Sun T, Zhu J, Yang X, Wang S (2006) Growth of miniature pig parotid cells on biomaterials in vitro. Arch Oral Biol 51(5):351–358

Tran SD, Wang J, Bandyopadhyay BC, Redman RS, Dutra A, Pak E, Swaim WD, Gerstenhaber JA, Bryant JM, Zheng C, Goldsmith CM, Kok MR, Wellner RB, Baum BJ (2005) Primary culture of polarized human salivary epithelial cells for use in developing an artificial salivary gland. Tissue Eng 11(1-2):172–181

Tran SD, Liu Y, Xia D, Maria OM, Khalili S, Wang RW-J, Quan V-H, Hu S, Seuntjens J (2013) Paracrine Effects of Bone Marrow Soup Restore Organ Function, Regeneration, and Repair in Salivary Glands Damaged by Irradiation. PLoS One 8(4):e61632

Tucker AS, Miletich I (2010) Salivary glands; Development, adaptations, and Disease. Karger, London, UK

Vissink A, Mitchell JB, Baum BJ, Limesand KH, Jensen SB, Fox PC, Elting LS, Langendijk JA, Coppes RP, Reyland ME (2010) Clinical management of salivary gland hypofunction and xerostomia in head-and-neck cancer patients: Successes and barriers. Int J Radiat Oncol Biol Phys 78:983–991

Voutetakis A, Bossis I, Kok MR, Zhang W, Wang J, Cotrim AP, Zheng C, Chiorini JA, Nieman LK, Baum BJ (2005) Salivary glands as a potential gene transfer target for gene therapeutics of some monogenetic endocrine disorders. J Endocrinol 185(3):363–372

Wei C, Larsen M, Hoffman MP, Yamada KM (2007) Self-Organization and Branching Morphogenesis of Primary Salivary Epithelial Cells. Tissue Eng 13(4):721–735

Wu SM, Hochedlinger K (2011) Harnessing the potential of induced pluripotent stem cells for regenerative medicine. Nat Cell Biol 13(5):497–505

Yan X, Qin H, Qu C, Tuan RS, Shi S, Huang GT (2010) iPS cells reprogrammed from human mesenchymal-like stem/progenitor cells of dental tissue origin. Stem Cells Dev 19(4):469–480

Yoo C, Vines JB, Alexander G, Murdock K, Hwang P, Jun HW. (2014) Adult stem cells and tissue engineering strategies for salivary gland regeneration: a review. Biomater Res 18:9.

Chapter 8
Functional Lacrimal Gland Regeneration

Masatoshi Hirayama, Kazuo Tsubota, and Takashi Tsuji

Abstract The lacrimal gland is an indispensable organ to secrete tear for maintaining a homeostatic microenvironment for healthy ocular surfaces. The lacrimal gland develops from its organ germ at ocular epithelium via epithelial and mesenchymal interactions into mature secretory gland structure, which consists of acini, ducts, myoepithelial cells, and peripheral tissues such as nerves. Dysfunction of the lacrimal gland leads to dry eye disease, which is one of the prevalent eye disorders involving ocular discomfort, significant visual disturbances, and a reduced quality of life. Current clinical therapies for dry eye disease are artificial tear eye drops, but they are transient and palliative approach. To restore functions of the lacrimal gland, lacrimal tissue stem cells have been identified for regenerative therapeutic approaches. Fully functional organ replacement such as for tooth and hair follicles has also been developed as a novel three-dimensional organ regeneration using stem cell manipulation—named the organ germ method. Recently, we successfully demonstrated fully functional bioengineered lacrimal gland replacements after transplanting a bioengineered organ germ. This study was a significant advance in possible lacrimal gland organ replacement as a next-generation regenerative therapy for dry eye disease. In this review, we summarize recent progress in lacrimal regeneration research and the development of bioengineered lacrimal gland organ replacement therapy.

Keywords Bioengineered lacrimal gland • Organ germ method • Cell manipulation • Transplantation • Lacrimal gland germ • Lacrimal glands • Tear • Dry eye disease • Regenerative medicine • Organ regeneration

---

M. Hirayama • K. Tsubota
Department of Ophthalmology, Keio University School of Medicine, Shinjuku-ku, Tokyo 160-8582, Japan

T. Tsuji (✉)
Laboratory for Organ Regeneration, RIKEN Center for Developmental Biology, 2-2-3, Minatojima-mimamimschi, Chuo-ku, Kobe, Hyogo 650-0047, Japan
e-mail: t-tsuji@cdb.riken.jp

T. Tsuji (ed.), *Organ Regeneration Based on Developmental Biology*,
DOI 10.1007/978-981-10-3768-9_8

8.1 Introduction

Regenerative medicine to restore organ function has been developed as realizable therapy due to perpetual advance of our understanding of developmental biology, stem cell biology, and tissue engineering (Brockes and Kumar 2005; Atala 2005). In various organs, the concept of regenerative medicine, including stem cell transplantation of enriched tissue-derived stem cells and activation therapy of tissue stem cell using cytokine, has been clinically applied to treat damaged tissues (Copelan 2006; Segers and Lee 2008; Wang et al. 2003; Takahashi et al. 1998). Advances in tissue engineering technologies enable to make 2D tissue regeneration such as a cultivated cell sheet transplant procedure (Ge et al. 2015; Ito et al. 2014; Sawa and Miyagawa 2013; Takagi et al. 2014). Current ophthalmology, including corneal limbal stem cell transplantation surgery and cultivated cell sheets of corneal epithelial cell or oral mucosal epithelial cell, has accomplished clinical evidence to reconstruct the ocular surface for severe ocular surface disorders by integrating the concept of regenerative medicine (Tsubota et al. 1999; Shimazaki et al. 2002; Nishida et al. 2004; Hirayama et al. 2012). This methodology has expanded as pluripotent stem cell-derived retinal pigment epithelium cell sheet transplant to overcome age-related macular degeneration, the vision-threatening retinal disease (Kamao et al. 2014; Westenskow et al. 2014). Regenerative medicine has been expected to be an indispensable therapy for future ophthalmology.

Functional restoration of the lacrimal glands, which protect a healthy microenvironment of the ocular surface via secretion of tears, has attracted ophthalmologists as an effective way to be realized for improvement of visual functions (Mathers 2000; Hirayama et al. 2013b). Tear fluids contain water, electrolytes, and various tear proteins, and they promise the smoothness of the ocular surface and lid lubrication with cooperation of lipid secretion from periocular glands (Schechter et al. 2010; Mishima 1965; Seal et al. 1986). Acini of the lacrimal glands produce tear water and tear proteins, which are critical for physiological tear functions such as tear stability and anti-microbiotic effect. Produced fluids reach to the ocular surface all over through the lacrimal gland excretory duct. A stable ocular surface with homeostatic tear environment is fundamental to maintain visual functions.

Dysfunction of the lacrimal glands leads to dry eye disease (DED), which has become one of the big public health problems because it is a multifactorial disease, including aging, and environmentally induced such as use of visual display, immunological reaction, and trauma (Uchino et al. 2011; Uchino et al. 2008a, b; Lin et al. 2003; Paulsen et al. 2014). Severe DED decreases the quality of life due to significant visual disturbance from epithelial cell damage of the cornea, which can cause loss of vision (Tsubota et al. 1996; Dogru et al. 2008; Kaido et al. 2011). The first step to restore tear function for DED is replenishment therapy using artificial tear solution, which is mainly constituted of water, whereas it has conservative effect (Matsumoto et al. 2012; Johnson 1989; Ubels et al. 2004; Rangarajan et al. 2015). Therefore, an establishment of curable therapeutic approach to restore functions of the lacrimal glands has been required to redeem biological tear fluids (Hirayama et al. 2015).

This review summarized the physiological functions, development of the lacrimal glands, and attempts to restore functions of the lacrimal glands from published research. We detailed that there is potential for a novel, fully functional lacrimal gland regeneration as a next-generation regenerative medicine.

8.2 Physiological Function of Tears from the Lacrimal Glands

Tear secretion from the lacrimal glands is a critical factor for function and homeostatic microenvironment of the ocular surface (Mishima 1965; Lemp 1973). The lacrimal gland consists of the main lacrimal gland, which produces most of aqueous tears, and small accessory lacrimal glands (Schechter et al. 2010). The histological structure of mature lacrimal glands is constructed by an organized tubuloalveolar system with the acini, the ducts that are a pathway of fluid from the acini to a mucosal surface, and the myoepithelial cells that wrap the acini and early duct elements (Schechter et al. 2010). A tear film consisting of triple layers of lipid, aqueous, and mucin contributes to the ocular surface wetness, lid lubrication, and optical properties of the ocular surface (Lemp 1987; Yokoi et al. 2014). The tear aqueous layer is formed by secretion from the lacrimal glands and contains water and many tear proteins, including lactoferrin, for expression of biological functions such as moisturizing the ocular surface and antimicrobial activity (Hirayama et al. 2013b; Ohashi et al. 2006). The lacrimal gland and tear functions are essential for healthy ocular surface in protecting the ocular epithelium and visual function.

8.3 Organogenesis of the Lacrimal Glands

The lacrimal glands are anatomically classified as serous secretory glands, which are constituted of various cell types including acini, acini-enveloping myoepithelial cells, duct, and peripheral interstitial tissues including nerves (Schechter et al. 2010). This reasonable 3D histo-architecture for tear secretion is organized by a reciprocal epithelial and mesenchymal interaction during organogenesis process in embryonic stage, which is a developmental system commonly observed in other ectodermal organs such as teeth, hair follicle, and salivary glands (Takahashi et al. 1998; Makarenkova et al. 2000; Pispa and Thesleff 2003; Thesleff 2003; Tucker et al. 2004; Sharpe and Young 2005). After the induction of initial epithelial bud of the lacrimal gland on ocular surface epithelium, the lacrimal gland germ shows branching morphogenesis with stalk elongation and cleft formation (Kammandel et al. 1999). In murine, the lacrimal gland's bud occurs on embryonic day (ED) 13.5

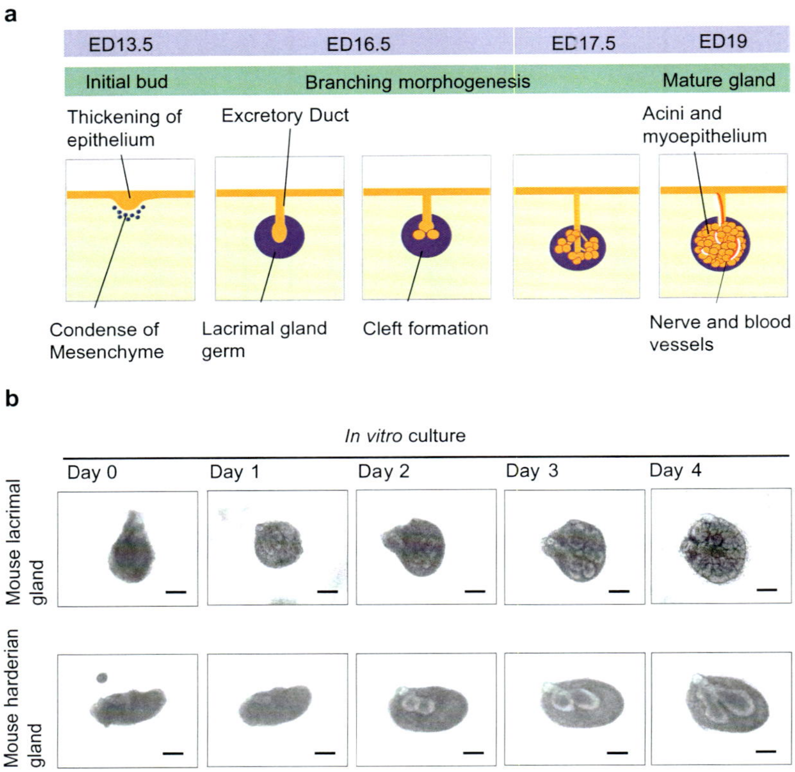

Fig. 8.1 Embryonic organogenesis of the lacrimal glands through epithelial-mesenchymal interactions. (**a**) Schematic representation of the lacrimal gland development. (**b**) Phase-contrast images of development of the lacrimal gland germ and harderian gland germ in vitro. Scale bars, 100 μm (Modified and reprinted from Hirayama et al[112])

as an invagination of the conjunctival epithelium to mesenchymal sac at the temporal region of the eye and then starts to proliferate rapidly and branch to form a lobular structure as the lacrimal gland germ (Fig. 8.1a, b) (Kammandel et al. 1999; Johnston et al. 1979). The fundamental structure of the lacrimal gland is achieved by ED 19. The lacrimal gland matures by the time the eyes open 1 week after birth and prepares to secrete tear to the ocular surface (Schechter et al. 2010; Franklin 1989). The harderian glands in mice, which secrete tear lipids to the ocular surface, originate from the nasal region of the conjunctival epithelium at ED 16 by similar course of branching as the lacrimal glands and are located behind the eye (Fig. 8.1b) (Payne 1994). In human, meibomian glands secrete lipids, whereas the harderian glands are absent or rudimentary (Kozak et al. 2007).

The development and differentiation of the lacrimal glands are regulated by many of the molecular signaling pathways underlying branching morphogenesis (Fig. 8.2) (Hogan 1999). *Pax6* expression is necessary to speculate embryonic

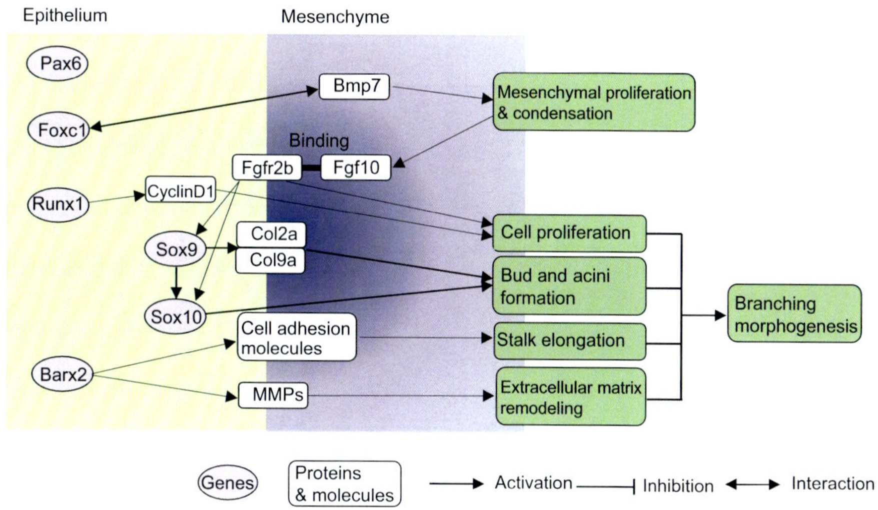

Fig. 8.2 Schematic illustration of signaling pathways involving development of the lacrimal glands. Bmp7 (bone morphogenetic protein 7), Fgf10, fibroblast growth factor 7, Fgfr2b, fgf10 receptor, Col (collagen), MMP (matrix metalloproteinase)

conjunctival epithelium into the lacrimal gland epithelium bud (Grindley et al. 1995; Hill et al. 1992). The temporal periorbital mesenchyme around the *Pax6*-expressing epithelium expresses fibroblast growth factor 10 (Fgf10), which plays an essential role in the induction of the lacrimal glands and the harderian glands (Makarenkova et al. 2000; De Moerlooze et al. 2000; Finch et al. 1989; Lu et al. 1999; Govindarajan et al. 2000; Ahrens and Schlosser 2005). The signaling of Fgf10 involves the regulation of branching morphogenesis, including the formation of acinar lobes, by activation of *Sox9-Sox10* pathways through Fgfr2 (Chen et al. 2014). Heparan sulfate, a sulfated glycosaminoglycan on cell surface, interacts with Fgf10-Fgfr2 signaling pathway to regulate the development of the lacrimal gland branching (Qu et al. 2011; Patel et al. 2007, 2008; Izvolsky et al. 2003). Fgfs also control *Six1* expression, which involves the induction of the ocular placode and the development of the lacrimal glands (El-Hashash et al. 2011; Ahrens and Schlosser 2005). *Barx2*, which expresses in epithelium of the lacrimal gland germ, regulates the expression of metalloproteinases for epithelial elongation and outgrowth (Tsau et al. 2011). Bone morphogenetic protein (Bmp) family such as Bmp7 promotes mesenchymal proliferation and condensation during branching morphogenesis (Dean et al. 2004). The exchange of signals between epithelium and mesenchyme modulates the progress and maturation of the lacrimal gland development.

8.4 Dry Eye Disease and Current Clinical Trials to Restore Lacrimal Gland Functions

DED is caused by a tear shortage due to lacrimal gland dysfunction, and it leads to ocular surface epithelial damage with ocular pain, impaired visual acuity, and reduction of the quality of life (Tsubota et al. 1996; Stern et al. 1998; Stern et al. 2004; The definition and classification of dry eye disease: report of the Definition and Classification Subcommittee of the International Dry Eye WorkShop (2007). Complete lack of tears, which is caused by chemical or heat burn and immunological disorders such as Stevens-Johnson syndrome and Sjogren syndrome, leads to corneal and conjunctival keratinization and then formed a skin-like surface (Gregory 2008; Ciralsky et al. 2013; Hyon et al. 2007; Ramos-Casals et al. 2010; Wagoner 1997; Fish and Davidson 2010). The only means of surgical intervention to prevent loss of vision is corneal transplantation, including artificial device-supported cornea such as keratoprosthesis devices using polymers or autologous dental tissue (Srikumaran et al. 2014; Tan et al. 2012; Liu et al. 2008). Usage of artificial tear solutions contributes to increase temporary tear volume on the ocular surface and tear stability (Barabino et al. 2014; Kaya et al. 2015; Schmidl et al. 2015). To supply tear proteins and biological subjects from the lacrimal gland, albumin containing tear solution and autologous serum are frequently used as a substitute (Shimmura et al. 2003; Higuchi et al. 2007; Kojima et al. 2008; Kojima et al. 2005). An ectopic transplantation of autologous labial salivary glands to the ocular conjunctiva has been trialed in human (Sant' Anna et al. 2012; Marinho et al. 2010). The results seem promising preliminary; however, a more extensive study under well-constructed clinical research for long term is expected (Sant' Anna et al. 2012). Recent advances in regenerative medicine and stem cell biology have shown another possibility in regenerating a functional lacrimal gland.

8.5 Investigation of Lacrimal Gland Stem/Progenitor Cells in Adult Tissue

Self-renew response after partial tissue injury by activation of tissue stem cells has been observed in secretory glands, including the salivary glands, pancreas, and mammary glands (Takahashi et al. 1998; Feng et al. 2009; Sumita et al. 2011; Lanzoni et al. 2015; Linnemann et al. 2015). A concept of cell injection therapy using purified tissue stem cells came to be considered as an established method to restore damaged tissue (Copelan 2006; Segers and Lee 2008). In the lacrimal glands, many researchers have tackled to elucidate existence and details of tissue stem cells. An induced partial damage of mouse lacrimal gland's tissue using cytokine such as IL-1 leads to inflammation and cell death; however, some processes of tissue repair occur (Zoukhri et al. 2007). In the damaged lacrimal glands, increase of nestin-positive mesenchymal cells and cells expressing stem cell markers such as ABGC2 and Sca-1 has been reported (You et al. 2011, 2012; Zoukhri 2010). Activation of mesenchymal stem cells and epithelial-mesenchymal transition is thought to involve this regenerative capacity of the lacrimal glands (You et al. 2012). In in vitro culture of human lacrimal gland cells, candidates of tissue stem cell expressing stem cell markers such as c-Kit, ABCG2, and ALDH1 have been determined (Tiwari et al. 2012). Tissue regeneration by tissue stem cell injection in adult tissues to recover lacrimal gland function is a critical subject of intense research because of the potential clinical benefits.

8.6 Functional Lacrimal Gland Organ Regeneration by Transplantation of a Bioengineered Organ Germ

Functional design of organs is built up through an embryonic developmental process in coordination with kinetics and interactions of various types of cells. Organ replacement regenerative therapy using a fully functional bioengineered organ reconstructed by using an in vitro 3D cell culture is one of the successful goals in regenerative medicine, which holds great promise for the replacement of dysfunctional organs after disease, injury, or aging (Atala 2005; Purnell 2008). Organ replacement by transplantation of bioengineered organ germ to restore the function of damaged/lost ectodermal organs such as teeth, hair follicle, salivary glands, and lacrimal glands has been established, thanks to the development of tissue engineering using 3D cell manipulation procedure called an organ germ method (Nakao et al. 2007; Ikeda and Tsuji 2008; Oshima et al. 2012; Ogawa and Tsuji 2015; Hirayama et al. 2013b). We have investigated a possibility of functional lacrimal gland organ regeneration by using our organ germ method (Hirayama et al. 2013a). The bioengineered lacrimal gland germ, which was generated by in vitro cell compartmentation of epithelial and mesenchymal cells from the lacrimal gland germ of ED 16.5 mice, successfully progressed its branching morphogenesis followed by epithelial elongation and cleft formation in organ culture (Fig. 8.3a). Bioengineered harderian gland germs were also developed by using the organ germ method.

8.6.1 Transplantation of Bioengineered Lacrimal Gland Germ with Duct Connection

Duct formation, which connects the lacrimal glands and the ocular surface, also develops during organogenesis, and its reconstruction is indispensable to secretion of the tear (Schechter et al. 2010). To achieve this, both of the bioengineered lacrimal gland germ and the bioengineered harderian gland germ were transplanted to an extra-orbital lacrimal gland-removed mouse. By using our thread-guided transplantation methods, the duct epithelium of bioengineered glands was connected to the recipient's lacrimal excretory duct (Toyoshima et al. 2012) (Fig. 8.3b,c). After engraftment with this method, the bioengineered lacrimal and harderian glands developed into the appropriate histo-architecture, including acini-expressing aquaporin-5 and myoepithelial cells, duct, and nerve fibers, by reproducing the developmental process in recipient mouse (Fig. 8.3d). The transplantation of bioengineered organ germ of lacrimal gland and harderian gland could achieve mature secretory gland structure in vivo.

8.6.2 Tear Secretion Ability of the Bioengineered Lacrimal Gland

Tear secretion is strictly regulated by neural stimulation to appropriately respond to the outer environment (Dartt 2004, 2009). Cooling stimulation to the ocular surface leads to tearing, which is activated through corneal thermoreceptors, a representative neural pathway for lacrimal gland function (Parra et al. 2010; Robbins et al. 2012). We proved the bioengineered lacrimal glands could have physiological function in coordination with the peripheral and central nervous system, because it secreted tears in response to the cooling stimulation to ocular surface. Acini of the lacrimal glands or harderian glands secrete not only aqueous tear but also tear components including tear proteins such as lactoferrin and lipids (Schechter et al. 2010).

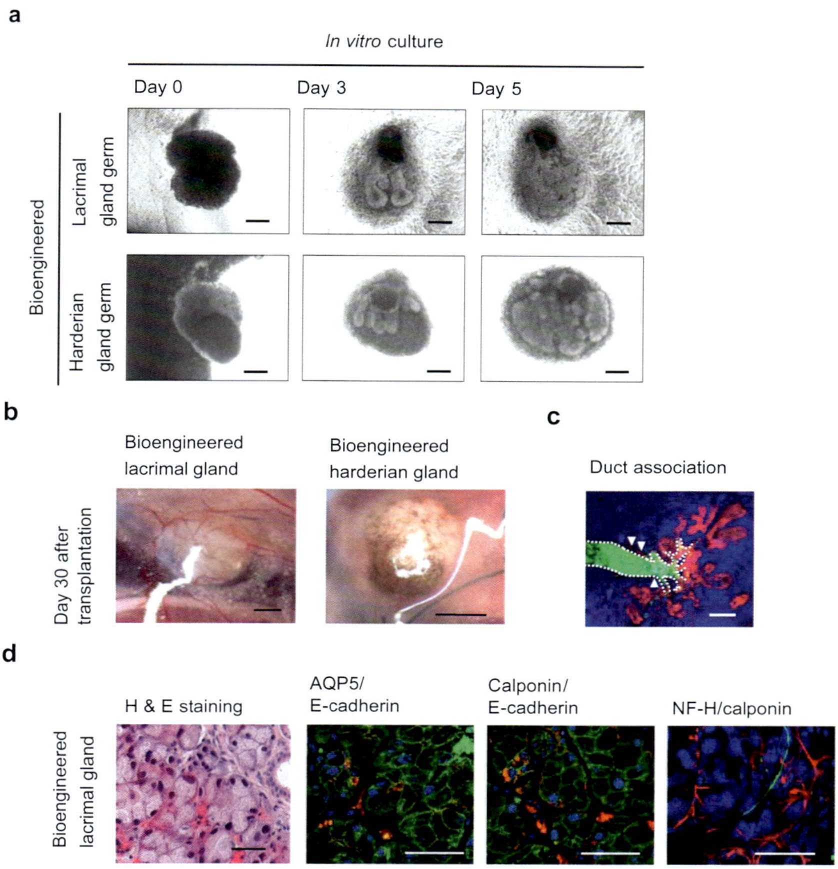

Fig. 8.3 Generation of bioengineered lacrimal gland germ using the organ germ method. Phase-contrast images of development of the bioengineered lacrimal gland germ (*upper line*) and bioengineered harderian gland germ (*lower line*). Scale bar, 100 μm. Photographs of the bioengineered lacrimal gland germ after transplantation. At 30 days after transplantation, the bioengineered lacrimal gland and bioengineered harderian gland were engrafted and developed (*right* and *left*; scale bar, 500 μm). Histological analysis of the duct association between the bioengineered lacrimal gland and recipient lacrimal excretory duct. Bioengineered lacrimal glands regenerated using DsRed transgenic mouse-derived epithelial cells (*red*), duct association between the bioengineered lacrimal gland and the recipient mouse (*arrowhead*). FITC-gelatin (*green*) which was injected from the recipient lacrimal excretory duct could successfully reach to the bioengineered lacrimal gland (*yellow*). DAPI (*blue*) and the excretory duct (*dotted line*) are shown. Scale bars, 100 μm. Histological and immunohistochemical analysis of the bioengineered lacrimal gland after transplantation. The HE staining revealed that the bioengineered lacrimal gland achieved a mature secretory gland structure including acini and duct (*left*; scale bar, 50 μm). Aquaporin-5 is red and E-cadherin is green in the center-left panel. Calponin is red and E-cadherin is green in the center-right panel. Calponin is red, neurofilament-H (NF-H) is green, and DAPI is blue in the right panel. Scale bars, 50 μm (Modified and reprinted from Hirayama et al.[112])

These play an important role for increasing tear stability, wound healing, and antibacterial effects (Mishima 1965; Chen et al. 2014; Goto et al. 2011). To substitute physiological function of tear proteins, many therapies including albumin eye drops and autologous serum eye drops have been attempted in current clinical therapies for severe lacrimal gland dysfunction (Shimmura et al. 2003; Kojima et al. 2008). Tears from the bioengineered lacrimal glands have a biological merit because they contained major tear proteins, such as lactoferrin. The bioengineered harderian glands could secrete tear lipids to ocular surface. Our results demonstrated that transplanted bioengineered glands can replace appropriate tear components.

8.6.3 Protection Effect of Ocular Surface by the Bioengineered Lacrimal Gland

The goal of bioengineered lacrimal gland organ replacement is to protect the ocular surface from DED. The impaired area on the ocular surface in DED shows corneal punctate staining and corneal epithelial changes including thinning and stromal fibroblast activation, which were observed in a DED mouse model with an extra-orbital lacrimal gland defect (Kaido et al. 2011). Our bioengineered lacrimal glands could prevent these pathological changes in the ocular surface and maintain a healthy status equivalent to those of normal mice (Fig. 8.4a,b). Our results indicate that functional replacements of the bioengineered lacrimal gland would be an attractive therapeutic strategy for severe DED.

8.7 Future Directions of Lacrimal Gland Regenerative Medicine

Complete substitution of physiological tear function is one of the goals to achieve stable ocular surface and overcome DED (Hirayama et al. 2013b). Current biological analysis of tears and lacrimal glands elucidated a role of key molecules such as mucin and improved the quality of artificial tear solution (Hamano 1998; Shigemitsu et al. 2002; Corrales et al. 2011). On the other hand, technological advances in stem cell biology, developmental biology, and tissue engineering have provided evidence to realize lacrimal gland regenerative therapy clinically. Especially in treatment for severe DED, the restoration of the lacrimal gland function including tissue stem cell transplantation and lacrimal gland organ regeneration has been expected as a curable therapeutic approach (Tsubota et al. 1996; Hirayama et al. 2015). Our bioengineered lacrimal glands showed a proof of concept for bioengineered organ replacement to functionally restore the lacrimal gland. To progress this concept clinically, it requires evidence of efficacy of bioengineered organ transplantation for disease-specific environment such as inflammation and immunological disorders

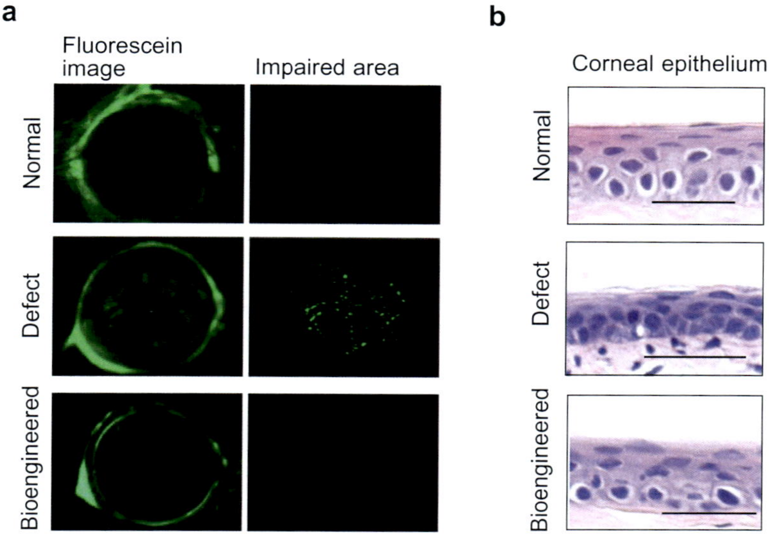

Fig. 8.4 Tear secretion and ocular surface protection for the bioengineered lacrimal gland. Representative images of the corneal surface of a normal lacrimal gland (*upper*), a lacrimal gland-defective mouse (*center*), and a bioengineered lacrimal gland-engrafted mouse (*lower*). The punctate staining area by fluorescein showed impaired area on the corneal surface. Scale bar, 1 mm (Modified and reprinted from Hirayama et al.[112])
Representative microscopic images of the corneal epithelium including a normal mouse (*upper*), lacrimal gland-defective mouse (center), and bioengineered lacrimal gland-transplanted mouse (*lower*) are shown. Chronic dry eye status in lacrimal gland-defective mouse induced corneal thickening as shown in the center panel, whereas these changes were not observed in the bioengineered lacrimal gland-transplanted mouse. Scale bars, 25 μm. (Modified and reprinted from Hirayama et al[112])

(Tiwari et al. 2012). Another demand for regenerating the lacrimal gland would be the use of available cell sources including pluripotent stem cells (Yoshida and Yamanaka 2011; Okano and Yamanaka 2014; Tsumaki et al. 2015). Application to 3D organ regeneration using embryonic stem cells and induced pluripotent stem cells to regenerate lacrimal glands would be powerful tools not only to realize functional lacrimal gland organ replacement but also to investigate developmental and pathological process of the lacrimal gland function (Okano 2010, 2012; Okano and Yamanaka 2014). Bioengineered organ regeneration is expected to be the next-generation therapeutic strategy of regenerative medicine.

Acknowledgments This work was partially supported by a Grant-in-Aid for Scientific Research A (to T.T.) from the Ministry of Education, Culture, Sports, Science and Technology, Japan.

Conflicts of Interest The authors declare no conflict of interest.

References

Ahrens K, Schlosser G (2005) Tissues and signals involved in the induction of placodal Six1 expression in *Xenopus laevis*. Dev Biol 288(1):40–59. doi:10.1016/j.ydbio.2005.07.022

Atala A (2005) Tissue engineering, stem cells and cloning: current concepts and changing trends. Expert Opin Biol Ther 5(7):879–892. doi:10.1517/14712598.5.7.879

Barabino S, Rolando M, Nardi M, Bonini S, Aragona P, Traverso CE (2014) The effect of an artificial tear combining hyaluronic acid and tamarind seeds polysaccharide in patients with moderate dry eye syndrome: a new treatment for dry eye. Eur J Ophthalmol 24(2):173–178. doi:10.5301/ejo.5000355

Brockes JP, Kumar A (2005) Appendage regeneration in adult vertebrates and implications for regenerative medicine. Science 310(5756):1919–1923. doi:10.1126/science.1115200

Chen Z, Huang J, Liu Y, Dattilo LK, Huh SH, Ornitz D, Beebe DC (2014) FGF signaling activates a Sox9-Sox10 pathway for the formation and branching morphogenesis of mouse ocular glands. Development 141(13):2691–2701. doi:10.1242/dev.108944

Ciralsky JB, Sippel KC, Gregory DG (2013) Current ophthalmologic treatment strategies for acute and chronic Stevens-Johnson syndrome and toxic epidermal necrolysis. Curr Opin Ophthalmol 24(4):321–328. doi:10.1097/ICU.0b013e3283622718

Copelan EA (2006) Hematopoietic stem-cell transplantation. N Engl J Med 354(17):1813–1826. doi:10.1056/NEJMra052638

Corrales RM, Narayanan S, Fernandez I, Mayo A, Galarreta DJ, Fuentes-Paez G, Chaves FJ, Herreras JM, Calonge M (2011) Ocular mucin gene expression levels as biomarkers for the diagnosis of dry eye syndrome. Invest Ophthalmol Vis Sci 52(11):8363–8369. doi:10.1167/iovs.11-7655

Dartt DA (2004) Dysfunctional neural regulation of lacrimal gland secretion and its role in the pathogenesis of dry eye syndromes. The ocular surface 2(2):76–91

Dartt DA (2009) Neural regulation of lacrimal gland secretory processes: relevance in dry eye diseases. Prog Retin Eye Res 28(3):155–177. doi:10.1016/j.preteyeres.2009.04.003

De Moerlooze L, Spencer-Dene B, Revest JM, Hajihosseini M, Rosewell I, Dickson C (2000) An important role for the IIIb isoform of fibroblast growth factor receptor 2 (FGFR2) in mesenchymal-epithelial signalling during mouse organogenesis. Development 127(3):483–492

Dean C, Ito M, Makarenkova HP, Faber SC, Lang RA (2004) Bmp7 regulates branching morphogenesis of the lacrimal gland by promoting mesenchymal proliferation and condensation. Development 131(17):4155–4165. doi:10.1242/dev.01285

Dogru M, Kaido M, Matsumoto Y, Tsubota K (2008) Dry eye disease and its implications on visual function. Archivos de la Sociedad Espanola de Oftalmologia 83(1):3–6

El-Hashash AH, Al Alam D, Turcatel G, Rogers O, Li X, Bellusci S, Warburton D (2011) Six1 transcription factor is critical for coordination of epithelial, mesenchymal and vascular morphogenesis in the mammalian lung. Dev Biol 353(2):242–258. doi:10.1016/j.ydbio.2011.02.031

Feng J, van der Zwaag M, Stokman MA, van Os R, Coppes RP (2009) Isolation and characterization of human salivary gland cells for stem cell transplantation to reduce radiation-induced hyposalivation. Radiother Oncol J Eur Soc Ther Radiol Oncol 92(3):466–471. doi:10.1016/j.radonc.2009.06.023

Finch PW, Rubin JS, Miki T, Ron D, Aaronson SA (1989) Human KGF is FGF-related with properties of a paracrine effector of epithelial cell growth. Science 245(4919):752–755

Fish R, Davidson RS (2010) Management of ocular thermal and chemical injuries, including amniotic membrane therapy. Curr Opin Ophthalmol 21(4):317–321. doi:10.1097/ICU.0b013e32833a8da2

Franklin RM (1989) The ocular secretory immune system: a review. Curr Eye Res 8(6):599–606

Ge Y, Gong YY, Xu Z, Lu Y, Fu W (2015) The application of sheet technology in cartilage tissue engineering. Tissue Eng B Rev. doi:10.1089/ten.TEB.2015.0189

Goto E, Dogru M, Sato EA, Matsumoto Y, Takano Y, Tsubota K (2011) The sparkle of the eye: the impact of ocular surface wetness on corneal light reflection. Am J Ophthalmol 151(4):691–696. doi:10.1016/j.ajo.2010.09.027e691

Govindarajan V, Ito M, Makarenkova HP, Lang RA, Overbeek PA (2000) Endogenous and ectopic gland induction by FGF-10. Dev Biol 225(1):188–200. doi:10.1006/dbio.2000.9812

Gregory DG (2008) The ophthalmologic management of acute Stevens-Johnson syndrome. Ocul Surf 6(2):87–95

Grindley JC, Davidson DR, Hill RE (1995) The role of Pax-6 in eye and nasal development. Development 121(5):1433–1442

Hamano T (1998) Dry eye treatment with eye drops that stimulate mucin production. Adv Exp Med Biol 438:965–968

Higuchi A, Ueno R, Shimmura S, Suematsu M, Dogru M, Tsubota K (2007) Albumin rescues ocular epithelial cells from cell death in dry eye. Curr Eye Res 32(2):83–88. doi:10.1080/02713680601147690

Hill RE, Favor J, Hogan BL, Ton CC, Saunders GF, Hanson IM, Prosser J, Jordan T, Hastie ND, van Heyningen V (1992) Mouse small eye results from mutations in a paired-like homeobox-containing gene. Nature 355(6362):750. doi:10.1038/355750a0

Hirayama M, Satake Y, Higa K, Yamaguchi T, Shimazaki J (2012) Transplantation of cultivated oral mucosal epithelium prepared in fibrin-coated culture dishes. Invest Ophthalmol Vis Sci 53(3):1602–1609. doi:10.1167/iovs.11-7847

Hirayama M, Ogawa M, Oshima M, Sekine Y, Ishida K, Yamashita K, Ikeda K, Shimmura S, Kawakita T, Tsubota K, Tsuji T (2013a) Functional lacrimal gland regeneration by transplantation of a bioengineered organ germ. Nat Commun 4:2497. doi:10.1038/ncomms3497

Hirayama M, Oshima M, Tsuji T (2013b) Development and prospects of organ replacement regenerative therapy. Cornea 32(Suppl 1):S13–S21. doi:10.1097/ICO.0b013e3182a18e6c

Hirayama M, Tsubota K, Tsuji T (2015) Bioengineered lacrimal gland organ regeneration in vivo. J Funct Biomater 6(3):634–649. doi:10.3390/jfb6030634

Hogan BL (1999) Morphogenesis. Cell 96(2):225–233

Hyon JY, Lee YJ, Yun PY (2007) Management of ocular surface inflammation in Sjogren syndrome. Cornea 26(9 Suppl 1):S13–S15. doi:10.1097/ICO.0b013e31812f6782

Ikeda E, Tsuji T (2008) Growing bioengineered teeth from single cells: potential for dental regenerative medicine. Expert Opin Biol Ther 8(6):735–744. doi:10.1517/14712598.8.6.735

Ito M, Shichinohe H, Houkin K, Kuroda S (2014) Application of cell sheet technology to bone marrow stromal cell transplantation for rat brain infarct. J Tissue Eng Regen Med. doi:10.1002/term.1920

Izvolsky KI, Shoykhet D, Yang Y, Yu Q, Nugent MA, Cardoso WV (2003) Heparan sulfate-FGF10 interactions during lung morphogenesis. Dev Biol 258(1):185–200

Johnson TL (1989) Ophthalmic solution, the ocular surface, and a unique therapeutic artificial tear formulation. Am J Ophthalmol 107(6):687

Johnston MC, Noden DM, Hazelton RD, Coulombre JL, Coulombre AJ (1979) Origins of avian ocular and periocular tissues. Exp Eye Res 29(1):27–43

Kaido M, Matsumoto Y, Shigeno Y, Ishida R, Dogru M, Tsubota K (2011) Corneal fluorescein staining correlates with visual function in dry eye patients. Invest Ophthalmol Vis Sci 52(13):9516–9522. doi:10.1167/iovs.11-8412

Kamao H, Mandai M, Okamoto S, Sakai N, Suga A, Sugita S, Kiryu J, Takahashi M (2014) Characterization of human induced pluripotent stem cell-derived retinal pigment epithelium cell sheets aiming for clinical application. Stem Cell Rep 2(2):205–218. doi:10.1016/j.stemcr.2013.12.007

Kammandel B, Chowdhury K, Stoykova A, Aparicio S, Brenner S, Gruss P (1999) Distinct cis-essential modules direct the time-space pattern of the Pax6 gene activity. Dev Biol 205(1):79–97. doi:10.1006/dbio.1998.9128

Kaya S, Schmidl D, Schmetterer L, Witkowska KJ, Unterhuber A, Aranha Dos Santos V, Baar C, Garhofer G, Werkmeister RM (2015) Effect of hyaluronic acid on tear film thickness as

assessed with ultra-high resolution optical coherence tomography. Acta Ophthalmol 93(5):439–443. doi:10.1111/aos.12647
Kojima T, Ishida R, Dogru M, Goto E, Matsumoto Y, Kaido M, Tsubota K (2005) The effect of autologous serum eyedrops in the treatment of severe dry eye disease: a prospective randomized case-control study. Am J Ophthalmol 139(2):242–246. doi:10.1016/j.ajo.2004.08.040
Kojima T, Higuchi A, Goto E, Matsumoto Y, Dogru M, Tsubota K (2008) Autologous serum eye drops for the treatment of dry eye diseases. Cornea 27(Suppl 1):S25–S30. doi:10.1097/ICO.0b013e31817f3a0e
Kozak I, Bron AJ, Kucharova K, Kluchova D, Marsala M, Heichel CW, Tiffany JM (2007) Morphologic and volumetric studies of the meibomian glands in elderly human eyelids. Cornea 26(5):610–614. doi:10.1097/ICO.0b013e318041f0d2
Lanzoni G, Cardinale V, Carpino G (2015) The hepatic, biliary and pancreatic network of stem/progenitor cells niches in humans: A new reference frame for disease and regeneration. Hepatology. doi:10.1002/hep.28326
Lemp MA (1973) Surfacing the precorneal tear film. Ann Ophthalmol 5(7):819–826
Lemp MA (1987) Tear film: new concepts and implications for the management of the dry eye. Trans New Orleans Acad Ophthalmol 35:53–64
Lin PY, Tsai SY, Cheng CY, Liu JH, Chou P, Hsu WM (2003) Prevalence of dry eye among an elderly Chinese population in Taiwan: the Shihpai Eye Study. Ophthalmology 110(6):1096–1101. doi:10.1016/S0161-6420(03)00262-8
Linnemann JR, Miura H, Meixner LK, Irmler M, Kloos UJ, Hirschi B, Bartsch HS, Sass S, Beckers J, Theis FJ, Gabka C, Sotlar K, Scheel CH (2015) Quantification of regenerative potential in primary human mammary epithelial cells. Development 142(18):3239–3251. doi:10.1242/dev.123554
Liu C, Hille K, Tan D, Hicks C, Herold J (2008) Keratoprosthesis surgery. Dev Ophthalmol 41:171–186. doi:10.1159/000131088
Lu W, Luo Y, Kan M, McKeehan WL (1999) Fibroblast growth factor-10. A second candidate stromal to epithelial cell andromedin in prostate. J Biol Chem 274(18):12827–12834
Makarenkova HP, Ito M, Govindarajan V, Faber SC, Sun L, McMahon G, Overbeek PA, Lang RA (2000) FGF10 is an inducer and Pax6 a competence factor for lacrimal gland development. Development 127(12):2563–2572
Marinho DR, Burmann TG, Kwitko S (2010) Labial salivary gland transplantation for severe dry eye due to chemical burns and Stevens-Johnson syndrome. Ophthal Plast Reconstr Surg 26(3):182–184. doi:10.1097/IOP.0b013e3181b8c3ad
Mathers WD (2000) Why the eye becomes dry: a cornea and lacrimal gland feedback model. CLAO J Off Publ Contact Lens Assoc Ophthalmol Inc 26(3):159–165
Matsumoto Y, Ohashi Y, Watanabe H, Tsubota K, Diquafosol Ophthalmic Solution Phase 2 Study G (2012) Efficacy and safety of diquafosol ophthalmic solution in patients with dry eye syndrome: a Japanese phase 2 clinical trial. Ophthalmology 119(10):1954–1960. doi:10.1016/j.ophtha.2012.04.010
Mishima S (1965) Some physiological aspects of the precorneal tear film. Arch Ophthalmol 73:233–241
Nakao K, Morita R, Saji Y, Ishida K, Tomita Y, Ogawa M, Saitoh M, Tomooka Y, Tsuji T (2007) The development of a bioengineered organ germ method. Nat Methods 4(3):227–230. doi:10.1038/nmeth1012
Nishida K, Yamato M, Hayashida Y, Watanabe K, Yamamoto K, Adachi E, Nagai S, Kikuchi A, Maeda N, Watanabe H, Okano T, Tano Y (2004) Corneal reconstruction with tissue-engineered cell sheets composed of autologous oral mucosal epithelium. N Engl J Med 351(12):1187–1196. doi:10.1056/NEJMoa040455
Ogawa M, Tsuji T (2015) Reconstitution of a bioengineered salivary gland using a three-dimensional cell manipulation method. Curr Protoc Cell Biol / editorial board, Juan S Bonifacino [et al.] 66:19 17 11–19 17 13. doi:10.1002/0471143030.cb1917s66

Ohashi Y, Dogru M, Tsubota K (2006) Laboratory findings in tear fluid analysis. Clin Chim Acta Int J Clin Chem 369(1):17–28. doi:10.1016/j.cca.2005.12.035
Okano H (2010) Investigation of regeneration of damaged CNS and neurological diseases mechanisms using iPS cells. Rinsho Shinkeigaku Clin Neurol 50(11):887
Okano H (2012) Investigation of pathophysiology of Parkinson's disease using iPS cells technology. Rinsho Shinkeigaku Clin Neurol 52(11):899
Okano H, Yamanaka S (2014) iPS cell technologies: significance and applications to CNS regeneration and disease. Mol Brain 7:22. doi:10.1186/1756-6606-7-22
Oshima M, Ogawa M, Yasukawa M, Tsuji T (2012) Generation of a bioengineered tooth by using a three-dimensional cell manipulation method (organ germ method). Methods Mol Biol 887:149–165. doi:10.1007/978-1-61779-860-3_14
Parra A, Madrid R, Echevarria D, del Olmo S, Morenilla-Palao C, Acosta MC, Gallar J, Dhaka A, Viana F, Belmonte C (2010) Ocular surface wetness is regulated by TRPM8-dependent cold thermoreceptors of the cornea. Nat Med 16(12):1396–1399. doi:10.1038/nm.2264
Patel VN, Knox SM, Likar KM, Lathrop CA, Hossain R, Eftekhari S, Whitelock JM, Elkin M, Vlodavsky I, Hoffman MP (2007) Heparanase cleavage of perlecan heparan sulfate modulates FGF10 activity during ex vivo submandibular gland branching morphogenesis. Development 134(23):4177–4186. doi:10.1242/dev.011171
Patel VN, Likar KM, Zisman-Rozen S, Cowherd SN, Lassiter KS, Sher I, Yates EA, Turnbull JE, Ron D, Hoffman MP (2008) Specific heparan sulfate structures modulate FGF10-mediated submandibular gland epithelial morphogenesis and differentiation. J Biol Chem 283(14):9308–9317. doi:10.1074/jbc.M709995200
Paulsen AJ, Cruickshanks KJ, Fischer ME, Huang GH, Klein BE, Klein R, Dalton DS (2014) Dry eye in the beaver dam offspring study: prevalence, risk factors, and health-related quality of life. Am J Ophthalmol 157(4):799–806. doi:10.1016/j.ajo.2013.12.023
Payne AP (1994) The harderian gland: a tercentennial review. J Anat 185(Pt 1):1–49
Pispa J, Thesleff I (2003) Mechanisms of ectodermal organogenesis. Dev Biol 262(2):195–205
Purnell B (2008) New release: the complete guide to organ repair. Introduction. Science 322(5907):1489. doi:10.1126/science.322.5907.1489
Qu X, Carbe C, Tao C, Powers A, Lawrence R, van Kuppevelt TH, Cardoso WV, Grobe K, Esko JD, Zhang X (2011) Lacrimal gland development and Fgf10-Fgfr2b signaling are controlled by 2-O- and 6-O-sulfated heparan sulfate. J Biol Chem 286(16):14435–14444. doi:10.1074/jbc.M111.225003
Ramos-Casals M, Tzioufas AG, Stone JH, Siso A, Bosch X (2010) Treatment of primary Sjogren syndrome: a systematic review. JAMA 304(4):452–460. doi:10.1001/jama.2010.1014
Rangarajan R, Kraybill B, Ogundele A, Ketelson HA (2015) Effects of a hyaluronic Acid/Hydroxypropyl guar artificial tear solution on protection, recovery, and lubricity in models of corneal epithelium. J Ocul Pharmacol Ther Off J Assoc Ocul Pharmacol Ther 31(8):491–497. doi:10.1089/jop.2014.0164
Robbins A, Kurose M, Winterson BJ, Meng ID (2012) Menthol activation of corneal cool cells induces TRPM8-mediated lacrimation but not nociceptive responses in rodents. Invest Ophthalmol Vis Sci 53(11):7034–7042. doi:10.1167/iovs.12-10025
Sant' Anna AE, Hazarbassanov RM, de Freitas D, Gomes JA (2012) Minor salivary glands and labial mucous membrane graft in the treatment of severe symblepharon and dry eye in patients with Stevens-Johnson syndrome. Br J Ophthalmol 96(2):234–239. doi:10.1136/bjo.2010.199901
Sawa Y, Miyagawa S (2013) Cell sheet technology for heart failure. Curr Pharm Biotechnol 14(1):61–66
Schechter JE, Warren DW, Mircheff AK (2010) A lacrimal gland is a lacrimal gland, but rodent's and rabbit's are not human. Ocul Surf 8(3):111–134
Schmidl D, Schmetterer L, Witkowska KJ, Unterhuber A, dos Santos VA, Kaya S, Nepp J, Baar C, Rosner P, Werkmeister RM, Garhofer G (2015) Tear film thickness after treatment with artificial

tears in patients with moderate dry eye disease. Cornea 34(4):421–426. doi:10.1097/ICO.0000000000000358

Seal DV, McGill JI, Mackie IA, Liakos GM, Jacobs P, Goulding NJ (1986) Bacteriology and tear protein profiles of the dry eye. Br J Ophthalmol 70(2):122–125

Segers VF, Lee RT (2008) Stem-cell therapy for cardiac disease. Nature 451(7181):937–942. doi:10.1038/nature06800

Sharpe PT, Young CS (2005) Test-tube teeth. Sci Am 293(2):34–41

Shigemitsu T, Shimizu Y, Ishiguro K (2002) Mucin ophthalmic solution treatment of dry eye. Adv Exp Med Biol 506(Pt A):359–362

Shimazaki J, Aiba M, Goto E, Kato N, Shimmura S, Tsubota K (2002) Transplantation of human limbal epithelium cultivated on amniotic membrane for the treatment of severe ocular surface disorders. Ophthalmology 109(7):1285–1290

Shimmura S, Ueno R, Matsumoto Y, Goto E, Higuchi A, Shimazaki J, Tsubota K (2003) Albumin as a tear supplement in the treatment of severe dry eye. Br J Ophthalmol 87(10):1279–1283

Srikumaran D, Munoz B, Aldave AJ, Aquavella JV, Hannush SB, Schultze R, Belin M, Akpek EK (2014) Long-term outcomes of boston type 1 keratoprosthesis implantation: a retrospective multicenter cohort. Ophthalmology 121(11):2159–2164. doi:10.1016/j.ophtha.2014.05.030

Stern ME, Beuerman RW, Fox RI, Gao J, Mircheff AK, Pflugfelder SC (1998) The pathology of dry eye: the interaction between the ocular surface and lacrimal glands. Cornea 17(6):584–589

Stern ME, Gao J, Siemasko KF, Beuerman RW, Pflugfelder SC (2004) The role of the lacrimal functional unit in the pathophysiology of dry eye. Exp Eye Res 78(3):409–416

Sumita Y, Liu Y, Khalili S, Maria OM, Xia D, Key S, Cotrim AP, Mezey E, Tran SD (2011) Bone marrow-derived cells rescue salivary gland function in mice with head and neck irradiation. Int J Biochem Cell Biol 43(1):80–87. doi:10.1016/j.biocel.2010.09.023

Takagi S, Shimizu T, Kuramoto G, Ishitani K, Matsui H, Yamato M, Okano T (2014) Reconstruction of functional endometrium-like tissue in vitro and in vivo using cell sheet engineering. Biochem Biophys Res Commun 446(1):335–340. doi:10.1016/j.bbrc.2014.02.107

Takahashi S, Schoch E, Walker NI (1998) Origin of acinar cell regeneration after atrophy of the rat parotid induced by duct obstruction. Int J Exp Pathol 79(5):293–301

Tan A, Tan DT, Tan XW, Mehta JS (2012) Osteo-odonto keratoprosthesis: systematic review of surgical outcomes and complication rates. Ocul Surf 10(1):15–25. doi:10.1016/j.jtos.2012.01.003

The definition and classification of dry eye disease: report of the Definition and Classification Subcommittee of the International Dry Eye WorkShop (2007). Ocul Surf 5 (2):75–92

Thesleff I (2003) Epithelial-mesenchymal signalling regulating tooth morphogenesis. J Cell Sci 116(Pt 9):1647–1648

Tiwari S, Ali MJ, Balla MM, Naik MN, Honavar SG, Reddy VA, Vemuganti GK (2012) Establishing human lacrimal gland cultures with secretory function. PLoS One 7(1):e29458. doi:10.1371/journal.pone.0029458

Toyoshima KE, Asakawa K, Ishibashi N, Toki H, Ogawa M, Hasegawa T, Irie T, Tachikawa T, Sato A, Takeda A, Tsuji T (2012) Fully functional hair follicle regeneration through the rearrangement of stem cells and their niches. Nat Commun 3:784. doi:10.1038/ncomms1784

Tsau C, Ito M, Gromova A, Hoffman MP, Meech R, Makarenkova HP (2011) Barx2 and Fgf10 regulate ocular glands branching morphogenesis by controlling extracellular matrix remodeling. Development 138(15):3307–3317. doi:10.1242/dev.066241

Tsubota K, Satake Y, Shimazaki J (1996) Treatment of severe dry eye. Lancet 348(9020):123

Tsubota K, Satake Y, Kaido M, Shinozaki N, Shimmura S, Bissen-Miyajima H, Shimazaki J (1999) Treatment of severe ocular-surface disorders with corneal epithelial stem-cell transplantation. N Engl J Med 340(22):1697–1703. doi:10.1056/NEJM199906033402201

Tsumaki N, Okada M, Yamashita A (2015) iPS cell technologies and cartilage regeneration. Bone 70:48–54. doi:10.1016/j.bone.2014.07.011

Tucker AS, Headon DJ, Courtney JM, Overbeek P, Sharpe PT (2004) The activation level of the TNF family receptor, Edar, determines cusp number and tooth number during tooth development. Dev Biol 268(1):185–194. doi:10.1016/j.ydbio.2003.12.019

Ubels JL, Clousing DP, Van Haitsma TA, Hong BS, Stauffer P, Asgharian B, Meadows D (2004) Pre-clinical investigation of the efficacy of an artificial tear solution containing hydroxypropyl-guar as a gelling agent. Curr Eye Res 28(6):437–444. doi:10.1080/02713680490503787

Uchino M, Dogru M, Uchino Y, Fukagawa K, Shimmura S, Takebayashi T, Schaumberg DA, Tsubota K (2008a) Japan Ministry of Health study on prevalence of dry eye disease among Japanese high school students. Am J Ophthalmol 146(6):925–929 . doi:10.1016/j.ajo.2008.06.030e922

Uchino M, Schaumberg DA, Dogru M, Uchino Y, Fukagawa K, Shimmura S, Satoh T, Takebayashi T, Tsubota K (2008b) Prevalence of dry eye disease among Japanese visual display terminal users. Ophthalmology 115(11):1982–1988. doi:10.1016/j.ophtha.2008.06.022

Uchino M, Nishiwaki Y, Michikawa T, Shirakawa K, Kuwahara E, Yamada M, Dogru M, Schaumberg DA, Kawakita T, Takebayashi T, Tsubota K (2011) Prevalence and risk factors of dry eye disease in Japan: Koumi study. Ophthalmology 118(12):2361–2367. doi:10.1016/j.ophtha.2011.05.029

Wagoner MD (1997) Chemical injuries of the eye: current concepts in pathophysiology and therapy. Surv Ophthalmol 41(4):275–313

Wang X, Foster M, Al-Dhalimy M, Lagasse E, Finegold M, Grompe M (2003) The origin and liver repopulating capacity of murine oval cells. Proc Natl Acad Sci U S A 100(Suppl 1):11881–11888. doi:10.1073/pnas.1734199100

Westenskow PD, Kurihara T, Friedlander M (2014) Utilizing stem cell-derived RPE cells as a therapeutic intervention for age-related macular degeneration. Adv Exp Med Biol 801:323–329. doi:10.1007/978-1-4614-3209-8_41

Yokoi N, Bron AJ, Georgiev GA (2014) The precorneal tear film as a fluid shell: the effect of blinking and saccades on tear film distribution and dynamics. Ocul Surf 12(4):252–266. doi:10.1016/j.jtos.2014.01.006

Yoshida Y, Yamanaka S (2011) iPS cells: a source of cardiac regeneration. J Mol Cell Cardiol 50(2):327–332. doi:10.1016/j.yjmcc.2010.10.026

You S, Kublin CL, Avidan O, Miyasaki D, Zoukhri D (2011) Isolation and propagation of mesenchymal stem cells from the lacrimal gland. Invest Ophthalmol Vis Sci 52(5):2087–2094. doi:10.1167/iovs.10-5686

You S, Avidan O, Tariq A, Ahluwalia I, Stark PC, Kublin CL, Zoukhri D (2012) Role of epithelial-mesenchymal transition in repair of the lacrimal gland after experimentally induced injury. Invest Ophthalmol Vis Sci 53(1):126–135. doi:10.1167/iovs.11-7893

Zoukhri D (2010) Mechanisms involved in injury and repair of the murine lacrimal gland: role of programmed cell death and mesenchymal stem cells. Ocul Surf 8(2):60–69

Zoukhri D, Macari E, Kublin CL (2007) A single injection of interleukin-1 induces reversible aqueous-tear deficiency, lacrimal gland inflammation, and acinar and ductal cell proliferation. Exp Eye Res 84(5):894–904. doi:10.1016/j.exer.2007.01.015

Chapter 9
Early Kidney Specification and Its Recapitulation by Pluripotent Stem Cells

Atsuhiro Taguchi and Ryuichi Nishinakamura

Abstract Successful generation of the kidney from pluripotent stem cells is challenging because of insufficient knowledge of the underlying developmental processes. In addition to the technical difficulties of physically examining the early stages of embryogenesis, the unique features surrounding the development of the kidney from three distinct primordia hamper a more complete understanding of the relevant cell fate acquisition mechanisms. We have recently addressed these issues by combining in vivo lineage tracing experiments and ex vivo directed differentiation culture systems. Our strategy has revealed the mechanism by which the kidney morphogenic field is patterned along the anteroposterior axis and also identified the key signals which promote posteriorization, specification, and maturation of nephron progenitors from pluripotent stem cells. Importantly, these newly identified biological insights have enabled the production of three-dimensional complex nephron structure from both mouse and human pluripotent stem cells, which would be a big progress toward the realization of kidney regenerative medicine.

Keywords Kidney • Regeneration • Development • Stem cell • Nephron progenitor • Wolffian duct • A-P patterning • Intermediate mesoderm

9.1 Introduction

The myriad types of cells that comprise a human develop from a single fertilized egg. Recent advancements in developmental biology have identified multiple growth factors responsible for cell-type specification during embryogenesis. Furthermore, multiple cell types have been successfully differentiated from pluripotent stem cells following the combination of these growth factors in vitro, demonstrating the sufficiency of extrinsic cues for cell fate acquisition (Murry and Keller 2008). However, directing differentiation to defined cell types requires fine optimization of the factor exposure concentrations, timing, duration, and combinations.

A. Taguchi • R. Nishinakamura (✉)
Department of Kidney Development, Institute of Molecular Embryology and Genetics, Kumamoto University, 2-2-1 Honjo, Kumamoto 860-0811, Japan
e-mail: ataguchi777@gmail.com; ryuichi@kumamoto-u.ac.jp

T. Tsuji (ed.), *Organ Regeneration Based on Developmental Biology*,
DOI 10.1007/978-981-10-3768-9_9

Therefore, a deeper understanding of the key signals required for lineage specification in vivo is essential for the induction of desired cell types. Presently, the precise mechanisms underlying kidney development during early embryogenesis are largely unknown (Dressler 2009; Taguchi et al. 2014; Taguchi and Nishinakamura 2014). This is in part attributable to the unique developmental processes that were acquired during evolution of the kidney. Here we review recent advances in the understanding of early kidney specification and its recapitulation by pluripotent stem cells. These advances have been obtained largely by the combination of in vivo biological and ex vivo directed stem cell differentiation approaches.

9.2 Kidney Function and Structure

The kidney filters blood to produce urine, thereby facilitating excretion of metabolic waste and extra fluid from the body. Renal arteries branch directly from the aorta, and the kidneys receive approximately one-fifth of resting cardiac output to promote filtration efficiency. This filtration function is accomplished by millions of individual filtration units—nephrons. Each nephron comprises a filtering apparatus—a renal corpuscle—and a connecting tubule unit, a nephric tubule. The nephric tubule balances the fluid volume, pH, and ion concentration of urine. The renal corpuscle contains characteristic arterioles which are covered by epithelial podocytes to form the glomerulus. Slit diaphragms form between the cellular foot processes of podocytes, serving as a molecular filtration barrier to prevent the leakage of blood proteins and cellular components. Each nephron is connected to the collecting ducts and urine then drains through the ureter to the bladder. This complex and sophisticated architecture is essential for the function of the kidney.

9.3 Overview of Kidney Organogenesis In Vivo

The kidney is a unique organ that developed during evolution when animals made the transition from water to land. In mammals, three distinct primordia—pronephros, mesonephros, and metanephros—form in an anterior-to-posterior direction along the body trunk during embryogenesis (Fig. 9.1). The pronephros is the most basic form and is found in mature primitive fish. The mesonephros is found in other fishes and amphibians. Both the pronephros and mesonephros appear transiently and then regress during mammalian gestation, though the functionality of these primitive rudiments varies among species (Saxen 1987). In mice, a rudimentary pronephros arises at embryonic day 8.5 (E8.5), and the mesonephros subsequently appears at E9.5–E10.0 posterior to the forelimb position. The metanephros then develops at E10.5 at the hind limb level of the body trunk.

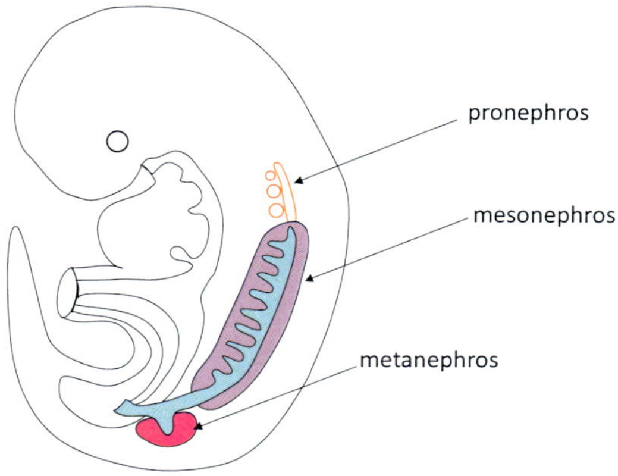

Fig. 9.1 Schematic of an E10.5 embryo. The kidney develops from three distinct primordia—pronephros, mesonephros, and metanephros—in an anterior-to-posterior direction

9.3.1 Interdependent Progenitor Interaction Is Required for Mature Kidney Architecture

The adult kidney is derived from the embryonic metanephros. The metanephros is anatomically divided into mesenchymal blastema—the metanephric mesenchyme (MM)—and epithelial blastema, the ureteric bud (UB). The MM contains at least three types of progenitors: nephron progenitors, stromal progenitors, and vascular endothelial cells (Fig. 9.2a). Nephron progenitors express the transcription factor *Six2* and give rise to most of the epithelial nephron components, including glomerular podocytes, proximal tubules, loop of Henle, and distal tubules (Self et al. 2006; Kobayashi et al. 2008; Osafune et al. 2006). Stromal progenitors express *Foxd1* and contribute to interstitial cells, smooth muscle cells, and mesangial cells (Humphreys et al. 2010, Kobayashi et al. 2014). Importantly, interactions among these cell populations enable the maintenance of a progenitor cell population at the periphery of the embryonic metanephros in a so-called nephrogenic zone and the continuous production of epithelialized nephron structures inside the layer (Fig. 9.2b). In the mouse, this serial generation of nephrons continues up to postnatal day 3, ultimately yielding tens of thousands of individual nephrons within a kidney.

9.3.2 Signals that Maintain the Ureteric Bud and Nephron Progenitors

Recent progress in the field of developmental biology has identified several key signals that are required for the maintenance and differentiation of nephron and ureteric progenitors (Fig. 9.2a). Glial cell line-derived neurotrophic factor (Gdnf) is

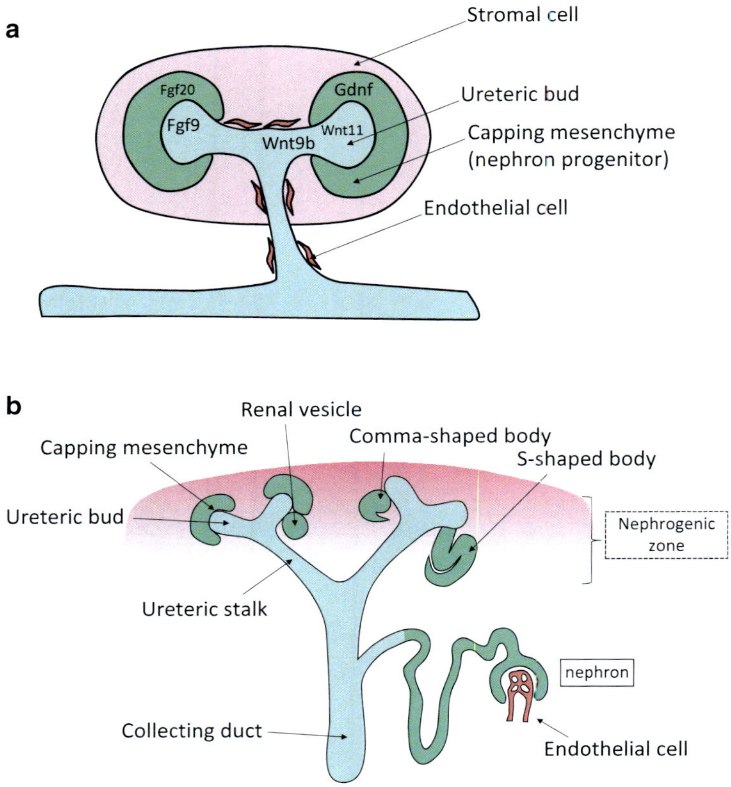

Fig. 9.2 Schematic of a developing metanephros. (**a**) The metanephric rudiment at E11.5 contains several progenitor types: stromal cells, capping mesenchyme, endothelial cells, and the ureteric bud. (**b**) Schematic of E15.5 metanephros. The capping mesenchyme (nephron progenitor) differentiates into an epithelial nephron by way of the renal vesicle, comma-shaped body, and S-shaped body. The peripheral kidney cortex, or "nephrogenic zone," contains an immature cell population, including nephron progenitors, ureteric bud tips, and cortical stromal progenitors (*pink-colored layer*)

secreted by nephron progenitor cells to signal via the tyrosine kinase receptor ret proto-oncogene (Ret) on the ureteric epithelia to enable dichotomous branching morphogenesis (Vega et al. 1996; Shakya et al. 2005). This signal network is positively coordinated by UB tip-derived Wnt11 in an autoregulatory feedback loop manner (Majumdar 2003; Kispert et al. 1996). Contrastingly, Fgf9 from the ureteric epithelium concomitant with the Fgf20-expressing nephron progenitor itself maintains the immature nephron progenitors on the surface of the ureteric tips (Barak et al. 2012). Bmp7 and canonical Wnt signaling have also been identified as the maintenance signals for nephron progenitors (Karner et al. 2011; Muthukrishnan et al. 2015). Additional to these interactions between the UB and capping mesenchyme, stromal cells are also important for UB branching and nephron differentiation (Das et al. 2013; Mendelsohn et al. 1999). Therefore, the interactions among the UB, metanephric nephron progenitors, and stromal progenitors are indispensable for the maintenance of an immature cell population in the nephrogenic zone during kidney organogenesis.

9.3.3 Wnt Triggers Self-Organizing Nephron Differentiation

A pioneering developmental biology study demonstrated that an inductive signal from the UB promotes differentiation of the MM to an epithelialized nephron structure (Grobstein 1953). Later, the embryonic dorsal spinal cord was identified as an alternative inducer that can trigger MM differentiation in an organ culture setting (Grobstein 1955). Other reports have since investigated the efficacy of multiple growth factors and cytokines in promoting the in vitro differentiation of MM, including LIF (Barasch et al. 1999), HGF (Woolf et al. 1995), bFGF (Perantoni et al. 1995), and TGFβ1 (Rogers et al. 1993).

An in vivo knockout study revealed that Wnt9b-driven Wnt4 upregulation within the nephron progenitor is an essential regulator of nephron differentiation (Stark et al. 1994; Carroll et al. 2005). The sufficiency of Wnt signaling for this process was shown by an in vitro study in which isolated MM was cocultured with Wnt-expressing feeder cells (Kispert et al. 1998; Carroll et al. 2005). Genetic stabilization of β-catenin induced the differentiation program but blocked the completion of MET (mesenchymal-to-epithelial transition), suggesting that canonical Wnt signaling must be switched off at an appropriate time (Park et al. 2007). Indeed, an in vitro study showed that transient application and subsequent removal of a chemical Wnt activator could induce formation of an epithelialized structure in the cultured MM (Kuure et al. 2007). Alternatively, others have provided evidence of a contribution from the Wnt-calcium pathway to the MET process rather than the canonical pathway (Tanigawa et al. 2011; Burn et al. 2011). Furthermore, ectopic activation of Notch signaling can drive MET in the absence of Wnt9b and Wnt4 in vitro (Boyle et al. 2011). Bmp7 signaling is also involved with this process by priming the $Cited1^{+}/Six2^{+}$ naïve capping mesenchyme into a $Cited1^{-}/Six2^{+}$ state, thereby facilitating receptivity to Wnt (Brown et al. 2013).

Intriguingly, after the Wnt-driven MET step, the subsequent conformational changes of nephron progenitors take place in a self-organizing manner. The epithelialized nephron progenitors, or renal vesicles, polarize proximo-distally under the influence of Notch-mediated patterning and elongate to form comma-shaped and subsequently S-shaped tubular structures (Cheng et al. 2003; Kopan et al. 2007) (Fig. 9.2b). These S-shaped bodies further elongate in a corticomedullary direction and form the loop of Henle presumably under the influence of Wnt-mediated signaling from the medullary stromal cells (Yu et al. 2009). Recent in vivo and in vitro studies have shown that the gradient of the Wnt signal may also play a role during the proximo-distal patterning of the nephron (Lindstrom et al. 2014; Schneider et al. 2015).

In summary, canonical Wnt signaling plays a central role in the initiation of nephron progenitor differentiation, whereas the subsequent processes—including MET and proximo-distal axis formation—are regulated in part by permissive signaling, enabling the self-organization of nephron structure construction. Importantly, however, the formation of complete nephron structures and their connection with the urinary exit tracts results from interdependent interactions between the nephron progenitor, stromal progenitor, and ureteric epithelium. Thus all components of the metanephric precursors are essential for the construction of well-organized kidney architecture.

9.4 Definition of Nephron Progenitors

The term “nephron progenitor” refers to cells with the ability to form a nephron. A genetic lineage tracing study utilizing a tamoxifen-inducible Cre mouse line found that Six2-positive cells in the MM develop into multiple types of nephron epithelial cells, including glomerular podocytes and nephric tubules—but not collecting ducts. Furthermore, transient pulse labeling of Six2-positive cells at E10.5–E11.5 found that they contributed to the E19.5 Six2-positive capping mesenchyme, indicating the capacity to self-renew the nephron progenitors (Kobayashi et al. 2008). However, a recent and more extensive study at a single-cell resolution identified stage-dependent aging and heterogeneity progression in the cellular population, suggesting the gradual loss of genuine stem cells (Chen et al. 2015). Other detailed analyses have revealed subpopulations or subdomains within the Six2-positive nephron progenitors, permitting classification of the population into Cited1^{+}/Six2^{+} and Cited1^{-}/Six2^{+} groups (Mugford et al. 2009; Brown et al. 2013). Nevertheless, expression of core transcriptional factors such as *Six2*, *Pax2*, *Wt1*, *Sall1*, *Eya1,* and *Osr1* is retained within the capping mesenchyme throughout nephrogenesis. Therefore, these genes are widely accepted as representative markers of nephron progenitors in mice (Dressler et al. 1993; Kreidberg et al. 1993; Nishinakamura et al. 2001; Xu et al. 1999; James et al. 2006). Notably, a recent study concerning the differential regulation of *SIX* transcription factors between mouse and human homologs in the mid-gestation-stage metanephros suggested the existence of differences in nephron progenitor programs between the species (O'Brien et al. 2016). Indeed, a previous report demonstrated expression of *PAX8* in the capping mesenchyme of the human embryonic metanephros, while *PAX8* expression in the mouse cap mesenchyme was restricted to differentiated pretubular aggregates or renal vesicles (Poleev et al. 1992). However, the limited accessibility of human fetal kidneys renders a comprehensive analysis of renal developmental stages difficult. Therefore, the criteria which validate genuine human nephron progenitors on the basis of representative markers remain incompletely understood.

9.5 Early Kidney Specification

All somatic tissues develop from the epiblast of the fertilized egg (E5.5 in mice). At E6.5 the embryo develops a primitive streak (PS) in the posterior region, in an area marked by the expression of transcription factor *T* (*Brachyury*) (Fig. 9.3a). The epiblast cells which migrate through the PS differentiate to form the nascent endoderm and mesoderm (Tam and Loebel 2007). As the kidney derives from mesodermal tissue, it is widely accepted that the kidney develops via the *T*-positive population. The *T*-positive nascent mesoderm becomes further specified into three subdomains along the dorsoventral axis: the paraxial mesoderm (PAM), intermediate mesoderm (IM), and a lateral plate mesoderm (LPM). These divisions first become apparent around E8.5 and are present until E9.5 in mice (Fig. 9.3b). The urogenital system—including all the kidney rudiments—is derived from IM. The IM further specifies along the anteroposterior (A-P) axis between E8.5 and E9.5, differentiating into the pro-, meso-, and metanephric anlagen in an anterior-to-posterior direction. A more complete understanding of the mechanisms underlying both dorsoventral patterning within the mesoderm and A-P patterning within the IM is therefore essential.

9.5.1 Positioning Within the Primitive Streak Influences Mesoderm Dorsoventral Patterning

Development of the nascent mesoderm occurs during the gastrulation stage (E6.5–E7.5 in mice) and is dependent on Nodal/activin, BMP, and WNT signaling (Tam and Loebel 2007). This combination of signaling pathways together with FGF establishes the primitive streak (Fig. 9.3a). The classic model explains germ layer specification based on the anterior-to-posterior gradient of Nodal/activin signal and the opposite concentration gradient of BMP. The anteriorly located cells within the primitive streak are exposed to a high concentration of Nodal/activin signal and develop into endoderm, whereas the posterior cells differentiate into mesoderm under the influence of BMP. Indeed, directed differentiation experiments employing pluripotent stem cells mimic findings that the combined actions of activin and BMP promote differentiation of the epiblast to mesoderm or endoderm fates (Gadue et al. 2006; Murry and Keller 2008).

Dye labeling-based lineage tracing experiment in chick embryos at a late gastrulation stage showed that the anterior cellular population within the primitive streak gives rise to the PAM, the posterior population to LPM, and the mid-streak cells to IM (Psychoyos and Stern 1996). Therefore, the signals that determine the A-P axis within the late primitive streak could specify the fate of future IM to some extent (James and Schultheiss 2003) (Fig. 9.3b).

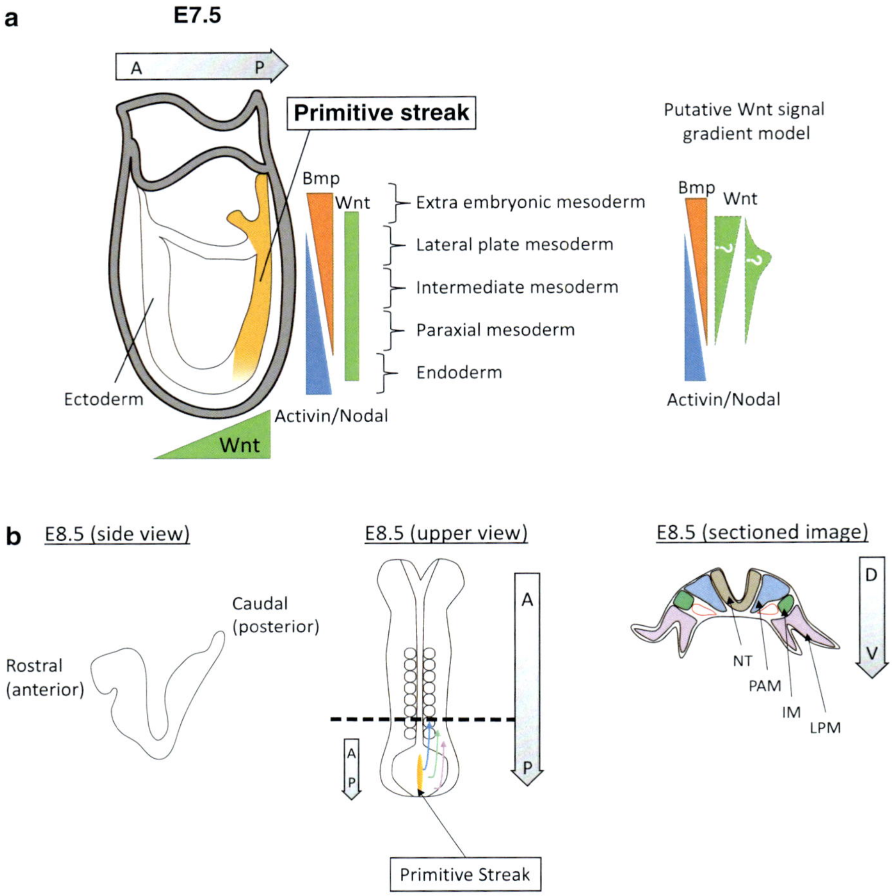

Fig. 9.3 Schematic of early lineage specification at E7.5 and E8.5. (**a**) Signal gradients within the primitive streak at E7.5. The anterior activin/Nodal signal and posterior Bmp signal together with Wnt signal patterns in the primitive streak (*left*). Hypothetical Wnt signal gradient within the primitive streak (*right*). (**b**) Schematic images of E8.5 embryo: side view (*left*), upper view (*middle*), and sectioned image (*right*). Continuous cell differentiation from the caudal immature population (primitive streak) produces the caudal body trunk. The anteroposterior position within the primitive streak largely determines the future dorsoventral patterning within the mesodermal tissues. The *dashed line* in the *middle panel* indicates the sectioned level of *right panel* image. *Arrows* indicate the direction of anterior (*A*)-to-posterior (*P*) axis (*middle panel*) or dorsal (*D*)-to-ventral (*V*) axis (*right panel*)

9.5.2 *Lateral Patterning of the IM During Post-migration*

In addition to the pre-patterning within the PS (before migration), previous studies proposed a patterning signal in the post-migration phase. Studies utilizing chick embryos demonstrated the importance of PAM-derived signals and an appropriate level of BMP signaling for subsequent IM marker gene expression following in vitro culture of the presomitic mesenchyme and in vivo ectopic activation of BMP signal within the PAM (Mauch et al. 2000; James and Schultheiss 2003; James and Schultheiss 2005). Another report has identified a putative role for Vg1/Nodal signal in the specification of IM (Fleming et al. 2013). These results indicate that factors from the neighboring PAM and LPM presumably signal to give the IM identity by establishing the dorsoventral signal gradients following migration from the primitive streak.

9.5.3 A-P Patterning Within the Intermediate Mesoderm

Because the distinct three types of kidney primordia form within the IM, it is essential to understand the A-P axis specification mechanism within the IM.

The initial A-P patterning within the IM becomes apparent at approximately E9.5 (Fig. 9.4a). Although the representative IM markers *Osr1* and *Wt1* are uniformly expressed in the IM at this stage, there exist distinct differences between the anterior and posterior regions of the body trunk. The anteriorly located IM at this stage (just beneath the forelimb level) exhibits a condensed structure—the nephrogenic cord—which is reminiscent of the capping mesenchyme in the metanephros (Fig. 9.4b). Indeed, they share a molecular signature with mature metanephric nephron progenitors: expression of *Pax2*, *Six2*, and *Gdnf* and the cell surface pattern Itga8^{+}/Pdgfra^{-} (Taguchi et al. 2014). Contrastingly, the posterior IM (PIM; located at the hind limb level) at this stage is negative for these markers, indicating that they are a newly formed immature population. Meanwhile, the PIM begins to express the posterior *Hox* genes including the *Hox10*, *Hox11*, and *Hox12* families, which are essential for metanephros development (Wellik et al. 2002; Mugford et al. 2008a). These findings suggest that the anteriorly located IM at this stage is the formerly developed progenitor for anterior kidney rudiments (mesonephros) and the PIM is the later-developing precursor for the posterior kidney anlage (metanephros).

Lineage tracing experiments using Six2-CreER mice identified that the *Six2*-positive anteriorly located IM at E9.5 contributes predominantly to the mesonephric nephrons but not to the metanephros. Accordingly, sorted Osr1- or Wt1-positive PIM cells at E9.5 could be induced to differentiate into metanephric nephron progenitors in vitro (Fig. 9.5). Here, the addition of retinoic acid, activin, and Bmp4 in the presence of a high concentration of Wnt agonist inhibited induction of nephron progenitors. Meanwhile, addition of Fgf9 and a low concentration of Wnt agonist synergistically promoted metanephric nephron progenitor induction from the Wt1^{+} PIM.

Taken together, the above findings suggest that anteriorly located IM at E9.5, which exhibits a mature signature pattern of nephron progenitors and predominantly contributes to the mesonephros, may not migrate to give rise to the metanephric nephron progenitors. Meanwhile, the PIM—which expresses posterior *Hox* genes—matures later to develop into the metanephric nephron progenitors (Figs. 9.5 and 9.6).

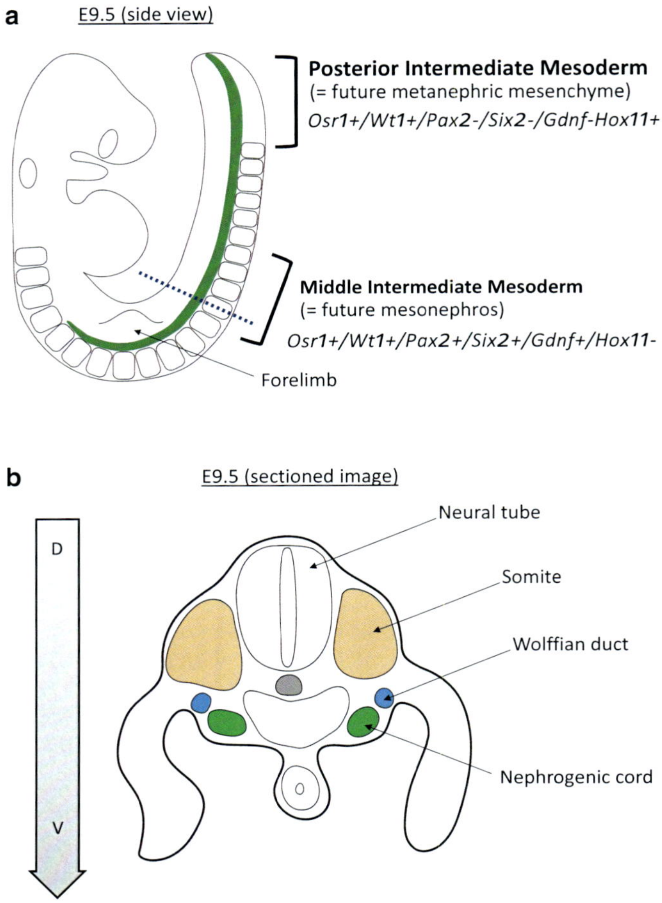

Fig. 9.4 Schematic of an E9.5 embryo. (**a**) Side view of an E9.5 embryo. *Green* indicates the intermediate mesoderm. The middle intermediate mesoderm is located in the anterior region of the body trunk and contributes to the mesonephros. The posterior intermediate mesoderm is located in the caudal region of the embryo and gives rise to the metanephric mesenchyme. The *dashed line* indicates the sectioned level of panel B. (**b**) Sectioned image of the E9.5 embryo at the middle intermediate mesoderm level. The middle intermediate mesoderm forms a condensate called the nephrogenic cord. *Arrow* indicates the direction of dorsal (*D*)-to-ventral (*V*) axis

Thus, similar to the process of somitogenesis, the IM develops in an anterior-to-posterior direction alongside caudal trunk elongation. Notably, the position of the anteriorly located IM is not at the most rostral part of the embryo, but is located in between the caudal forelimb edge and the rostral hind limb border level. Therefore, this region could be more appropriately referred to as a middle IM (Figs. 9.4, 9.5, and 9.6).

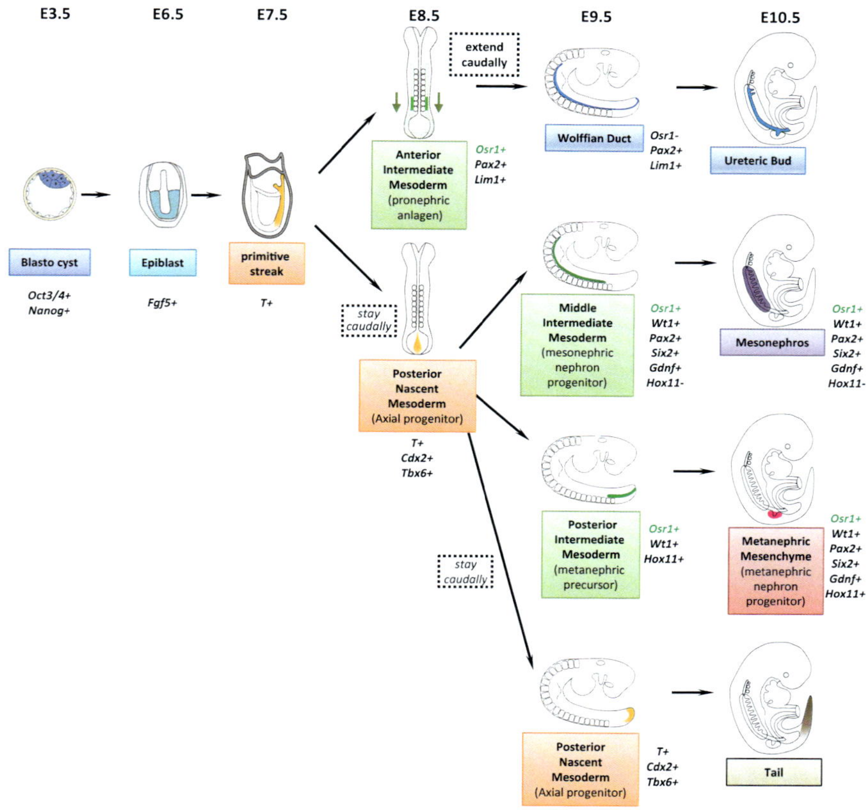

Fig. 9.5 Schematic of early kidney lineage specification. The transition timing from the T-positive immature population to the T-negative differentiated population determines the anteroposterior level in the body trunk

9.5.4 *Mechanism of A-P Axis Formation Along the Body Trunk Elongation*

During embryogenesis, formation of the rostral body trunk (the head and the anterior-most trunk region) occurs by the end of gastrulation (~E8.5), while elongation of the caudal body trunk continues in later embryogenesis (E8.5–E13.5) (Wilson et al. 2009). This axis elongation is attributed to the continuous supply of cells at the caudal end of the embryo, the so-called tail bud mesenchyme, which is marked by expression of *T* and *Cdx2* (Figs. 9.5 and 9.6).

The caudal immature population is maintained by Wnt and FGF signals and the suppression of retinoic acid signals by the degradation enzyme Cyp26s (cytochrome P450, family 26). Wnt directly promotes *T* expression, and *T* upregulates the expression of Wnt3a in a positive-feedback manner. Indeed, both *T* and *Wnt3a* knockout

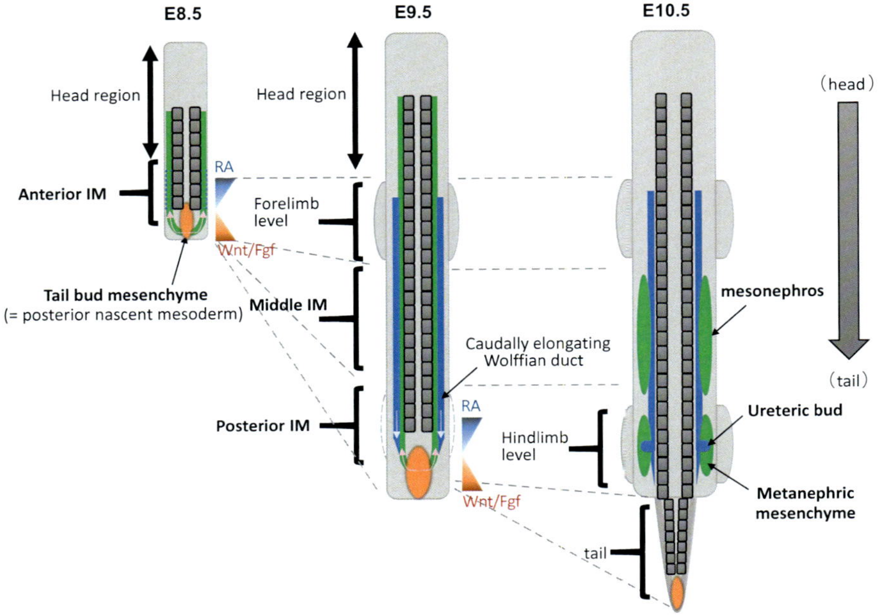

Fig. 9.6 Schematic of the body trunk elongation process during development. The T-positive tail bud mesenchyme persists in the caudal end of the embryo up to E13.5, leading to development of the caudal body trunk. The caudal immature cell population is maintained in a Wnt/Fgf-rich environment, whereas retinoic acid signaling promotes differentiation. *Green*, intermediate mesoderm; *blue*, Wolffian duct; *orange*, caudal immature population

mice exhibit caudal truncation of the embryo (Takada et al. 1994; Yamaguchi et al. 1999). Contrastingly, cells undergo differentiation by leaving the tail bud region and migrating to the former side, where they receive less Wnt/FGF signal and more retinoic acid signal. Such continuous recruitment of differentiated cells enables formation of the caudal body trunk, including somite formation. Therefore, the timing of exposure to differentiation signals determines the axial level within the caudal body trunk (Aulehla and Pourquie 2010) (Figs. 9.5 and 9.6).

Because the caudal body trunk includes both neural and mesodermal tissues, it is regarded as a bipotent progenitor, namely, the "axial progenitor" or "neuromesodermal progenitor." Recent in vivo studies have provided supporting evidence using genetic lineage tracing and genetic manipulation studies (Tzouanacou et al. 2009; Takemoto et al. 2011). These bipotent cells have been recently classified as a *T* and *Sox2* double-positive cell population (Henrique et al. 2015). However, the distinct origin of this bipotent population in the early gastrulation-stage embryo (E6.5–E7.5) is not known, and the contribution of this population to the metanephros is unclear.

9.5.5 *UB and MM Differentiate from a Caudal* T^+ *Population at Different Times*

A lineage tracing experiment utilizing Osr1-GFPCreER mice revealed that Osr1-positive cells at E6.5 give rise to a small part of the ureteric epithelium. Additionally, Osr1-positive cells from E7.5 to E8.5 contribute to the metanephros, and those present after E9.5 give rise to metanephros other than the ureteric epithelium (Mugford et al. 2008b). However, because *Osr1* is continuously expressed from the early IM (E8.5) to the late IM (E9.5), it remains unclear whether the bipotent progenitor cells which give rise to both the MM and UB exist.

We recently employed T-GFPCreER mice to show that T-positive cells at E6.5 and E7.5 give rise to both the MM and UB. Meanwhile, at E8.5 these cells contributed to the MM but not the UB. The T-positive cells at E9.5 contributed exclusively to the tail (Taguchi et al. 2014). Because expression of T is a marker of immature caudal tissue, our results indicate that the timing of differentiation from the T^+ population is different between the UB and the MM. Therefore, we concluded that the UB differentiates from the T^+ population between E7.5 and E8.5 to give rise to the anterior IM (the early-formed IM), and the MM differentiates from E8.5 to E9.5 T^+ cells and develops via the PIM (the late-formed IM) (Figs. 9.5 and 9.6).

The Wolffian duct (WD), also known as the pronephric or mesonephric duct, is the precursor of the UB, and its morphogenic process has been well studied in chicken and amphibians (Drawbridge et al. 2003). Previous reports have shown that the WD originates from anterior IM at the axial level of somites 8–10, and committed WD progenitors migrate toward the cloaca in an anterior-to-posterior direction. The equivalent WD progenitors in the mouse are proposed to be in the IM at the level of somites 6–8. These progenitors are identified by the well-conserved markers, *Osr1*, *Lim1*, *Pax2* (*Pax8*), and *Ret*.

Taken together, these findings indicate that at E8.5 precursors of the UB are located in the $Osr1^+/T^-$ population and are already segregated from precursors of the MM, which are localized in the T^+ posterior nascent mesoderm (Figs. 9.5 and 9.6). Additionally, we successfully induced metanephric nephron progenitors from a sorted E8.5 caudal T^+ population. To this end, we employed a high concentration of Wnt agonist in combination with Bmp4 to mimic the conditions within the posterior tail bud and showed maintenance of the $T^+/Cdx2^+$ state and induction of posterior *Hox* gene expression with time. Treatment with activin, Bmp4, an intermediate concentration of Wnt agonist, and retinoic acid synergistically induced a PIM-like population that expressed *Osr1*, *Wt1*, and *Hox11*. Importantly, the depletion of each single factor or administration of inhibitors hampered nephron progenitor induction, suggesting the requirement of all signals for this step. Finally, the maturation factors (Fgf9 + low Wnt) identified in the E9.5 PIM culture successfully induced the $Six2^+/Pax2^+/Gdnf^+$ metanephric nephron progenitors in vitro (Fig. 9.7).

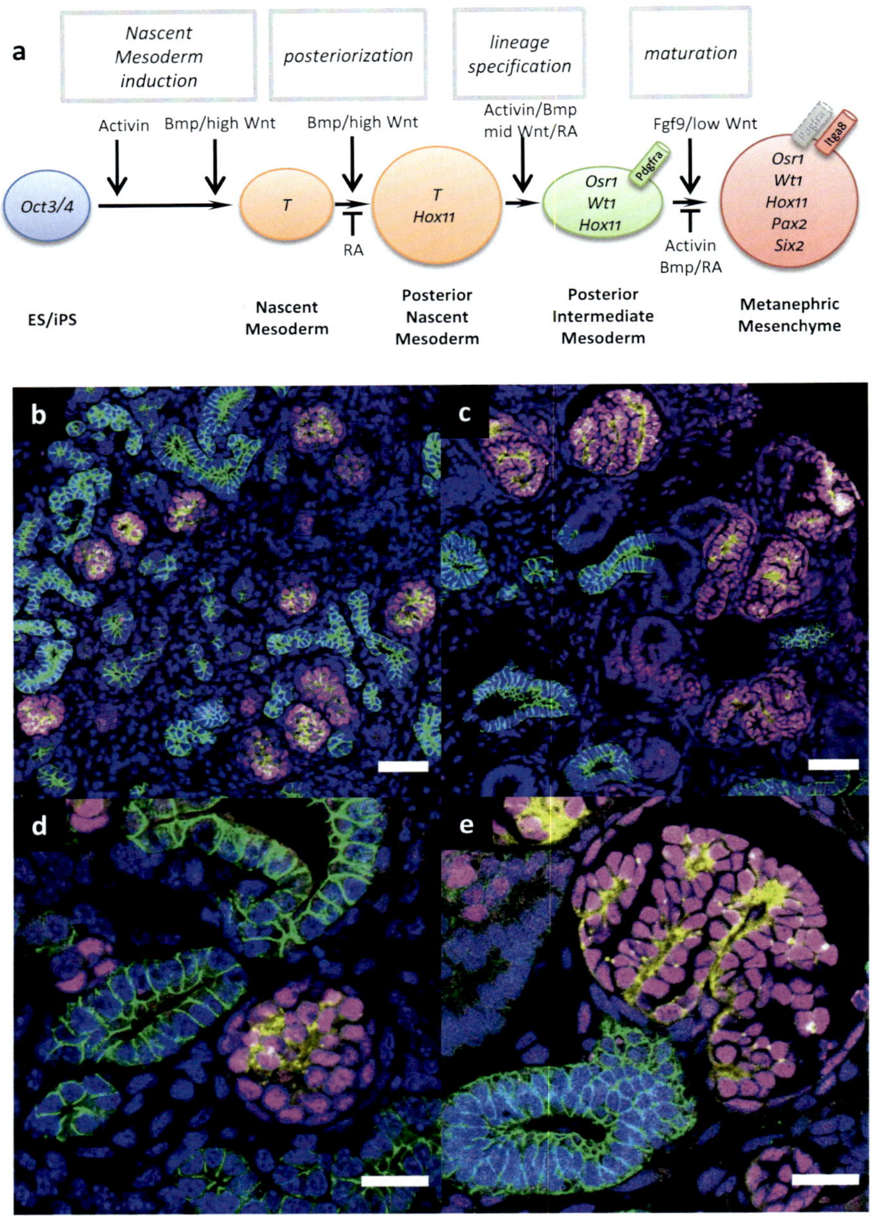

Fig. 9.7 In vitro reconstitution of three-dimensional kidney structures. (**a**) Model of directed differentiation of pluripotent stem cells toward the metanephric mesenchyme. (**b–e**) Development of three-dimensional kidney structures from mouse ES cells and human iPS cells. Immunostaining of induced nephron progenitors for the nephric tubule marker E-cadherin (*green*) and the glomerular markers nephrin (*yellow*) and Wt1 (*red*). (**b**, **c**) Lower magnification images of induced mouse (**b**) and human (**c**) kidney tissue. (**d**, **e**) Higher magnification images of induced mouse (**d**) and human (**e**) kidney tissue. Scale bars, 50 μm (**b**, **c**), 20 μm (**d**, **e**). ES cells, embryonic stem cells; iPS cells, induced pluripotent stem cells; *RA* retinoic acid

9.6 Induction of Metanephric Nephron Progenitors from Pluripotent Stem Cells

In contrast to conventionally accepted Nodal−/activin- and Bmp-driven patterning within the PS, we found that a high concentration of Wnt agonist combined with an optimal concentration of Bmp4 induced the $T^{+}/Cdx2^{+}$ posterior nascent mesoderm (or tail bud mesenchyme, late primitive streak). This caudal immature population should be further maintained by high concentrations of Wnt to posteriorize the population (see 1.4.5). Indeed, prolonged Wnt activation followed by exposure to the differentiation cocktail efficiently induced metanephric nephron progenitors from both mouse embryonic stem cells and human-induced pluripotent stem cells (Fig. 9.7). Importantly, the nephron progenitors derived from mouse ES cells express representative key transcription factors at similar levels to those observed in the embryonic mouse metanephric nephron progenitors.

In response, several groups investigating kidney induction have recently revised their original protocols and now employ a prolonged Wnt-driven strategy, although the duration of Wnt treatment slightly varies among protocols, from 4 (Takasato et al. 2015; Morizane et al. 2015) to 6 days (Taguchi et al. 2014). The subsequent conditions used to induce PIM and mature nephron progenitor development are also somewhat different among the protocols. Nevertheless, the shared stream of differentiation—exposure to a prolonged strong Wnt signal followed by attenuation of the Wnt signal and administration of Fgf9—could presumably facilitate programming into nephron progenitors to a certain extent. Importantly, however, the endogenous activity of each growth factor signal varies among cell lines and maintenance conditions. Therefore, the fine tuning of each growth factor signaling is indispensable to optimize the process in each laboratory.

Intriguingly, a recent study that aimed to induce caudal paraxial mesoderm reported the efficacy of initial Wnt treatment, concluding that the effect of Wnt signaling was the induction of the late primitive streak (Mendjan et al. 2014). Other studies that aimed to induce neuromesodermal progenitors—the bipotent precursors of caudal PAM and the neural tube—have shown the efficacy of initial prolonged Wnt and FGF treatment (Gouti et al. 2014; Turner et al. 2014; Henrique et al. 2015). Taken together, in vitro directed differentiation experiments may indicate the slope, or local accumulation of Wnt signals within the primitive streak confers the caudal identity in addition to the Nodal/BMP gradient (Fig. 9.3a).

9.7 Reconstitution of a Three-Dimensional Nephron by Self-Organizing Morphogenesis

Pluripotent stem cell-derived metanephric nephron progenitors can differentiate into three-dimensional nephron structures in a self-organizing manner under the influence of Wnt (see 1.2.4). Indeed, the conventional coculture method that

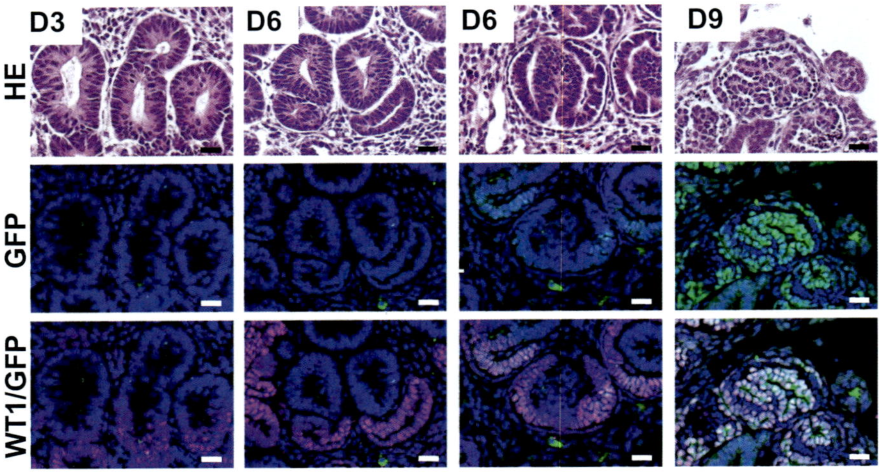

Fig. 9.8 Fluorescent visualization of human glomerular podocytes generated from NPHS1-GFP iPS cells. Histologic sections of glomeruli developing in vitro. Tissues at days 3, 6, and 9 after recombination with the spinal cord were analyzed. *Top panels*: hematoxylin-eosin (*HE*) staining. *Middle panels*: GFP (*green*) staining. *Bottom panels*: dual staining with GFP and WT1. Nuclei were stained with 4′9,6-diamidino-2-phenylindole (DAPI, *blue*). Scale bars, 20 μm

employs embryonic spinal cord and Wnt4-expressing 3T3 feeder cells efficiently differentiates both mouse and human PSC-derived nephron progenitors into nephron structures. These three-dimensional nephron structures include spherical glomerular podocytes and proximo-distally specified nephric tubules (Fig. 9.7). Intriguingly, such conformation changes take place in a similar sequence to those of the in vivo morphological processes. At day 3 of differentiation, renal vesicle-like structures were observed. At day 6, S-shaped bodies were specified into E-cadherin$^+$ presumptive distal tubules, cadherin 6$^+$ proximal tubules, and WT1$^+$ glomerular podocytes. Finally, capillary convolution takes place within the glomerular structure, later developing into the spherical renal corpuscle at approximately day 9 (Sharmin et al. 2015) (Fig. 9.8). Scanning electron microscopy identified the Bowman's capsule surrounding the glomerulus and primitive foot process-like and slit diaphragm-like structures in between the induced podocytes in vitro.

The podocytes are aligned along the type IV collagen-positive basement membrane. Additionally, an apicobasally polarized distribution of cell surface molecules is present, including apically distributed podocalyxin and basally localized nephrin, podocin, and neph1 (Fig. 9.9). Microarray gene expression analysis utilizing a newly generated nephrin-GFP knock-in iPSC identified well-conserved molecular signatures within mouse adult podocytes and human adult glomeruli. Such molecules include podocyte development-related transcriptional factors and cytokines

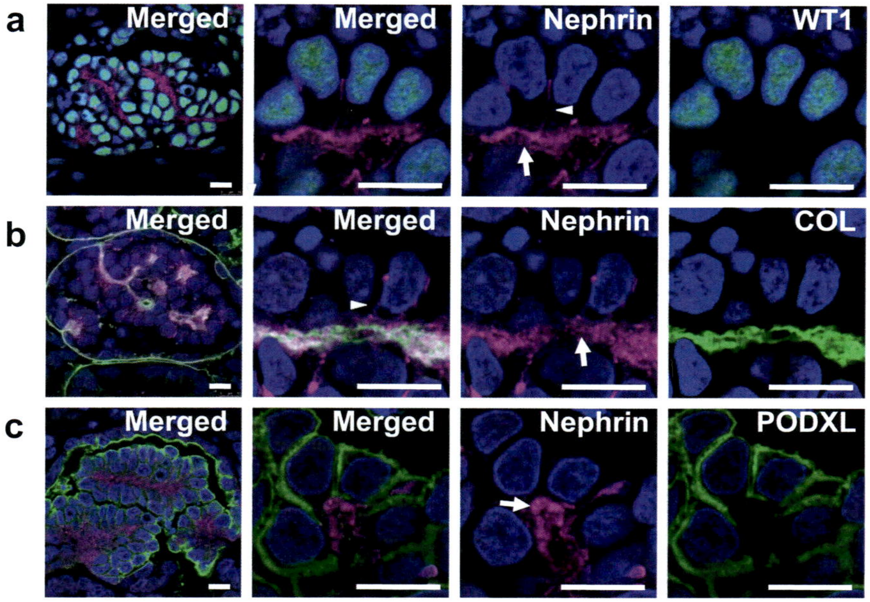

Fig. 9.9 Induced podocytes exhibit apicobasal polarity and basal localization of nephrin. (**a**) Nephrin (*magenta*) and WT1 (*green*) staining of the induced glomerulus at day 9. (**b**) Nephrin (*magenta*) and type IV collagen (COL, *green*) staining. (**c**) Nephrin (*magenta*) and podocalyxin (PODXL, *green*) staining. The *left* columns are at lower magnification to show a whole glomerulus. The *right* two columns are singly stained, while the *left* two columns represent merged images. *Arrows*: nephrin proteins localized to the basal domain; *arrowheads*: nephrin-positive dot-like or filamentous structures. Scale bars, 10 μm

(WT1, MAFB, FOXD1, TCF21, and VEGFA), typical slit diaphragm-related genes (NPHS1, NPHS2, CD2AP, CLIC5, and dendrin), basolateral adhesion molecules (claudin 5 and integrin α3), and causative genes for hereditary kidney diseases (phospholipase ε1 and Myh9) (Table 9.1).

Specific interactions between vascular endothelial cells and podocytes are essential for normal glomerular development. Indeed, the induced glomerular podocytes underwent further maturation following their transplantation to beneath the kidney capsule of immunodeficient mice. Here, host mouse vasculature integrated with the glomeruli of the transplanted tissue and the foot process and slit diaphragms that developed were more distinct than those formed in vitro. These findings indicate that human iPSC-derived glomerular podocytes matured further following transplantation, possibly under the influence of host vascular endothelial cells and circulation (Fig. 9.10).

Table 9.1 Genes common to iPS-derived podocytes in vitro, human glomeruli, and mouse podocytes in vivo

Gene Symbol	Gene name	iPS-derived podocytes	Human glomeruli	Mouse podocytes
NPHS1	Nephrosis 1, congenital, Finnish type (nephrin)	133.16	10.41	22.75
CLIC5	Chloride intracellular channel 5	120.01	6.48	9.10
NPHS2	Nephrosis 2, idiopathic, steroid resistant (podocin)	104.25	9.99	19.95
PODXL	Podocalyxin-like	77.43	9.34	26.60
CLDN5	Claudin 5	61.52	14.33	5.39
ITGA3	Integrin, alpha 3 (antigen CD49C, alpha 3 subunit of VLA-3 receptor)	54.54	9.33	9.98
CTGF	Connective tissue growth factor	36.95	8.31	4.48
MAFB	v-maf avian musculoaponeurotic fibrosarcoma oncogene homolog B	27.76	14.07	8.56
FOXD1	Forkhead box D1	26.92	3.87	18.67
SEMA3G	Sema domain, immunoglobulin domain (Ig), short basic domain, secreted, (semaphorin) 3G	23.97	31.46	6.33
WT1	Wilms tumor 1	22.01	14.02	19.15
DDN	Dendrin	21.91	9.91	4.25
BCAM	Basal cell adhesion molecule (Lutheran blood group)	21.07	2.04	2.71
TCF21	Transcription factor 21	18.95	31.46	14.60
MAFB	v-maf avian musculoaponeurotic fibrosarcoma oncogene homolog B	16.09	14.07	8.56
VEGFA	Vascular endothelial growth factor A	14.72	2.33	6.20
LAMA5	Laminin, alpha 5	14.05	2.27	2.32
VEGFA	Vascular endothelial growth factor A	13.78	2.33	6.20
SYNPO	Synaptopodin	9.62	19.26	7.40
MAGI2	Membrane-associated guanylate kinase, WW and PDZ domain containing 2	8.37	25.02	16.28
FOXC1	Forkhead box C1	6.53	6.92	2.71
PDPN	Podoplanin	6.07	8.99	17.51
MAGI2	Membrane-associated guanylate kinase, WW and PDZ domain containing 2	4.72	25.02	16.28
LAMB2	Laminin, beta 2 (laminin S)	4.55	5.00	3.54
SYNPO	Synaptopodin	4.51	19.26	7.40
MYH9	Myosin, heavy chain 9, non-muscle	4.47	4.38	4.07
LAMB2	Laminin, beta 2 (laminin S)	4.43	5.00	3.54
MYH9	Myosin, heavy chain 9, non-muscle	3.44	4.38	4.07
CD2AP	CD2-associated protein	3.29	1.58	3.52
LAMC1	Laminin, gamma 1 (formerly LAMB2)	2.98	2.69	2.29

(continued)

Table 9.1 (continued)

Gene Symbol	Gene name	iPS-derived podocytes	Human glomeruli	Mouse podocytes
CTNNAL1	Catenin (cadherin-associated protein), alpha-like 1	2.66	2.13	5.97
COL4A4	Collagen, type IV, alpha 4	2.57	1.60	4.42
ITGAV	Integrin, alpha V	2.43	4.59	3.16
COL4A5	Collagen, type IV, alpha 5	2.13	11.05	2.51
SYNPO	Synaptopodin	2.07	19.26	7.40

9.8 Applications Toward the Disease Modeling

It is now possible to effectively induce metanephric nephron progenitor cells from a stem cell population. Further, the three-dimensional structure of the kidney—including both the glomerulus and the renal tubule—can be reconstructed. Therefore, it is feasible that patient-derived iPS cells could be used to produce disease-specific renal tubules and podocytes that could be applied in the treatment of renal diseases and the development of novel drugs. Indeed, recent reports have described the partial reproduction of nephric tubule injury. For example, apoptosis or the expression of KIM1 has been detected following application of the nephrotoxic drug cisplatin to induced kidney tissue in vitro (Takasato et al. 2015, Morizane et al. 2015, Freedman et al. 2015). However, to fully replicate physiological disease dynamics, it may be essential to induce blood perfusion into the induced kidney tissue.

9.9 Future Directions Toward Creation of Functional Kidneys

The kidney functions by filtering blood to produce urine. Therefore, a continuous blood supply, a filtration unit (nephron), and a drainage system are essential for renal function. Generation of functional renal tissues will require replication of the physiological interactions that occur between innate kidney precursors, i.e., the UB, nephron progenitors, stromal cells, and the endothelia (see 1.2.1). Most reports to date have focused on the selective induction of nephron components via the nephron progenitors and have required coculture with Wnt-producing cells or administration of Wnt analogues to induce formation of epithelialized nephron structures. Because these differentiation signals deplete the supply of uninduced nephron progenitors, they disrupt the continuous propagation of nephron structures as seen in vivo (see 1.2.2). Additionally, induced nephron structures within the tissue are not connected to each other via the collecting duct system and hence cannot excrete urine outside

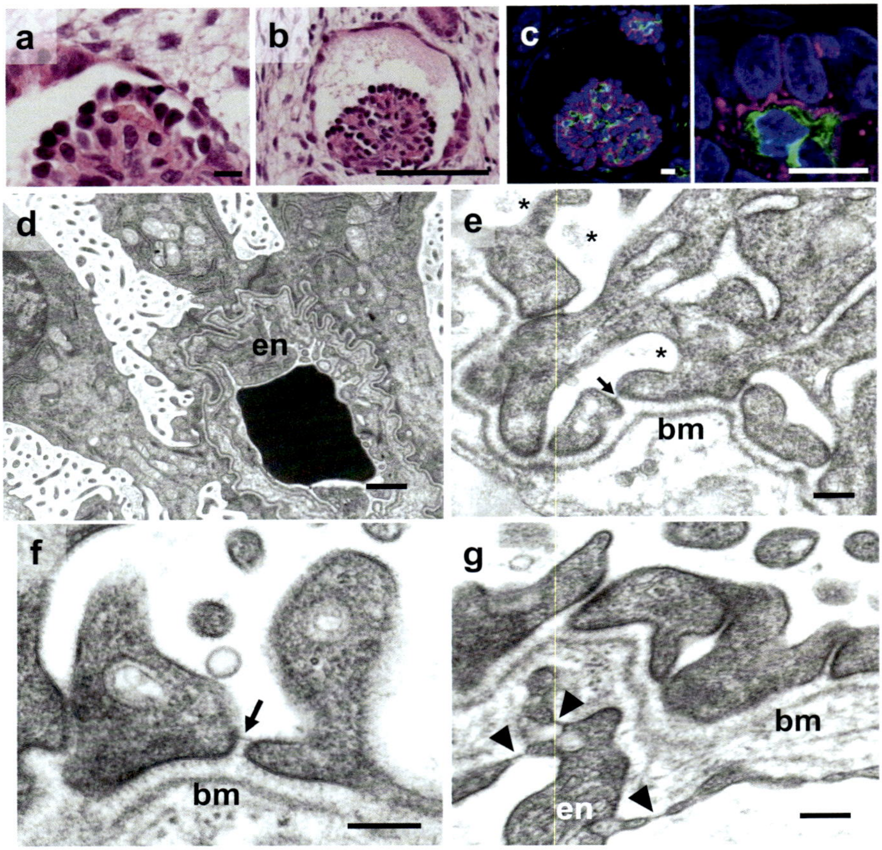

Fig. 9.10 Transplanted iPS cell-derived nephron progenitors form vascularized glomeruli. (**a**) Hematoxylin-eosin staining showing red blood cells in the induced glomeruli. (**b**) Hematoxylin-eosin staining showing the eosin-positive precipitates in the Bowman's space. (**c**) Staining of nephrin (*magenta*) and CD31 (*green*). *Right* panel shows the basal localization of nephrin. Scale bars: A, C–F, I, 100 μm; B, 1 μm; G, H, J, 10 μm. (**d**) Induced podocytes surrounding the vascular endothelial cells and extending multiple cell processes, shown by transmission electron microcopy. (**e** and **f**) Formation of slit diaphragm-like structures (*arrows*) between the cell processes of induced podocytes. Note the electron-dense substance in the Bowman's capsule (*asterisk*). (**g**) Fenestrated endothelial cells with diaphragms (*black arrowheads*). bm, Basement membrane derived from induced podocytes; en, endothelial cells. Scale bars: A, 1 μm; B, E, 0.5 μm; C, D, F, 0.2 μm

of the tissue when transplanted regardless of their potential filtration capacity (Sharmin et al. 2015). Although some reports have shown directed or simultaneous induction of UB lineage cells, they are not programmed enough to exhibit functionality (Xia et al. 2013; Takasato et al. 2015). Therefore, it is desirable to induce fully competent UB to recapitulate a functional kidney organoid. Furthermore, the stromal cell population within the kidney is essential for growth of the UB and the differentiation of nephron progenitors (Mendelsohn et al. 1999; Das et al. 2013). Because the stromal cells could be a mixed population from several distinct origins, further developmental examination should be undertaken to clarify early lineage specification processes.

Future investigations should examine how best to validate induced human metanephric nephron progenitors. Although current available protocols evaluate the quality of nephron progenitors based on epithelial nephron structure reconstitution capacity and marker genes expression, several problems with this approach exist. First, the precise quantitative gene expression profile of human fetal nephron progenitors is unknown, complicating the evaluation of induced nephron progenitors. We have evaluated the quantitative gene expression profile of murine ESC-derived nephron progenitors using a pure fetal population, and similar information from human equivalents is essential. One problem complicating this issue is the difficulty in discriminating between mesonephric and metanephric nephron progenitors. Because both populations differentiate into nephric tubules and glomeruli, they cannot be distinguished by morphological examination. Additionally, the widely used method of posterior *Hox* gene examination may be insufficient, because Hox expression is dose dependent and not all-or-none. Second, there is no reliable functional assay to quantitatively assess early nephron differentiation or late functional nephron maturation. The currently used chemical Wnt agonist-driven differentiation method varies between protocols, and it may be less effective in inducing differentiation of nephron progenitors when compared with conventional Wnt-producing cell-mediated stimulation. This also complicates attempts to compare induction protocols. Furthermore, the late developmental process by which the kidney matures into a functional organ is still poorly understood, and there is no method to recapitulate this process at present. Further evaluation of the capacity for induced nephron progenitor cells to produce a functional kidney organoid will require a more thorough understanding of the mechanisms underlying late-stage kidney organogenesis.

9.10 Concluding Remarks

Many years of scientific effort toward understanding kidney developmental processes have enabled the production of three-dimensional kidney tissue in a dish. This milestone is a product of the careful analysis and understanding of multiple biological mechanisms, including interdependent kidney organogenesis, self-organizing nephron induction, and early lineage specification. Although the currently available tissue remains immature and cannot replace the adult kidney functions, continued progress in the field of kidney regeneration may one day yield an era in which hemodialysis is no longer needed.

References

Aulehla A, Pourquie O (2010) Signaling gradients during paraxial mesoderm development. Cold Spring Harb Perspect Biol 2:a000869

Barak H, Huh SH, Chen S, Jeanpierre C, Martinovic J, Parisot M, Bole-Feysot C, Nitschke P, Salomon R, Antignac C, Ornitz DM, Kopan R (2012) FGF9 and FGF20 maintain the stemness of nephron progenitors in mice and man. Dev Cell 22:1191–1207

Barasch J, Yang J, Ware CB, Taga T, Yoshida K, Erdjument-Bromage H, Tempst P, Parravicini E, Malach S, Aranoff T, Oliver JA (1999) Mesenchymal to epithelial conversion in rat metanephros is induced by LIF. Cell 99:377–386

Boyle SC, Kim M, Valerius MT, Mcmahon AP, Kopan R (2011) Notch pathway activation can replace the requirement for Wnt4 and Wnt9b in mesenchymal-to-epithelial transition of nephron stem cells. Development 138:4245–4254

Brown AC, Muthukrishnan SD, Guay JA, Adams DC, Schafer DA, Fetting JL, Oxburgh L (2013) Role for compartmentalization in nephron progenitor differentiation. Proc Natl Acad Sci U S A 110:4640–4645

Burn SF, Webb A, Berry RL, Davies JA, Ferrer-Vaquer A, Hadjantonakis AK, Hastie ND, Hohenstein P (2011) Calcium/NFAT signalling promotes early nephrogenesis. Dev Biol 352:288–298

Carroll TJ, Park JS, Hayashi S, Majumdar A, Mcmahon AP (2005) Wnt9b plays a central role in the regulation of mesenchymal to epithelial transitions underlying organogenesis of the mammalian urogenital system. Dev Cell 9:283–292

Chen S, Brunskill EW, Potter SS, Dexheimer PJ, Salomonis N, Aronow BJ, HONG CI, Zhang T, Kopan R (2015) Intrinsic age-dependent changes and cell-cell contacts regulate nephron progenitor lifespan. Dev Cell 35:49–62

Cheng HT, Miner JH, Lin M, Tansey MG, Roth K, Kopan R (2003) Gamma-secretase activity is dispensable for mesenchyme-to-epithelium transition but required for podocyte and proximal tubule formation in developing mouse kidney. Development 130:5031–5042

Das A, Tanigawa S, Karner CM, Xin M, Lum L, Chen C, Olson EN, Perantoni AO, Carroll TJ (2013) Stromal-epithelial crosstalk regulates kidney progenitor cell differentiation. Nat Cell Biol 15:1035–1044

Drawbridge J, Meighan CM, Lumpkins R, Kite ME (2003) Pronephric duct extension in amphibian embryos: migration and other mechanisms. Dev Dyn 226:1–11

Dressler GR (2009) Advances in early kidney specification, development and patterning. Development 136:3863–3874

Dressler GR, Wilkinson JE, Rothenpieler UW, Patterson LT, Williams-Simons L, Westphal H (1993) Deregulation of Pax-2 expression in transgenic mice generates severe kidney abnormalities. Nature 362:65–67

Fleming BM, Yelin R, James RG, Schultheiss TM (2013) A role for Vg1/Nodal signaling in specification of the intermediate mesoderm. Development 140:1819–1829

Freedman BS, Brooks CR, Lam AQ, Fu H, Morizane R, Agrawal V, Saad AF, Li MK, Hughes MR, Werff RV, Peters DT, Lu J, Baccei A, Siedlecki AM, Valerius MT, Musunuru K, Mcnagny KM, Steinman TI, Zhou J, Lerou PH, Bonventre JV (2015) Modelling kidney disease with CRISPR-mutant kidney organoids derived from human pluripotent epiblast spheroids. Nat Commun 6:8715

Gadue P, Huber TL, Paddison PJ, Keller GM (2006) Wnt and TGF-beta signaling are required for the induction of an in vitro model of primitive streak formation using embryonic stem cells. Proc Natl Acad Sci U S A 103:16806–16811

Gouti M, Tsakiridis A, Wymeersch FJ, Huang Y, Kleinjung J, Wilson V, Briscoe J (2014) In vitro generation of neuromesodermal progenitors reveals distinct roles for wnt signalling in the specification of spinal cord and paraxial mesoderm identity. PLoS Biol 12:e1001937

Grobstein C (1953) Inductive epithelio-mesenchymal interaction in cultured organ rudiments of the mouse metanephros. Science 118:52–55

Grobstein C (1955) Inductive interaction in the development of the mouse metanephros. J Exp Zool 130:319–340
Henrique D, Abranches E, Verrier L, Storey KG (2015) Neuromesodermal progenitors and the making of the spinal cord. Development 142:2864–2875
Humphreys BD, Lin SL, Kobayashi A, Hudson TE, Nowlin BT, Bonventre JV, Valerius MT, Mcmahon AP, Duffield JS (2010) Fate tracing reveals the pericyte and not epithelial origin of myofibroblasts in kidney fibrosis. Am J Pathol 176:85–97
James RG, Schultheiss TM (2003) Patterning of the avian intermediate mesoderm by lateral plate and axial tissues. Dev Biol 253:109–124
James RG, Schultheiss TM (2005) Bmp signaling promotes intermediate mesoderm gene expression in a dose-dependent, cell-autonomous and translation-dependent manner. Dev Biol 288:113–125
James RG, Kamei CN, Wang Q, Jiang R, Schultheiss TM (2006) Odd-skipped related 1 is required for development of the metanephric kidney and regulates formation and differentiation of kidney precursor cells. Development 133:2995–3004
Karner CM, Das A, Ma Z, Self M, Chen C, Lum L, Oliver G, Carroll TJ (2011) Canonical Wnt9b signaling balances progenitor cell expansion and differentiation during kidney development. Development 138:1247–1257
Kispert A, Vainio S, Shen L, Rowitch DH, Mcmahon AP (1996) Proteoglycans are required for maintenance of Wnt-11 expression in the ureter tips. Development 122:3627–3637
Kispert A, Vainio S, Mcmahon AP (1998) Wnt-4 is a mesenchymal signal for epithelial transformation of metanephric mesenchyme in the developing kidney. Development 125:4225–4234
Kobayashi A, Valerius MT, Mugford JW, Carroll TJ, Self M, Oliver G, Mcmahon AP (2008) Six2 defines and regulates a multipotent self-renewing nephron progenitor population throughout mammalian kidney development. Cell Stem Cell 3:169–181
Kobayashi A, Mugford JW, Krautzberger AM, Naiman N, Liao J, Mcmahon AP (2014) Identification of a multipotent self-renewing stromal progenitor population during mammalian kidney organogenesis. Stem Cell Rep 3:650–662
Kopan R, Cheng HT, Surendran K (2007) Molecular insights into segmentation along the proximal-distal axis of the nephron. J Am Soc Nephrol 18:2014–2020
Kreidberg JA, Sariola H, Loring JM, Maeda M, Pelletier J, Housman D, Jaenisch R (1993) WT-1 is required for early kidney development. Cell 74:679–691
Kuure S, Popsueva A, Jakobson M, Sainio K, Sariola H (2007) Glycogen synthase kinase-3 inactivation and stabilization of beta-catenin induce nephron differentiation in isolated mouse and rat kidney mesenchymes. J Am Soc Nephrol 18:1130–1139
Lindstrom NO, Lawrence ML, Burn SF, Johansson JA, Bakker ER, Ridgway RA, Chang CH, Karolak MJ, Oxburgh L, Headon DJ, Sansom OJ, Smits R, Davies JA, Hohenstein P (2014) Integrated beta-catenin, BMP, PTEN, and Notch signalling patterns the nephron. elife 3:e04000
Majumdar A (2003) Wnt11 and Ret/Gdnf pathways cooperate in regulating ureteric branching during metanephric kidney development. Development 130:3175–3185
Mauch TJ, Yang G, Wright M, Smith D, Schoenwolf GC (2000) Signals from trunk paraxial mesoderm induce pronephros formation in chick intermediate mesoderm. Dev Biol 220:62–75
Mendelsohn C, Batourina E, Fung S, Gilbert T, Dodd J (1999) Stromal cells mediate retinoid-dependent functions essential for renal development. Development 126:1139–1148
Mendjan S, Mascetti VL, Ortmann D, Ortiz M, Karjosukarso DW, Ng Y, Moreau T, Pedersen RA (2014) NANOG and CDX2 pattern distinct subtypes of human mesoderm during exit from pluripotency. Cell Stem Cell 15:310–325
Morizane R, Lam AQ, Freedman BS, Kishi S, Valerius MT, Bonventre JV (2015) Nephron organoids derived from human pluripotent stem cells model kidney development and injury. Nat Biotechnol 33:1193–1200
Mugford JW, Sipila P, Kobayashi A, Behringer RR, Mcmahon AP (2008a) Hoxd11 specifies a program of metanephric kidney development within the intermediate mesoderm of the mouse embryo. Dev Biol 319:396–405

Mugford JW, Sipila P, Mcmahon JA, Mcmahon AP (2008b) Osr1 expression demarcates a multi-potent population of intermediate mesoderm that undergoes progressive restriction to an Osr1-dependent nephron progenitor compartment within the mammalian kidney. Dev Biol 324:88–98

Mugford JW, Yu J, Kobayashi A, Mcmahon AP (2009) High-resolution gene expression analysis of the developing mouse kidney defines novel cellular compartments within the nephron progenitor population. Dev Biol 333:312–323

Murry CE, Keller G (2008) Differentiation of embryonic stem cells to clinically relevant populations: lessons from embryonic development. Cell 132:661–680

Muthukrishnan SD, Yang X, Friesel R, Oxburgh L (2015) Concurrent BMP7 and FGF9 signalling governs AP-1 function to promote self-renewal of nephron progenitor cells. Nat Commun 6:10027

Nishinakamura R, Matsumoto Y, Nakao K, Nakamura K, Sato A, Copeland NG, Gilbert DJ, Jenkins NA, Scully S, Lacey DL, Katsuki M, Asashima M, Yokota T (2001) Murine homolog of SALL1 is essential for ureteric bud invasion in kidney development. Development 128:3105–3115

O'brien LL, Guo Q, Lee Y, Tran T, Benazet JD, Whitney PH, Valouev A, Mcmahon AP (2016) Differential regulation of mouse and human nephron progenitors by the six family of transcriptional regulators. Development 143:595–608

Osafune K, Takasato M, Kispert A, Asashima M, Nishinakamura R (2006) Identification of multipotent progenitors in the embryonic mouse kidney by a novel colony-forming assay. Development 133:151–161

Park JS, Valerius MT, Mcmahon AP (2007) Wnt/ -catenin signaling regulates nephron induction during mouse kidney development. Development 134:2533–2539

Perantoni AO, Dove LF, Karavanova I (1995) Basic fibroblast growth factor can mediate the early inductive events in renal development. Proc Natl Acad Sci U S A 92:4696–4700

Poleev A, Fickenscher H, Mundlos S, Winterpacht A, Zabel B, Fidler A, Gruss P, Plachov D (1992) PAX8, a human paired box gene: isolation and expression in developing thyroid, kidney and Wilms' tumors. Development 116:611–623

Psychoyos D, Stern CD (1996) Fates and migratory routes of primitive streak cells in the chick embryo. Development 122:1523–1534

Rogers SA, Ryan G, Purchio AF, Hammerman MR (1993) Metanephric transforming growth factor-beta 1 regulates nephrogenesis in vitro. Am J Phys 264:F996–1002

Saxen L (1987) Organogenesis of the kidney. Cambridge University Press, Cambridge

Schneider J, Arraf AA, Grinstein M, Yelin R, Schultheiss TM (2015) Wnt signaling orients the proximal-distal axis of chick kidney nephrons. Development 142:2686–2695

Self M, Lagutin OV, Bowling B, Hendrix J, Cai Y, Dressler GR, Oliver G (2006) Six2 is required for suppression of nephrogenesis and progenitor renewal in the developing kidney. EMBO J 25:5214–5228

Shakya R, Watanabe T, Costantini F (2005) The role of GDNF/Ret signaling in ureteric bud cell fate and branching morphogenesis. Dev Cell 8:65–74

Sharmin S, Taguchi A, Kaku Y, Yoshimura Y, Ohmori T, Sakuma T, Mukoyama M, Yamamoto T, Kurihara H, Nishinakamura R (2015) Human induced pluripotent stem cell-derived podocytes mature into vascularized glomeruli upon experimental transplantation. J Am Soc Nephrol 27(6):1778–1791

Stark K, Vainio S, Vassileva G, Mcmahon AP (1994) Epithelial transformation of metanephric mesenchyme in the developing kidney regulated by Wnt-4. Nature 372:679–683

Taguchi A, Nishinakamura R (2014) Nephron reconstitution from pluripotent stem cells. Kidney Int 87(5):894–900

Taguchi A, Kaku Y, Ohmori T, Sharmin S, Ogawa M, Sasaki H, Nishinakamura R (2014) Redefining the in vivo origin of metanephric nephron progenitors enables generation of complex kidney structures from pluripotent stem cells. Cell Stem Cell 14:53–67

Takada S, Stark KL, Shea MJ, Vassileva G, Mcmahon JA, Mcmahon AP (1994) Wnt-3a regulates somite and tailbud formation in the mouse embryo. Genes Dev 8:174–189

Takasato M, Er PX, Chiu HS, Maier B, Baillie GJ, Ferguson C, Parton RG, Wolvetang EJ, Roost MS, Chuva De Sousa Lopes SM, Little MH (2015) Kidney organoids from human iPS cells contain multiple lineages and model human nephrogenesis. Nature 526:564–568

Takemoto T, Uchikawa M, Yoshida M, Bell DM, Lovell-Badge R, Papaioannou VE, Kondoh H (2011) Tbx6-dependent Sox2 regulation determines neural or mesodermal fate in axial stem cells. Nature 470:394–398

Tam PP, Loebel DA (2007) Gene function in mouse embryogenesis: get set for gastrulation. Nat Rev Genet 8:368–381

Tanigawa S, Wang H, Yang Y, Sharma N, Tarasova N, Ajima R, Yamaguchi TP, Rodriguez LG, Perantoni AO (2011) Wnt4 induces nephronic tubules in metanephric mesenchyme by a non-canonical mechanism. Dev Biol 352:58–69

Turner DA, Hayward PC, Baillie-Johnson P, Rue P, Broome R, Faunes F, Martinez Arias A (2014) Wnt/beta-catenin and FGF signalling direct the specification and maintenance of a neuromesodermal axial progenitor in ensembles of mouse embryonic stem cells. Development 141:4243–4253

Tzouanacou E, Wegener A, Wymeersch FJ, Wilson V, Nicolas J-F (2009) Redefining the progression of lineage segregations during mammalian embryogenesis by clonal analysis. Dev Cell 17:365–376

Vega QC, Worby CA, Lechner MS, Dixon JE, Dressler GR (1996) Glial cell line-derived neurotrophic factor activates the receptor tyrosine kinase RET and promotes kidney morphogenesis. Proc Natl Acad Sci U S A 93:10657–10661

Wellik DM, Hawkes PJ, Capecchi MR (2002) Hox11 paralogous genes are essential for metanephric kidney induction. Genes Dev 16:1423–1432

Wilson V, Olivera-Martinez I, Storey KG (2009) Stem cells, signals and vertebrate body axis extension. Development 136:1591–1604

Woolf AS, Kolatsi-Joannou M, Hardman P, Andermarcher E, Moorby C, Fine LG, Jat PS, Noble MD, Gherardi E (1995) Roles of hepatocyte growth factor/scatter factor and the met receptor in the early development of the metanephros. J Cell Biol 128:171–184

Xia Y, Nivet E, Sancho-Martinez I, Gallegos T, Suzuki K, Okamura D, Wu MZ, Dubova I, Esteban CR, Montserrat N, Campistol JM, Izpisua Belmonte JC (2013) Directed differentiation of human pluripotent cells to ureteric bud kidney progenitor-like cells. Nat Cell Biol 15:1507–1515

Xu PX, Adams J, Peters H, Brown MC, Heaney S, Maas R (1999) Eya1-deficient mice lack ears and kidneys and show abnormal apoptosis of organ primordia. Nat Genet 23:113–117

Yamaguchi TP, Takada S, Yoshikawa Y, Wu N, Mcmahon AP (1999) T (Brachyury) is a direct target of Wnt3a during paraxial mesoderm specification. Genes Dev 13:3185–3190

Yu J, Carroll TJ, Rajagopal J, Kobayashi A, Ren Q, Mcmahon AP (2009) A Wnt7b-dependent pathway regulates the orientation of epithelial cell division and establishes the cortico-medullary axis of the mammalian kidney. Development 136:161–171

Chapter 10
Generation of Gastrointestinal Organoids Derived from Human Pluripotent Stem Cells

Jorge O. Múnera and James M. Wells

Abstract Over the past two decades, the stem cell field has developed methods to differentiate human pluripotent stem cells (hPSCs) into specific tissue types. These studies have been largely driven by developmental biologists who have identified pathways and tissue-specific markers that can be used to direct the differentiation of hPSCs. Furthermore, the identification of Lgr5+ adult stem cells in the mouse small intestine led to the development of protocols to grow these stem cells into self-organizing, self-renewing, multicellular "organoids." PSC-derived and adult organoids derived from human samples now allow researchers to study cell lineage processes and model complex cell-cell interactions. Both PSC-derived gastrointestinal organoids and those generated from freshly excised tissue have many properties of gastrointestinal physiology. However, PSC-derived gastrointestinal organoids are generated through a stepwise differentiation that largely mimics gastrointestinal development and is therefore a good system to study congenital defects of the human GI tract. Moreover, PSC-derived organoids are complex and contain mesodermal cell types comprising smooth muscle and subepithelial myofibroblasts. Lastly, developmentally inspired approaches have been used to tissue engineer human PSC-derived organoids with a functional enteric nervous system. In this chapter we describe the development of both the intestine and stomach. We then describe how pathways identified by developmental studies can be used to direct the differentiation of pluripotent stem cells into human intestinal organoids (HIOs) and human gastric organoids (HGOs). In addition, we discuss potential applications of these systems for studying human gastrointestinal development and disease and in engineering GI tissues for eventual transplantation-based therapies.

Keywords Definitive endoderm • Foregut • Midgut • Hindgut • Human pluripotent stem cells • Gastric organoids • Intestinal organoids

---

J.O. Múnera • J.M. Wells (✉)
Division of Developmental Biology Cincinnati Children's Hospital Research Foundation, 3333 Burnet Avenue, Cincinnati, OH 45229, USA
e-mail: jorge.munera@cchmc.org; james.wells@cchmc.org

T. Tsuji (ed.), *Organ Regeneration Based on Developmental Biology*,
DOI 10.1007/978-981-10-3768-9_10

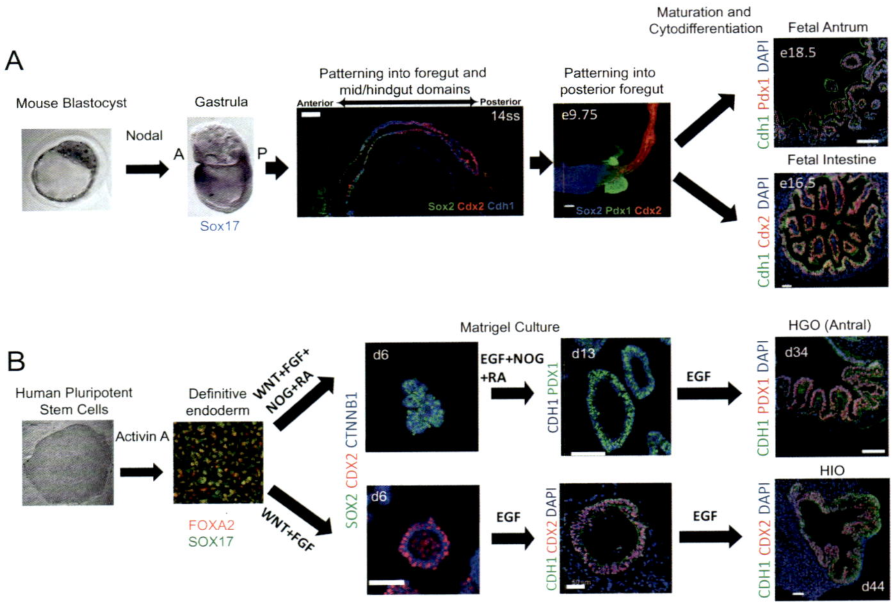

Fig. 10.1 Mouse gastrointestinal development time course from blastocyst stage through fetal antrum and fetal small intestine (**a**). Time course of human gastrointestinal organoid development from human pluripotent stem cells through human gastrointestinal organoids (**b**)

10.1 Establishment of Endoderm and Patterning into Foregut, Midgut, and Hindgut Domains

The definitive endoderm, one of the primary germ layers that are established during gastrulation, gives rise to the epithelium that lines the entire gastrointestinal tract (Spence et al. (2011a). Wnt and Nodal signaling are required at various stages of gastrulation. Wnt signaling is required for gastrulation, and loss of beta-catenin, a key downstream effector of canonical Wnt signaling, results in embryos that fail to gastrulate (Haegel et al. 1995). Wnt signaling is also required for the maintenance of definitive endoderm, and knocking out beta-catenin in embryonic endoderm (including DE) results in its conversion to pre-cardiac mesoderm (Lickert et al. 2002). Mouse Nodal mutants fail to form or maintain the primitive streak and fail to form the definitive endoderm (Brennan et al. 2001).

The definitive endoderm is first present as a flat sheet of cells, which through the process of gut tube morphogenesis transforms into a primitive gut tube (Fig. 10.1a). Formation of the gut tube is initiated at the anterior and posterior ends of the embryo, where the endoderm folds over forming the foregut and hindgut. The folding of the foregut and hindgut tube continues until they meet in the midgut, forming a primitive gut tube. Establishing the molecular patterns in the endoderm happens throughout these stages of development and is governed by four major signaling pathways, Wnt,

Fgf, retinoic acid (RA), and Bmp. In Xenopus and mouse embryos, Wnt/β-catenin signaling promotes posterior gene expression in endoderm while at the same time inhibiting anterior gene expression in the posterior (McLin et al. 2007; Sherwood et al. 2011; Zorn and Wells 2009). Bmp, Fgf, and RA all have been reported to promote posterior pattern of endoderm between the late gastrula and early somite stages of embryonic development in several vertebrate species (Bayha et al. 2009; Dessimoz et al. 2006; Huang et al. 1998; Kumar et al. 2003; Niederreither et al. 2003; Rankin et al. 2011; Roberts et al. 1995; Stafford and Prince 2002; Tiso et al. 2002; Wang et al. 2006; Wells and Melton 2000). However, it has been difficult to identify how these pathways act in concert to pattern the endoderm in the context of the embryo. This requires temporal and spatial manipulation of multiple signaling pathways.

The foregut gives rise to the esophagus, trachea, stomach, lungs, thyroid, liver, biliary system, and pancreas. The midgut gives rise to the small intestine, and the hindgut gives rise to the large intestine. The early regionalization of the gastrointestinal tract into foregut and midgut/hindgut domains can be observed by the region-specific expression of transcription factors (San Roman and Shivdasani 2011). At early somite stages, Sox2 is expressed in the foregut, while Cdx2 is expressed in the midgut/hindgut. At slightly later stages, a third marker, Pdx1, further delineates presumptive organ boundaries into Sox2+ anterior foregut, Sox2+/Pdx1+ posterior foregut (antrum), Pdx1+/Cdx2+ anterior intestine (duodenum), and Cdx2+ small and large intestine. Although Sox2 and Cdx2 expression domains are thought to mark foregut and midgut/hindgut, respectively, it should be noted that (1) Sox2+/Cdx2+ cells are present in e8.75–9.25 embryos (Sherwood et al. 2009a).

10.2 Intestinal Development

The intestine is derived from the midgut/hindgut tissue which expresses the homeobox transcription factor Cdx2 as early as e7.5 (Beck et al. 1995; Sherwood et al. 2009b, 2011). Cdx2 is critical for intestinal development, and when Cdx2 is conditionally knocked out in the endoderm using Foxa3-Cre, animals fail to form a colon, and the posterior small intestine is converted into stratified squamous tissue which molecularly and morphologically resembles the esophagus (Gao et al. 2009). The pathways which specify the midgut and hindgut from the common Cdx2 intestinal domain remain to be determined although studies from model organisms suggest that Wnt and Bmp signaling are involved (Kumar et al. 2003; Roberts et al. 1995; Sherwood et al. 2011; Tiso et al. 2002; Wills et al. 2008). Following patterning into midgut and hindgut domains, from e9.5 to e13.5, the intestine increases in length and circumference and undergoes a series of morphological transitions (Fig. 10.1a). At e12.5 the intestine displays a pseudostratified morphology which transitions to a simple cuboidal epithelium (Spence et al. 2011a; Wells and Spence 2014). At e14.5 the intestinal tract undergoes cytodifferentiation and villus formation through coordinated formation of mesenchymal clusters marked by Pdgfrα (Shyer et al. 2015;

Walton et al. 2012, 2016). The process of villi formation is dependent on BMP and Sonic Hedgehog signaling as clusters of Pdgfrα + mesenchymal cells form at sites of high activity of these pathways. Villus morphogenesis leads to the compartmentalization of progenitor domains to the intervillous regions, a process which requires Shh and Bmp signaling (Shyer et al. 2015; Walton et al. 2012, 2016). Following cytodifferentiation, goblet cells and enteroendocrine cells can be detected in the small intestine and colon, while Paneth cells are not detected in the small intestine until after birth (Fordham et al. 2013; Mustata et al. 2013). The progenitors of the intervillous regions are dependent on the Wnt effector TCF4, and mice deficient in this transcription factor display loss progenitors by birth (Korinek et al. 1998). Although most developmental studies have focused on small intestinal development, development of the colon involves similar transitions with the exception of villus formation. At birth, the small intestine is comprised of villi with immature crypts, while the colon is comprised of a flat epithelium with immature crypts.

During the first 2 weeks after birth in mice, the intervillous regions of the small intestine and the base of crypts in the colon begin to invade the submucosa to establish the adult stem cell domain. At 2 weeks Paneth cells, which support the intestinal stem cells that will sustain self-renewal of the small intestine throughout the lifetime of the animal, are present in the base of the crypts (Sato et al. 2011b). Unlike the small intestine which contains Paneth cells, the colon and its stem cell niche are dependent on other cell types including cKit + goblet cells and Reg4+ cells deep crypt secretory cells (Rothenberg et al. 2012; Sasaki et al. 2016). In contrast to mice, humans are born with crypts (Montgomery et al. 1999) and Paneth cells, although colonization by microbes helps to mature Paneth cells. In humans, intestinal maturation occurs in utero such that human enterocytes at mid-gestation resemble those of suckling rodents. At 22 weeks the human intestine resembles that of fully weaned rodents (Montgomery et al. 1999). The developmental processes outlined above culminate in the development of the most proliferative organ in the body which undergoes self-renewal every 4–5 days in mice and every week in humans.

10.3 Generation of Three-Dimensional Intestinal Organoids from Human Pluripotent Stem Cells

In the past few years, the identification of adult intestinal stem cell markers and the ability to isolate and culture these cells as organoids have led to significant advances in our understanding of intestinal regeneration and gastrointestinal cancer (Barker et al. 2007, 2009; Snippert et al. 2010) (Jung et al. 2011; Sato et al. 2009, 2011a, b), but they are not well suited to studies of developmental processes. However, a separate approach was developed that directed the differentiation of human pluripotent stem cells into human intestinal organoids (HIOs) using a process that recapitulates intestinal development. In this approach the Nodal mimetic Activin A is used to induce differentiation of pluripotent stem cells into definitive endoderm

(D'Amour et al. 2005). Subsequent exposure of definitive endoderm to high levels of WNT3A and FGF4 induces patterning and morphogenesis into midgut/hindgut spheroids that express CDX2. Once formed, these midgut/hindgut spheroids can be grown in three-dimensional culture under conditions that favor intestinal growth giving rise to HIOs (Sato et al. 2009; Spence et al. 2011b). Moreover, growth of these spheroids closely recapitulates developmental events that occur in vivo. Midgut/hindgut spheroids transition from a simple cuboidal epithelium into a pseudostratified epithelium, which then undergoes cytodifferentiation and transitions into a polarized epithelium which contains zones of differentiation and proliferation. When compared to developing mouse intestine, these intestinal organoid cultures undergo strikingly similar transitions (Spence et al. 2011a) and contain all of the cell types found in the fetal gut (Fig. 10.1b). Furthermore, HIOs contain discreet epithelial domains that express the stem cell markers LGR5 and ASCL2, suggesting that early stages of intestinal stem cell development occur normally in organoid cultures.

Pluripotent stem cell-derived intestinal organoids also have another unique characteristic which is the presence of mesenchyme. Addition of FGF4 during the midgut/hindgut morphogenesis process allows the expansion of Brachyury-expressing Cdx2+ mesenchymal cells that codevelop with the intestinal epithelium. Follow-up studies also revealed that mesenchyme containing HIOs grow in the absence of Noggin and R-spondin, factors that are necessary for the maintenance of epithelial organoids (Watson et al. 2014). This suggests that the mesenchyme present in the HIOs provides endogenous factors that promote growth and maintenance of the epithelium. Conversely, the mesenchymal cells grow in the absence of exogenous factors which could promote the growth of these cells such as FGF9 and SHH. This suggests that the epithelium is capable of generating factors which support the growth of the mesenchyme. How this epithelial-mesenchymal cross talk drives growth and maturation of HIOs remains an open question.

It should be noted that human intestinal organoids can be generated through 3D cultures of aggregated 2D midgut/hindgut endoderm derived from PSCs (Fordham et al. 2013; Tamminen et al. 2015). Exposure of definitive endoderm to the GSK3B inhibitor CHIR99021, unlike WNT3A, allows generation of CDX2-positive endoderm in the absence of FGF4. These culture aggregates undergo self-organization and form organoids similar to those generated from induced morphogenesis. This suggests that once patterned, midgut/hindgut tissue retains the capacity to self-organize in the presence of correct environmental cues such as the extracellular matrix components and growth factors present in Matrigel.

10.4 Gastric Development

After patterning of the foregut and midgut/hindgut into distinct Sox2+ and Cdx2+ domains, respectively, the foregut domain is further patterned into anterior and posterior domains, the latter of which will give rise to the stomach (Zorn and Wells 2009).

The stomach is then further subdivided into the proximal stomach that will give rise to the forestomach and corpus and the distal stomach which will give rise to the antrum which abuts the duodenum. Little is known about the specification and patterning of the stomach; however, as with other organs, this appears to require coordinated epithelial-mesenchymal signaling. In one example, the mesenchymal homeobox protein Barx1 is required for normal stomach development, and Barx1-null embryos have ectopic Cdx2 expression in the distal stomach remnant (Kim et al. 2005). Knockdown of Barx1 in gastric mesenchyme cultures results in reduced expression of secreted frizzled-related proteins (Sfrps), known inhibitors of Wnt signaling. This suggests that inhibition of Wnt in the distal stomach is required to prevent ectopic Cdx2 expression. The authors further demonstrate using Wnt reporter mice, that Wnt signaling is low in the distal stomach and high in the fundus and forestomach, suggesting that there is a Wnt-repressive mechanism patterning the antrum.

Several additional pathways have been implicated in anterior-posterior patterning of the stomach including retinoic acid, Indian hedgehog, and Sonic hedgehog (Ramalho-Santos et al. 2000; Wang et al. 2006). Mutants in the retinoic acid synthesizing enzyme Raldh2 display reduced expression of Pea3 and Nkx2–5 which are mesenchymal transcription factors expressed in the hindstomach. In addition mesenchymal transcription factors such as Bapx1 (Nkx3.2) and COUP-TFII are also required for proper stomach morphogenesis and patterning (Takamoto et al. 2005; Verzi et al. 2009). Bapx1 mutants' stomach displays a truncated antral region, while COUP-TFII displays altered hindstomach epithelial patterning as well as perturbed smooth muscle differentiation. Furthermore several epithelial transcription factors regulate various aspects of stomach patterning including Hnf1β, Pdx1, and Gata4 (Haumaitre et al. 2005; Jacobsen et al. 2002; Larsson et al. 1996). Hnf1β and Pdx1 regulate patterning of the hindstomach, while Gata4 is required for proper cytodifferentiation of both the corpus and antrum. How these signaling pathways and transcription factors are integrated and coordinated remains an open question. For a more extensive review of stomach development, we refer the reader to this review (Kim and Shivdasani 2016).

Similar to the intestine, the stomach undergoes growth and morphogenesis and transitions (Fig. 10.1a) from a pseudostratified epithelium to a glandular epithelium comprised of simple columnar epithelial cells (McCracken et al. 2014). The corpus contains five distinct differentiated cell types, surface pit mucous cells which are located at the luminal surface and secrete mucus (predominantly Muc5AC), mucous neck cells (Muc6), chief cells which secrete pepsinogen, parietal cells which secrete HCL, tuft cells (unknown function), and endocrine cells which secrete hormones. In contrast, the antrum lacks chief cells and parietal cells but contains both types of mucous cells and G cells, which are endocrine cells that secrete the hormone gastrin and are restricted to the antrum. The epithelium of the corpus is maintained by a Mist1-expressing cell under homeostatic conditions, while the antrum is maintained by Lgr5+ stem cells located at the base of the glands (Barker et al. 2010; Hayakawa et al. 2015).

10.5 Generation of Three-Dimensional Gastric Tissue from Human Pluripotent Stem Cells

Exposure of definitive endoderm to high levels of Wnt and FGF promotes a posterior/Cdx2+ fate and induces gut tube morphogenesis, which results in the formation of midgut/hindgut tubelike structures. Evidence from model organisms suggests that BMP signaling is also required for posterior patterning; however, the relationship between WNT, BMP, and FGF in promoting a posterior fate was undetermined. To investigate the involvement of BMP in the WNT/FGF driven posteriorization of definitive endoderm, McCracken and colleagues determined the involvement of endogenous BMP signaling in promoting a posterior endoderm fate. Inhibition of endogenous BMP signaling through the addition of NOGGIN, while at the same time activating WNT and FGF, results in the formation of three-dimensional spheroids which lack CDX2 expression and instead express SOX2. This suggests that endoderm posteriorization is BMP dependent, but that morphogenesis is BMP independent. SOX2 expressing foregut spheroids could be further patterned into posterior foregut by the addition of retinoic acid. Once formed, these posterior foregut spheroids grow in three-dimensional culture and give rise to human gastric organoids (HGOs). HGOs express SOX2 and PDX1 suggesting that they are antral in nature (McCracken et al. 2014). Growth of spheroids into HGOs closely resembles developmental transitions that occur in vivo (Fig. 10.1b). Posterior foregut spheroids transition from a simple cuboidal epithelium into a pseudostratified epithelium and then into a glandular epithelium. This corresponds with cytodifferentiation and patterning of glands into luminal domains with surface mucous cells and glandular neck cells similar to the developing mouse stomach. In addition, HGOs have gastrin-expressing G cells, a hallmark of the antrum. Lastly, HGOs contain discreet epithelial domains that express the stem cell marker LGR5 and the progenitor marker SOX9, suggesting that early stages of gastric stem cell development occurred normally in these organoid cultures.

10.6 Applications of HIOs and HGOs

HIOs and HGOs offer an unprecedented opportunity for studying human gastrointestinal development and disease (Fig. 10.2). The utility of HIOs and HGOs is particularly important for studying aspects of human development and human disease which cannot be recapitulated in model organisms. The accessibility of HIOs and HGOs allows the interrogation of signaling pathways using recombinant proteins or small molecules. In addition, hPSCs are amenable to genetic manipulation including shRNA-mediated knockdown and lentiviral-mediated doxycycline-inducible gene expression, BAC transgenesis, and CRISPR-CAS9-mediated gene editing. This allows for generation of HIOs and HGOs with a vast array of tools. In this

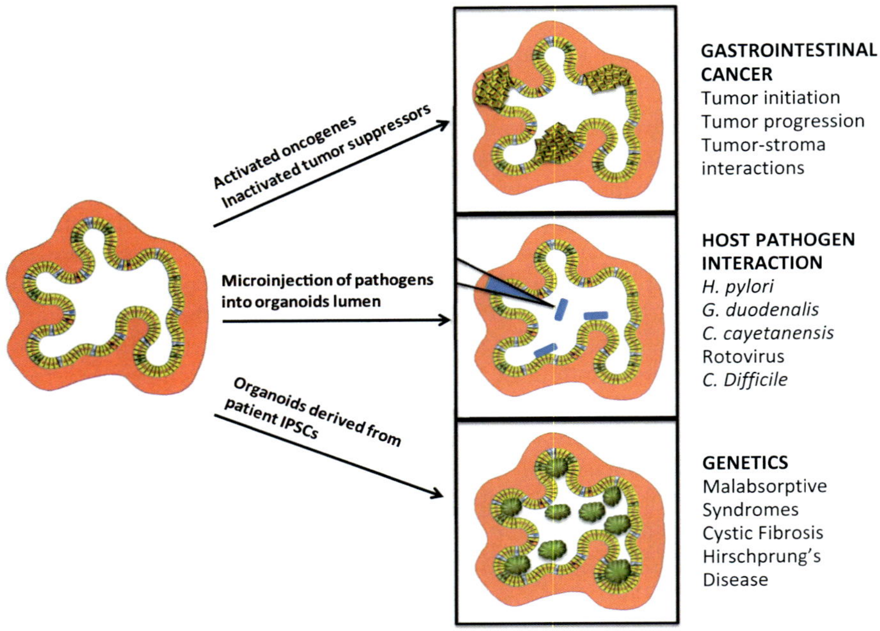

Fig. 10.2 Potential uses of human gastrointestinal organoid systems

section we discuss how organoids can be used to model human gastrointestinal development and disease.

HIOs have been used to study the role of the transcription factors in the development of enteroendocrine cells (Du et al. 2012; Spence et al. 2011b). By overexpressing human Neurogenin-3 using adenovirus, Spence and colleagues were able to demonstrate that NEUROG3 was sufficient to induce the differentiation of intestinal enteroendocrine cells. This approach was improved by the incorporation of a lentiviral-mediated doxycycline-inducible system for gene overexpression (McCracken et al. 2014). This Dox-inducible system offers the ability to control the timing and dosage of transgene expression. Using such a system, HIOs have been used to demonstrate that overexpression of a dominant-negative form of FOXO1 is sufficient to convert a subset of enteroendocrine cells into insulin-expressing cells (Bouchi et al. 2014). There are now several approaches to study loss of function in HIOs. Lentiviral-mediated shRNA-mediated knockdown of NEUROG3 and ARX demonstrated the requirement of these factors for normal development of human intestinal enteroendocrine cells (Du et al. 2012; Spence et al. 2011b). CRISPR-CAS9-mediated knockout of NEUROG3 in human embryonic stem cells has been used to demonstrate the requirement of this transcription factor in endocrine pancreas development (McGrath et al. 2015). In addition, BAC-mediated transgenesis has been used to generate an LGR5-GFP and NEUROG3-GFP reporter lines that mark intestinal epithelial progenitors and endocrine cells in organoids (McCracken et al. 2014; Watson et al. 2014).

HIOs and HGOs have also been used for studying infectious disease and inflammatory responses. HIOs can be infected in order to grow and expand rotaviruses. HGOs have been used to model *Helicobacter pylori* infection (McCracken et al. 2014). HGOs infected with *H. pylori* display increased proliferation and CagA-dependent phosphorylation of C-met. Another study showed that HIOs express TNFα in response to hypoxia and that this effect is mediated by EPAS1 (Xue et al. 2013). HIOs have also been shown to maintain viable *Clostridium difficile* capable of generating the toxins TcdA and TcdB (Leslie et al. 2015). This allows modeling of epithelial barrier disruption caused by these bacterial toxins (Leslie et al. 2015). HIOs and HGOs could potentially be used to model other infectious diseases and study the impact of commensal bacteria on disease progression and epithelial function. Lastly, generation of HIOs and HGOs from patients can be used to study putative disease-causing mutations during GI development or the pathogenesis of GI diseases.

HIOs and HGOs lack an enteric nervous system which is responsible for peristalsis in vivo (Heanue and Pachnis 2007). The incorporation of enteric nerve cells is complicated since these cells are derived from neural crest cells, which originate at a distant site in the embryo and then migrate into the mesoderm of the gastrointestinal tract. However, incorporation of an ENS into HIOs has recently been achieved, resulting in intestinal tissues capable of peristalsis (Workman et al. 2017). By combining vagal neural crest cells with HIOs, neural crest cells were able to incorporate into the mesenchyme of organoids. Transplantation of these chimeric HIOs allowed maturation of the mesenchyme into smooth muscle which was innervated by neural crest-derived cells. To demonstrate the translational potential of this approach, this system was used to model a genetic form of Hirschsprung's disease.

10.7 Limitations of HIOs and HGOs

Although HIOs and HGOs hold great promise as a tool for gastrointestinal research, these systems are not without their limitations. For example, HIOs are fetal in nature and lack expression of brush border enzymes like alkaline phosphatase, sucrase-isomaltase, and lactase, thus limiting their use for modeling absorptive diseases (Finkbeiner et al. 2015; Watson et al. 2014). However, following transplantation in vivo, HIOs undergo remarkable maturation and robustly express brush border and transport proteins. Moreover, in vivo-grown HIOs form villi, stem cell-containing crypts, and submucosal and myenteric muscle layers. These findings demonstrate that HIOs derived in vitro contain the progenitor cells capable of giving rise to more mature and functional intestinal tissue and that a close examination of pathways activated by in vivo transplantations may identify pathways which could be manipulated to improve maturation in vitro.

HGOs consistently form glandular units which appear to reflect early postnatal mouse development (McCracken et al. 2014). However, long-term growth of HGOs has remained a challenge, and growth in vivo has not yet been demonstrated. Identification of long-term culture conditions and in vivo growth will likely depend on expanding our knowledge base on the pathways that regulate embryonic growth and maturation of the stomach in model organisms. For example, there is little known about pathways that pattern the embryonic corpus/fundus or about what controls differentiation of fundic cell types such as the acid-producing parietal cells. Consequently, efforts to generate human fundic organoids that contain functional parietal cells have not yet been successful. Given the significant differences between the human and mouse stomach, developing human gastric organoid systems is critical for modeling human-specific processes.

HIOs and HGOs are also limited by the cell types that can be generated within the organoid. The blood vessels, immune cells, lymphatics, and an enteric nervous system are absent in HIOs. The lack of enteric nerves makes HIOs and HGOs unsuitable for gastrointestinal motility studies. In addition, these cells may contribute to maturation of intestinal tissue. There are several examples of successful efforts to incorporate additional cell types into PSC-derived organoids. Addition of human umbilical cord vascular endothelial cells (HUVECs) and mesenchymal stem cells to PSC-derived liver progenitors resulted in morphogenesis and development of embryonic liver tissue (Takebe et al. 2013, 2014). Similar successes have been achieved with other organ primordia including the kidney and intestine (Takebe et al. 2015) suggesting this may be possible to achieve with HIOs and HGOs. Finally, incorporation of immune cells should allow the development of complex models of inflammation in the gastrointestinal tract.

10.8 Future Directions

The development of methods to generate intestinal and gastric organoids, either through the directed differentiation of PSCs or from adult organ tissues, has been a seminal advance into human research. HIO and HGO technology will also benefit from new technologies such as high-throughput culture systems (for cell interaction and transcription analysis) and microfluidics platforms which can be used to mimic the flow of luminal contents through the intestine (Gracz et al. 2015; Ingber 2016). HIOs and HGOs have been used to model gastrointestinal pathogens, but these systems may also be applicable to other diseases. HIOs and HGOs derived from patients could be used to model congenital diseases which lead to malformations in the stomach or intestines. For example, HIOs could be derived from patients with congenital short bowel diseases in order to identify pathways which can be manipulated through drug treatment. TGFα mutations can be introduced into HGOs to study Ménétrier's disease and the mechanism of disease progression. In vivo-transplanted HIOs could also be used to study rare malabsorption syndromes as well as to study mutations which affect intestinal barrier function.

HIOs and HGOs could also be used to study metaplasias of the stomach and intestine. Metaplasia, the conversion of one differentiated cell type into another, is considered a precancerous lesion in the stomach and intestine (Quinlan et al. 2007). Pyloric metaplasia of the ileum is common in patients suffering from ileocolonic Crohn's disease, with one study reporting a 73% incidence (Soucy et al. 2012). Intestinal metaplasia of the stomach is common in patients with *H. pylori*-associated gastritis (Correa et al. 2010). Examination of pathways which regulate specification of HIOs versus HGOs could shed insight into how metaplasia occurs. In addition, HIOs and HGOs could be used to determine the role of inflammation in metaplasia by treating HIOs with inflammatory cytokines such as TNFα and IL-8. This could elucidate developmental mechanisms which are reactivated by prolonged chronic inflammation. The availability of tools and the accessibility of HIOs and HGOs will allow for countless studies on the development and pathogenesis of the gastrointestinal tract.

References

Barker N, van Es JH, Kuipers J, Kujala P, van den Born M, Cozijnsen M, Haegebarth A, Korving J, Begthel H, Peters PJ et al (2007) Identification of stem cells in small intestine and colon by marker gene Lgr5. Nature 449:1003–1007

Barker N, Ridgway RA, van Es JH, van de Wetering M, Begthel H, van den Born M, Danenberg E, Clarke AR, Sansom OJ, Clevers H (2009) Crypt stem cells as the cells-of-origin of intestinal cancer. Nature 457:608–611

Barker N, Huch M, Kujala P, van de Wetering M, Snippert HJ, van Es JH, Sato T, Stange DE, Begthel H, van den Born M et al (2010) Lgr5(+ve) stem cells drive self-renewal in the stomach and build long-lived gastric units in vitro. Cell Stem Cell 6:25–36

Bayha E, Jorgensen MC, Serup P, Grapin-Botton A (2009) Retinoic acid signaling organizes endodermal organ specification along the entire antero-posterior axis. PLoS One 4:e5845

Beck F, Erler T, Russell A, James R (1995) Expression of Cdx-2 in the mouse embryo and placenta: possible role in patterning of the extra-embryonic membranes. Dev Dyn Off Publ Am Assoc Anat 204:219–227

Bouchi R, Foo KS, Hua H, Tsuchiya K, Ohmura Y, Sandoval PR, Ratner LE, Egli D, Leibel RL, Accili D (2014) FOXO1 inhibition yields functional insulin-producing cells in human gut organoid cultures. Nat Commun 5:4242

Brennan J, Lu CC, Norris DP, Rodriguez TA, Beddington RSP, Robertson EJ (2001) Nodal signalling in the epiblast patterns the early mouse embryo. Nature 411:965–969

Correa P, Piazuelo MB, Wilson KT (2010) Pathology of gastric intestinal metaplasia: clinical implications. Am J Gastroenterol 105:493–498

D'Amour KA, Agulnick AD, Eliazer S, Kelly OG, Kroon E, Baetge EE (2005) Efficient differentiation of human embryonic stem cells to definitive endoderm. Nat Biotechnol 23:1534–1541

Dessimoz J, Opoka R, Kordich JJ, Grapin-Botton A, Wells JM (2006) FGF signaling is necessary for establishing gut tube domains along the anterior-posterior axis in vivo. Mech Dev 123:42–55

Du A, McCracken KW, Walp ER, Terry NA, Klein TJ, Han A, Wells JM, May CL (2012) Arx is required for normal enteroendocrine cell development in mice and humans. Dev Biol 365:175–188

Finkbeiner SR, Hill DR, Altheim CH, Dedhia PH, Taylor MJ, Tsai YH, Chin AM, Mahe MM, Watson CL, Freeman JJ et al (2015) Transcriptome-wide analysis reveals hallmarks of human intestine development and maturation In Vitro and In Vivo. Stem Cell Rep 4(6):1140–1155

Fordham RP, Yui S, Hannan NR, Soendergaard C, Madgwick A, Schweiger PJ, Nielsen OH, Vallier L, Pedersen RA, Nakamura T et al (2013) Transplantation of expanded fetal intestinal progenitors contributes to colon regeneration after injury. Cell Stem Cell 13:734–744

Gao N, White P, Kaestner KH (2009) Establishment of intestinal identity and epithelial-mesenchymal signaling by Cdx2. Dev Cell 16:588–599

Gracz AD, Williamson IA, Roche KC, Johnston MJ, Wang F, Wang Y, Attayek PJ, Balowski J, Liu XF, Laurenza RJ et al (2015) A high-throughput platform for stem cell niche co-cultures and downstream gene expression analysis. Nat Cell Biol 17:340–349

Haegel H, Larue L, Ohsugi M, Fedorov L, Herrenknecht K, Kemler R (1995) Lack of beta-catenin affects mouse development at gastrulation. Development 121:3529–3537

Haumaitre C, Barbacci E, Jenny M, Ott MO, Gradwohl G, Cereghini S (2005) Lack of TCF2/vHNF1 in mice leads to pancreas agenesis. Proc Natl Acad Sci U S A 102:1490–1495

Hayakawa Y, Ariyama H, Stancikova J, Sakitani K, Asfaha S, Renz BW, Dubeykovskaya ZA, Shibata W, Wang H, Westphalen CB et al (2015) Mist1 expressing gastric stem cells maintain the normal and neoplastic gastric epithelium and are supported by a perivascular stem cell niche. Cancer Cell 28:800–814

Heanue TA, Pachnis V (2007) Enteric nervous system development and Hirschsprung's disease: advances in genetic and stem cell studies. Nat Rev Neurosci 8:466–479

Huang D, Chen SW, Langston AW, Gudas LJ (1998) A conserved retinoic acid responsive element in the murine Hoxb-1 gene is required for expression in the developing gut. Development (Cambridge, England) 125:3235–3246

Ingber DE (2016) Reverse engineering human pathophysiology with organs-on-chips. Cell 164:1105–1109

Jacobsen CM, Narita N, Bielinska M, Syder AJ, Gordon JI, Wilson DB (2002) Genetic mosaic analysis reveals that GATA-4 is required for proper differentiation of mouse gastric epithelium. Dev Biol 241:34–46

Jung P, Sato T, Merlos-Suarez A, Barriga FM, Iglesias M, Rossell D, Auer H, Gallardo M, Blasco MA, Sancho E et al (2011) Isolation and in vitro expansion of human colonic stem cells. Nat Med 17:1225–1227

Kim TH, Shivdasani RA (2016) Stomach development, stem cells and disease. Development 143:554–565

Kim BM, Buchner G, Miletich I, Sharpe PT, Shivdasani RA (2005) The stomach mesenchymal transcription factor Barx1 specifies gastric epithelial identity through inhibition of transient Wnt signaling. Dev Cell 8:611–622

Korinek V, Barker N, Moerer P, van Donselaar E, Huls G, Peters PJ, Clevers H (1998) Depletion of epithelial stem-cell compartments in the small intestine of mice lacking Tcf-4. Nat Genet 19:379–383

Kumar M, Jordan N, Melton D, Grapin-Botton A (2003) Signals from lateral plate mesoderm instruct endoderm toward a pancreatic fate. Dev Biol 259:109–122

Larsson LI, Madsen OD, Serup P, Jonsson J, Edlund H (1996) Pancreatic-duodenal homeobox 1 -role in gastric endocrine patterning. Mech Dev 60:175–184

Leslie JL, Huang S, Opp JS, Nagy MS, Kobayashi M, Young VB, Spence JR (2015) Persistence and toxin production by Clostridium difficile within human intestinal organoids result in disruption of epithelial paracellular barrier function. Infect Immun 83:138–145

Lickert H, Kutsch S, Kanzler B, Tamai Y, Taketo MM, Kemler R (2002) Formation of multiple hearts in mice following deletion of beta-catenin in the embryonic endoderm. Dev Cell 3:171–181

McCracken KW, Cata EM, Crawford CM, Sinagoga KL, Schumacher M, Rockich BE, Tsai YH, Mayhew CN, Spence JR, Zavros Y et al (2014) Modelling human development and disease in pluripotent stem-cell-derived gastric organoids. Nature 516:400–404

McGrath PS, Watson CL, Ingram C, Helmrath MA, Wells JM (2015) The basic Helix-Loop-Helix transcription factor NEUROG3 is required for development of the human endocrine pancreas. Diabetes 64:2497–2505

McLin VA, Rankin SA, Zorn AM (2007) Repression of Wnt/beta-catenin signaling in the anterior endoderm is essential for liver and pancreas development. Development (Cambridge, England) 134:2207–2217

Montgomery RK, Mulberg AE, Grand RJ (1999) Development of the human gastrointestinal tract: twenty years of progress. Gastroenterology 116:702–731

Mustata RC, Vasile G, Fernandez-Vallone V, Strollo S, Lefort A, Libert F, Monteyne D, Perez-Morga D, Vassart G, Garcia MI (2013) Identification of Lgr5-independent spheroid-generating progenitors of the mouse fetal intestinal epithelium. Cell Rep 5:421–432

Niederreither K, Vermot J, Le Roux I, Schuhbaur B, Chambon P, Dolle P (2003) The regional pattern of retinoic acid synthesis by RALDH2 is essential for the development of posterior pharyngeal arches and the enteric nervous system. Development (Cambridge, England) 130:2525–2534

Quinlan JM, Colleypriest BJ, Farrant M, Tosh D (2007) Epithelial metaplasia and the development of cancer. Biochim Biophys Acta 1776:10–21

Ramalho-Santos M, Melton DA, McMahon AP (2000) Hedgehog signals regulate multiple aspects of gastrointestinal development. Development 127:2763–2772

Rankin SA, Kormish J, Kofron M, Jegga A, Zorn AM (2011) A gene regulatory network controlling hhex transcription in the anterior endoderm of the organizer. Dev Biol 351:297–310

Roberts DJ, Johnson RL, Burke AC, Nelson CE, Morgan BA, Tabin C (1995) Sonic hedgehog is an endodermal signal inducing Bmp-4 and hox genes during induction and regionalization of the chick hindgut. Development 121:3163–3174

Rothenberg ME, Nusse Y, Kalisky T, Lee JJ, Dalerba P, Scheeren F, Lobo N, Kulkarni S, Sim S, Qian D et al (2012) Identification of a cKit(+) colonic crypt base secretory cell that supports Lgr5(+) stem cells in mice. Gastroenterology 142(1195–1205):e1196

San Roman AK, Shivdasani RA (2011) Boundaries, junctions and transitions in the gastrointestinal tract. Exp Cell Res 317:2711–2718

Sasaki N, Sachs N, Wiebrands K, Ellenbroek SIJ, Fumagalli A, Lyubimova A, Begthel H, van den Born M, van Es JH, Karthaus WR et al (2016) Reg4+ deep crypt secretory cells function as epithelial niche for Lgr5+ stem cells in colon. Proc Natl Acad Sci U S A 113:E5399–E5407

Sato T, Vries RG, Snippert HJ, van de Wetering M, Barker N, Stange DE, van Es JH, Abo A, Kujala P, Peters PJ et al (2009) Single Lgr5 stem cells build crypt-villus structures in vitro without a mesenchymal niche. Nature 459:262–265

Sato T, Stange DE, Ferrante M, Vries RG, Van Es JH, Van den Brink S, Van Houdt WJ, Pronk A, Van Gorp J, Siersema PD et al (2011a) Long-term expansion of epithelial organoids from human colon, adenoma, adenocarcinoma, and Barrett's epithelium. Gastroenterology 141:1762–1772

Sato T, van Es JH, Snippert HJ, Stange DE, Vries RG, van den Born M, Barker N, Shroyer NF, van de Wetering M, Clevers H (2011b) Paneth cells constitute the niche for Lgr5 stem cells in intestinal crypts. Nature 469:415–418

Sherwood RI, Chen T-YA, Melton DA (2009a) Transcriptional dynamics of endodermal organ formation. Dev Dyn 238:29–42

Sherwood RI, Chen TY, Melton DA (2009b) Transcriptional dynamics of endodermal organ formation. Dev Dyn Off Publ Am Assoc Anat 238:29–42

Sherwood RI, Maehr R, Mazzoni EO, Melton DA (2011) Wnt signaling specifies and patterns intestinal endoderm. Mech Dev 128:387–400

Shyer AE, Huycke TR, Lee C, Mahadevan L, Tabin CJ (2015) Bending gradients: how the intestinal stem cell gets its home. Cell 161:569–580

Snippert HJ, van der Flier LG, Sato T, van Es JH, van den Born M, Kroon-Veenboer C, Barker N, Klein AM, van Rheenen J, Simons BD et al (2010) Intestinal crypt homeostasis results from neutral competition between symmetrically dividing Lgr5 stem cells. Cell 143:134–144

Soucy G, Wang HH, Farraye FA, Schmidt JF, Farris AB, Lauwers GY, Cerda SR, Dendrinos KG, Odze RD (2012) Clinical and pathological analysis of colonic Crohn's disease, including a subgroup with ulcerative colitis-like features. Mod Pathol 25:295–307

Spence JR, Lauf R, Shroyer NF (2011a) Vertebrate intestinal endoderm development. Dev Dyn Off Publ Am Assoc Anat 240:501–520
Spence JR, Mayhew CN, Rankin SA, Kuhar MF, Vallance JE, Tolle K, Hoskins EE, Kalinichenko VV, Wells SI, Zorn AM et al (2011b) Directed differentiation of human pluripotent stem cells into intestinal tissue in vitro. Nature 470:105–109
Stafford D, Prince VE (2002) Retinoic acid signaling is required for a critical early step in zebrafish pancreatic development. Curr Biol 12:1215–1220
Takamoto N, You L-R, Moses K, Chiang C, Zimmer WE, Schwartz RJ, DeMayo FJ, Tsai M-J, Tsai SY (2005) COUP-TFII is essential for radial and anteroposterior patterning of the stomach. Development (Cambridge, England) 132:2179–2189
Takebe T, Sekine K, Enomura M, Koike H, Kimura M, Ogaeri T, Zhang RR, Ueno Y, Zheng YW, Koike N et al (2013) Vascularized and functional human liver from an iPSC-derived organ bud transplant. Nature 499:481–484
Takebe T, Zhang RR, Koike H, Kimura M, Yoshizawa E, Enomura M, Koike N, Sekine K, Taniguchi H (2014) Generation of a vascularized and functional human liver from an iPSC-derived organ bud transplant. Nat Protoc 9:396–409
Takebe T, Enomura M, Yoshizawa E, Kimura M, Koike H, Ueno Y, Matsuzaki T, Yamazaki T, Toyohara T, Osafune K et al (2015) Vascularized and complex organ buds from diverse tissues via mesenchymal cell-driven condensation. Cell Stem Cell 16:556–565
Tamminen K, Balboa D, Toivonen S, Pakarinen MP, Wiener Z, Alitalo K, Otonkoski T (2015) Intestinal commitment and maturation of human pluripotent stem cells is independent of exogenous FGF4 and R-spondin1. PLoS One 10:e0134551
Tiso N, Filippi A, Pauls S, Bortolussi M, Argenton F (2002) BMP signalling regulates anteroposterior endoderm patterning in zebrafish. Mech Dev 118:29–37
Verzi MP, Stanfel MN, Moses KA, Kim BM, Zhang Y, Schwartz RJ, Shivdasani RA, Zimmer WE (2009) Role of the homeodomain transcription factor Bapx1 in mouse distal stomach development. Gastroenterology 136:1701–1710
Walton KD, Kolterud A, Czerwinski MJ, Bell MJ, Prakash A, Kushwaha J, Grosse AS, Schnell S, Gumucio DL (2012) Hedgehog-responsive mesenchymal clusters direct patterning and emergence of intestinal villi. Proc Natl Acad Sci U S A 109:15817–15822
Walton KD, Whidden M, Kolterud A, Shoffner SK, Czerwinski MJ, Kushwaha J, Parmar N, Chandhrasekhar D, Freddo AM, Schnell S et al (2016) Villification in the mouse: Bmp signals control intestinal villus patterning. Development 143:427–436
Wang Z, Dolle P, Cardoso WV, Niederreither K (2006) Retinoic acid regulates morphogenesis and patterning of posterior foregut derivatives. Dev Biol 297:433–445
Watson CL, Mahe MM, Munera J, Howell JC, Sundaram N, Poling HM, Schweitzer JI, Vallance JE, Mayhew CN, Sun Y et al (2014) An in vivo model of human small intestine using pluripotent stem cells. Nat Med 20:1310–1314
Wells JM, Melton DA (2000) Early mouse endoderm is patterned by soluble factors from adjacent germ layers. Development (Cambridge, England) 127:1563–1572
Wells JM, Spence JR (2014) How to make an intestine. Development (Cambridge, England) 141:752–760
Wills A, Dickinson K, Khokha M, Baker JC (2008) Bmp signaling is necessary and sufficient for ventrolateral endoderm specification in Xenopus. Dev Dyn Off Publ Am Assoc Anat 237:2177–2186
Workman MJ, Mahe MM, Trisno S. Poling HM (2017) Engineered human pluripotent-stem-cell-derived intestinal tissues with a functional enteric nervous system. Nat Med. http://www.nature.com/nm/journal/v23/n1/full/nm.4233.html
Xue X, Ramakrishnan S, Anderson E, Taylor M, Zimmermann EM, Spence JR, Huang S, Greenson JK, Shah YM (2013) Endothelial PAS domain protein 1 activates the inflammatory response in the intestinal epithelium to promote colitis in mice. Gastroenterology 145:831–841
Zorn AM, Wells JM (2009) Vertebrate endoderm development and organ formation. Annu Rev Cell Dev Biol 25:221–251

Chapter 11
Recapitulating Development to Generate Kidney Organoid Cultures

Melissa H. Little, Minoru Takasato, Joanne Y.-C. Soo, and Thomas A. Forbes

Abstract Evidence of the successful recreation of kidney cell types via the directed differentiation of human pluripotent stem cells adds the kidney to the list of tissues for which this approach to tissue regeneration is becoming feasible. Arguably, the kidney is one of the most complex organs to recapitulate in vitro with the final adult organ containing more than 25 distinct cell types playing a diverse number of distinct functional roles. This organ also has a heavy reliance upon the architectural accuracy of the component cellular structures as well as their appropriate co-localisation and interaction with surrounding cell types. Here we will focus on the generation of complex multicellular kidney organoids in vitro and examine both the rationale for the protocol developed, the advantages of a complex kidney organoid and the short- and long-term applications of such an approach.

Keywords Kidney development • Pluripotent stem cell • Directed differentiation • Organoid • Mesodermal patterning

M.H. Little (✉) • J.Y.-C. Soo
Murdoch Childrens Research Institute, Flemington Rd, Parkville, VIC 3052, Australia

Department of Paediatrics, University of Melbourne, Parkville, VIC 3010, Australia
e-mail: melissa.little@mcri.edu.au

M. Takasato
Murdoch Childrens Research Institute, Flemington Rd, Parkville, VIC 3052, Australia

RIKEN Center for Developmental Biology, Kobe, Hyogo 650-0045, Japan

T.A. Forbes
Murdoch Childrens Research Institute, Flemington Rd, Parkville, VIC 3052, Australia

Department of Paediatrics, University of Melbourne, Parkville, VIC 3010, Australia

Royal Children's Hospital, Flemington Rd, Parkville, VIC 3052, Australia

T. Tsuji (ed.), *Organ Regeneration Based on Developmental Biology*,
DOI 10.1007/978-981-10-3768-9_11

Abbreviations

BMP	Bone morphogenetic protein
CM	Cap mesenchyme
EE	Embryonic ectoderm
FGF	Fibroblast growth factor
GBM	Glomerular basement membrane
GDNF	Glial derived neutrophilic factor
hESC	Human embryonic stem cell
IM	Intermediate mesoderm
iPSC	Induced pluripotent stem cell
MM	Metanephric mesenchyme
PS	Primitive streak RA retinoic acid
UB	Ureteric bud

11.1 Introduction

The human kidney is a complex organ that serves a variety of physiological roles, most of which are essential to sustain life. The functional unit of the kidney, the nephron, consists of a glomerulus that filters the blood into a complex, highly organised tubule. The glomerulus is a capillary bundle supported by a mesangial stroma and covered by a specialised epithelium of podocyte foot processes. The glomerular capillary endothelium, the podocyte foot processes and the intervening glomerular basement membrane (GBM) form the glomerular filtration barrier which filters the blood with a precise selectivity. The tubule consists of multiple segments, including the proximal tubule (segments 1–3), loop of Henle and distal tubule. Each segment of the tubule contains unique combinations of specialised epithelial cell types capable of sensing the composition of the filtrate; responding to molecular signals from other organs; processing this information to dynamically regulate electrolyte, water and acid-base homeostasis; and excreting toxic, nitrogenous products of normal metabolism (Fig. 11.1). The kidney also serves important endocrine functions, as the primary site of erythropoietin production and 1α hydroxylation (and hence activation) of vitamin D precursors (Lacombe et al. 1991; Nykjaer et al. 1999).

Developmentally, the kidney is the third excretory organ to form in the human embryo (the metanephros) and the only one to persist into postnatal life (Little 2015). Nephrons arise from a self-renewing population of nephron progenitor cells called the cap mesenchyme (Little and McMahon 2012). In contrast to mice and rats where nephrogenesis occurs postnatally, the human kidney generates all of its nephrons by 36-week gestation (Hinchliffe et al. 1991). Individual nephron endowment is highly variable (between 0.2 and 2.7×10^6 nephrons per person) and positively correlates with birth weight and gestational age (Keller et al. 2003; Hoy et al. 2003; Hughson et al. 2003; McNamara et al. 2008). Nephron loss with age can be

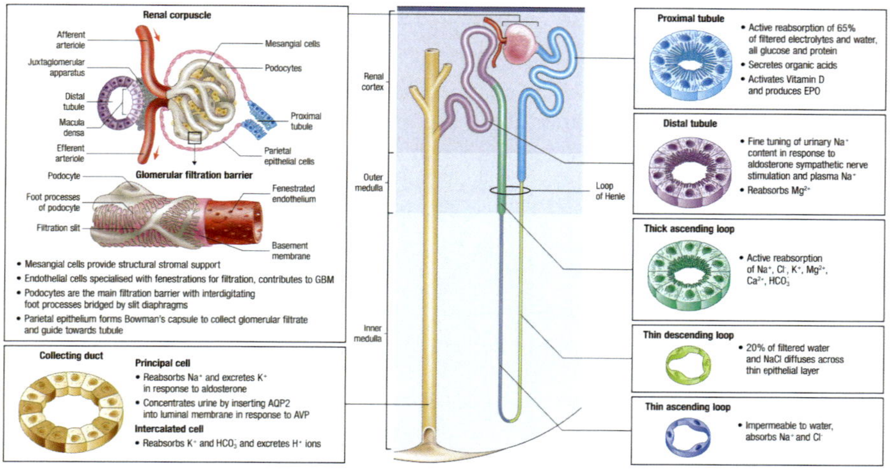

Fig. 11.1 Structure of the kidney. A single nephron illustrating the complex anatomy and architecture of the different segments and cell types and a description of their varied, specialised functions. GBM, glomerular basement membrane; AQP2, aquaporin 2; AVP, arginine vasopressin; EPO, erythropoietin

accelerated by acute or chronic kidney diseases, and while tubular epithelial cells can respond to elicit repair following mild-moderate acute tubular injury (Smeets et al. 2013; Berger et al. 2014), there is no persisting postnatal nephron progenitor cell population to regenerate new filtration units (Hartman et al. 2007; Rumballe et al. 2011). End-stage kidney disease occurs when approximately 90% of functional filtration is lost and results in considerable morbidity, mortality and healthcare expenditure. Despite advances in dialysis and kidney transplantation, these renal replacement therapies remain expensive and resource-limited interventions with significant complications.

The concept of regeneration of renal tissue is appealing in both research and clinical nephrology; however, the structural and functional complexity of the kidney makes it arguably one of the most complex tissues to regenerate in vitro. Many protocols have been published towards this aim (Wang et al. 2013; Takasato et al. 2014; Taguchi et al. 2014; Lam et al. 2014a; Freedman et al. 2015; Morizane et al. 2015), including a recent protocol that successfully derived kidney organoids from iPSCs containing ureteric epithelial collecting ducts, nephron, stromal and vascular progenitors as well as early nephrons which segment into proximal and distal tubules and immature glomeruli (Takasato et al. 2015). In this chapter, we discuss the current understanding of mammalian kidney development and the application of this knowledge to the formulation of protocols for the differentiation of human pluripotent stem cells (hPSCs) to multicellular kidney organoids. We discuss the potential applications and limitations of kidney organoids in the research of human kidney development, drug nephrotoxicity screening and kidney injury, repair and disease modelling. Finally, we discuss progress towards clinical regenerative therapies arising from these advances.

11.1.1 Formation of the Mammalian Kidney

Studies into the development of the mammalian kidney, largely performed in mouse, have provided much information about the progenitor populations present during foetal development that give rise to the distinct cell types of the kidney. In mammals, it is the metanephric kidneys which persist after birth, while the pronephroi and mesonephroi degenerate during foetal development. These metanephric kidneys are structurally distinct in that rather than being comprised of a series of simple tubules draining into the nephric duct, they form via the interaction between an epithelial side branch of the nephric duct, called the ureteric bud (Fig. 11.2). This ureteric bud (UB) forms as a swelling of the nephric duct in line with the hindlimbs. The UB then grows towards a domain of mesoderm, the metanephric mesenchyme (MM), expressing the growth factor glial-derived neurotrophic factor (GDNF). In response to GDNF from the MM, the UB upregulates RET and GFRα1, co-receptors for this ligand (Costantini and Kopan 2010). Once the UB reaches the MM, this ligand-receptor interaction triggers feed-forward signalling in which the UB begins to dichotomously branch within the MM while the MM is induced to upregulate GDNF to further support this process (Costantini and Kopan 2010). Ret signalling induces cell proliferation and migration within the tips of the branching ureteric tree as well as producing Wnt ligands that support the surrounding self-renewal of the

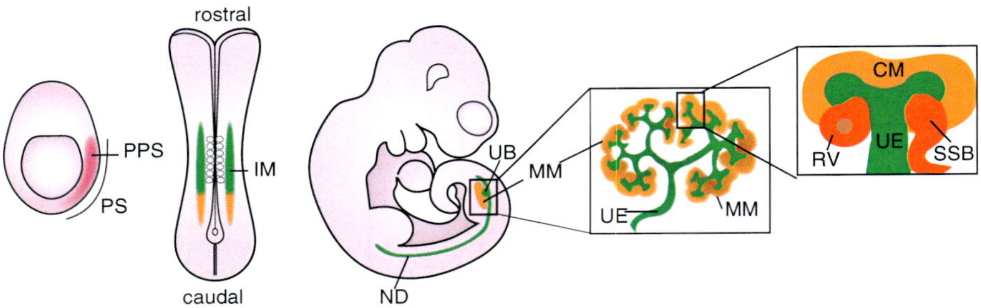

Fig. 11.2 Embryology of the kidney. The permanent kidney in mammals, the metanephros, is mesodermal in origin. The early embryo undergoes gastrulation to form the primitive streak (PS) which gives rise to the endoderm and mesoderm germ layers. The posterior end of the primitive streak (PPS) is the region from which the mesoderm of the embryo forms. Along the trunk of the embryo, the paraxial, intermediate and lateral plate mesoderm arises. The anterior end of the intermediate mesoderm (IM) gives rise to the nephric duct (ND) which extends along the embryo as it forms. A more posterior region of the IM gives rise to the metanephric mesenchyme (MM). This population induces the formation of a side branch of the ND. This is the ureteric bud (UB) which grows towards the MM and begins to branch as a ureteric epithelium (UE). The branching UE will mature into the collecting ducts of the kidney. MM cells closest to the tips of the branching UE will form the cap mesenchyme (CM). This has been shown to be the progenitor population for all of the cells of the nephron. Induction of nephron formation via a mesenchyme to epithelial transition first forms an epithelial renal vesicle (RV) which patterns and segments to form the S-shaped body (SSB) from which the nephron elongates and matures. The surrounding MM contributes to the interstitium/stroma of the kidney, some of which contributes to the perivascular and mesangial cells. It is thought that at least a portion of the vasculature also arises from the MM

mesoderm most closely associated with the tips, referred to as the cap mesenchyme (CM). This feedback loop restricts the domain of GDNF-Ret signalling to the very periphery of the expanding organ to form what is referred to as the nephrogenic zone. Lineage tracing using the CM marker, SIX2, has established that all epithelial cell types within the nephron arise from the CM (Kobayashi et al. 2008). The CM has also been shown to represent a self-renewing progenitor population critical for driving organ growth by supporting ongoing ureteric branching but also able to commit to nephron formation. The balance between self-renewal as CM and commitment to nephron formation via a mesenchyme to epithelial transition must be tightly regulated, although it remains incompletely understood. What is known is that signals from the adjacent ureteric tip, which represents the location of all progenitors for ongoing collecting duct morphogenesis, support CM maintenance as well as commitment (Kopan et al. 2014). Low levels of canonical Wnt signalling (Karner et al. 2011) together with FGF signalling (Barak et al. 2012) support self-renewal, while increased canonical Wnt signalling (Carroll et al. 2005) and subsequent non-canonical Wnt signalling (Tanigawa et al. 2011; Burn et al. 2011) are required for nephron formation. The remaining mesenchyme represents a population of stromal cells regarded as generating cells of the perivasculature (Kobayashi et al. 2014). This mesenchyme, or an earlier intermediate mesodermal population, is also likely to give rise to the vasculogenic endothelial progenitors that will contribute to the formation of the capillaries within the kidney. There is also a thought to be a contribution to the formation of kidney endothelium that arises via angiogenic ingrowth.

Almost as soon as a nephron is formed, there is immediate patterning and segmentation that occurs (Georgas et al. 2009). The cells closest to the ureteric tip as the nephron forms will become the distal segments of the nephron, while those further from the ureteric epithelium will form the proximal elements, including the glomeruli and proximal tubules. There is a declining gradient of canonical Wnt signalling from UB to glomerulus (Lindström et al. 2015), as well as distinctions in the cell adhesion proteins and transcription factors marking each segment. While it has been proposed that there are distinct progenitor cell types for different segments of the nephron (Barker et al. 2012; Rinkevich et al. 2014), the focus with respect to the recreation of the kidney has been the formation of the cap mesenchyme (nephron progenitor) and ureteric tip (collecting duct progenitor). Both of these cell types are known to arise from the intermediate mesoderm of the elongating trunk. Both also appear to be lost around (Hartman et al. 2007; Rumballe et al. 2011) or prior (Hinchliffe et al. 1991) to birth in mouse and humans, respectively, explaining why the formation of new nephrons is not observed during postnatal life in mammals. Indeed, what regulates this cessation of nephrogenesis is not understood although it has been the subject of intense investigation. Recent studies suggest that these populations are not transcriptionally uniform across morphogenesis, with both UB and CM showing declining rates of cell cycle (Short et al. 2014). However, studies into whether cessation is hard wired within CM cells suggest that placing a late CM cell into an early CM will allow such a cell to continue to self-renew in vitro longer than anticipated (Chen et al. 2015a). In addition, several studies have now developed defined culture conditions reported to be sufficient to maintain a CM phenotype in vitro away from the niche (Brown et al. 2015; Tanigawa et al. 2016). This may be important for the successful engineering of replacement organs from stem cells.

Almost all studies on mammalian kidney morphogenesis have been performed using rodent models, primarily mice. While mouse is not human, it would appear that many molecular pathways in kidney morphogenesis are conserved between these two species based upon the similarity of phenotype between gene knockouts in mouse and mutations identified in human patients (Olbrich et al. 2003; Jain 2009; Chatterjee et al. 2012; Hynes et al. 2014). More recent studies have confirmed the conserved expression of key genes during early human kidney development (O'Brien et al. 2016), including the conservation of key transcription factors within the cap mesenchyme and ureteric tip. Perhaps the most compelling evidence for a conservation between human and mouse have been recent studies in which the directed differentiation of hPSCs, based upon existing knowledge of mesodermal patterning and metanephric morphogenesis in the mouse, has successfully resulted in the generation of complex multicellular organoids containing segmenting and functionalising nephrons within a renal interstitium (Freedman et al. 2015; Takasato et al. 2015; Morizane et al. 2015). Expression profiling would suggest congruence between gene expression in trimester 1 human foetal kidney and pluripotent stem cell-derived kidney organoids, further supporting a conservation of key morphogenetic steps between these species. What we will focus on in the rest of this chapter is our knowledge of kidney development in the mouse, how this has been harnessed to generate kidney organoids and what applications are available using this approach.

11.2 Key Developmental Steps for Generating Kidney Organoids from Human Pluripotent Stem Cells

During embryogenesis, the pluripotent epiblast gives rise to three germ layers, ectoderm, mesoderm and endoderm, from which all tissues are formed. Successful protocols for the directed differentiation of hPSCs, including both human embryonic stem cells and induced pluripotent stem cells, attempt to recreate the induction events that occur during embryogenesis required for the desired cell type/tissue type, generally delivering this in a stepwise fashion via the addition or inhibition of specific signalling pathways. For directed differentiation to the kidney, this requires an understanding of the events involved in moving from pluripotency to mesoderm and from here to the kidney. As the molecular nature of these events has not been dissected in human, the field draws on our understanding of embryogenesis in the mouse.

11.2.1 Induction of the Primitive Streak During Development and In Vitro

The generation of both endodermal and mesodermal tissues commences with gastrulation, the process by which cells of the epiblast go through an epithelial-to-mesenchymal (EMT) transition to form the primitive streak (PS) (Fig. 11.2). In the mouse at 6.5 days post coitum, the anterior visceral endoderm forms on the distal epiblast and then migrates to the anterior side where it produced inhibitors of nodal, BMP and Wnt signalling. This results in the formation of the PS on the opposite side of the epiblast in response to activity of these same pathways (Perea-Gomez et al. 2002). Expression of BMP4 by the extraembryonic ectoderm (EE) adjacent to the posterior PS activates *Wnt3* expression (Ben-Haim et al. 2006), resulting in canonical β-catenin signalling within PS cells that in turn results in the expression of *Nodal*, *Mixl1*, *Eomes* and *Brachyury* (*T*) (Funa et al. 2015). Conversely, nodal signalling is essential for PS formation via upregulation of *Bmp4* in the EE. Both mesoderm and endoderm arise from the PS with fate determined by the position of cells along the streak. While *Mixl1* is expressed along the PS, expression is higher in the anterior PS which will form endoderm, this region also expressing *Sox17*, *Foxa2* and *Eomes*. Conversely, the posterior PS is marked by *T*, *Meox1* and *KDR*. This fate is patterned by the Bmp4/nodal gradient along the axis of the PS which is established via the presence of the BMP antagonists, *chordin* and *noggin*, in the anterior PS and the node counteracting the BMP4 secreted from the EE near the posterior PS (Bachiller et al. 2000). Activation of nodal/activin receptors upregulates the transcription of *Mixl1* (Hart et al. 2002; Pereira et al. 2012) determining endodermal fate.

Protocols for the induction of PS have been key for the generation of a variety of both endodermal and mesodermal tissues from hPSCs. Based on this understanding of early patterning, most employ combinations of nodal signalling (usually mimicked by activin A), BMP signalling (usually via the addition of recombinant BMP4) and canonical Wnt signalling (either via the addition of recombinant Wnt3a or inhibitors of GSK3β such as CHIR99021 or BIO) (Davis et al. 2008; Sumi et al. 2008; Burridge et al. 2012). Anterior PS has been induced via inhibition of BMP signalling using noggin, while posterior PS patterning is favoured via inhibition of nodal signalling. Mesodermal tissues such as cardiomyocyte are often initiated using a combination of activin A and BMP4, sometimes with FGF2. We have shown that varying the relative concentrations of BMP4 and activin A can alter specification from anterior (low BMP4, high activin A) to posterior (high BMP4, low Activin A) PS based on relative changes in *Sox17* versus *T* expression (Takasato et al. 2014). Indeed, induction of posterior PS from human iPSC was shown to be as effective using canonical Wnt activation alone (CHIR99021) (Takasato et al. 2014). In the mouse system, based on evidence of a later role for activin A signalling (Jackson et al. 2010), Taguchi et al. (2014) induced a T-positive 'nascent mesoderm' state using BMP4 and high Wnt signalling.

11.2.2 Inducing the Intermediate Mesoderm During Development and In Vitro

The trunk mesoderm of the vertebrate embryo is formed by migration of cells from the posterior PS. This mesoderm extends along the trunk of the embryo and is patterned from the middle of the embryo outwards (mediolaterally) into three regions: the paraxial, intermediate and lateral plate mesoderm. It is the intermediate mesoderm (IM), marked by *Osr1*, *Lhx1* and *Pax2*, that gives rise to the kidneys and the indeterminate gonad (Fig. 11.2). Hence, to generate this particular mesodermal region requires an understanding of how this mediolateral patterning occurs. Three growth factor pathways are involved in this process: BMP, nodal and FGF signalling (Takasato and Little 2015). High BMP signalling specifies lateral plate mesoderm as this region, as well as the overlying surface ectoderm, produces BMP4 (Obara-Ishihara et al. 1999; James and Schultheiss 2005). Inhibition of BMP signalling, specified in the embryo via the production of the BMP inhibitor, noggin, by the notochord, is required for paraxial mesoderm fate (Wijgerde et al. 2005). It is therefore proposed that the IM develops under an intermediate level of BMP signalling. Similarly, inhibition of nodal signalling via the production of the inhibitor Cer1 from the paraxial mesoderm is critical for paraxial mesoderm specification (Biben et al. 1998). In the absence of Cer1, there is an expansion of the IM (Fleming et al. 2013). Finally, FGF9 is produced by the paraxial and intermediate mesoderm in the caudal embryo, within this more restricted to IM in the trunk region (Colvin et al. 1999).

The trunk of the embryo extends from the head to the tail across time; hence there are also distinct differences in specification within all mesodermal domains across time and along this rostro-caudal (head-to-tail) axis. Indeed, three distinct pairs of excretory organs form along this axis (pronephros, mesonephros and metanephros) with the metanephroi representing the most caudal of these organs. A common component to all three excretory organs is the epithelium of the nephric duct. This initially forms early, and hence from anterior IM, and then extends along the embryo as the trunk elongates (Takasato et al. 2014; Taguchi et al. 2014; Takasato and Little 2015). Conversely, the cells contributing to the MM arise later and in a more caudal location. As a result, cells of the MM are subjected to prolonged canonical Wnt signalling within the caudal embryo and are also protected from retinoic acid signalling via the expression of *Cyp26a*. Clearly, to recreate a kidney, both anterior and posterior IM derivatives are required.

Attempts to recreate IM in vitro from hPSCs initially focussed on the generation of $OSR1^+$ mesenchymal populations (Mae et al. 2013). However, after initial posterior PS patterning, we have reported the spontaneous formation of $OSR1^+$ $FOXF1^+$ lateral plate mesoderm (Takasato et al. 2014), highlighting the non-specificity of OSR1 as an IM marker. Instead, it has been shown that FGF signalling, using either FGF2 or FGF9 with or without retinoic acid (RA), results in the induction of $PAX2^+$ $LHX1^+$ IM cells (Lam et al. 2014b). This suggested in human that FGF9 signalling is sufficient to specify IM even in the absence of BMP suppression. To test the hypothesis that rostro-caudal patterning could also be regulated in vitro, we varied the duration of CHIR99021-mediated specification between 2 and 5 days before the addition of FGF9 (Takasato et al. 2015). A shorter duration of canonical Wnt induction, particularly with the addition of RA, patterned posterior PS and then anterior IM, as evidenced by the formation of a GATA3$^+$ epithelium representative of nephric duct, whereas prolonged Wnt signalling, particularly in the presence of an inhibitor of RA signalling, increased the expression of the MM transcription factor, HOXD11, suggesting more posterior IM (Takasato et al. 2015). Of note, an intermediate duration of canonical induction prior to FGF9 generated cultures within which there was evidence for both anterior and posterior IM cell types, and it is from these populations that kidney organoids were generated.

11.2.3 Generation of Kidney Organoids from Human Pluripotent Stem Cells

As described above, the final permanent kidney in the mammal arises from the reciprocal interactions between the UB that branches from the nephric duct, a derivative of the anterior IM, and the MM, which is derived from the posterior IM (Fig. 11.2). In mouse, lineage tracing using *Eya1* confirms that while the mesenchyme of the metanephroi and the caudal region of the mesonephroi arise from an EYA1$^+$ population, the UB itself arises from an earlier OSR1$^+$ population (Xu et al. 2014). Despite this, after induction of posterior PS and then IM, we have observed the simultaneous formation of an epithelial population displaying markers of nephric duct/UB (PAX2, GATA3, ECAD1, DBL) within a mesenchyme displaying expression of MM/CM (SIX2, WT1, PAX2, HOXD11) (Takasato et al. 2014). While both of these cell types arose in monolayer culture in defined media, with time the cells within the dish reorganised into 3D clusters in which an epithelium was surrounded by mesenchyme spontaneously forming CDH6$^+$ early nephrons (Takasato et al. 2014).

During normal kidney development, morphogenesis is a highly dynamic process with cells migrating with respect to each other and in response to growth factor and cell-cell-based signals (Short et al. 2014; Combes et al. 2016). This process occurs in most tissues during development and is referred to as self-organisation. Individual mesenchymal cells within each CM/ureteric tip swarm around the tip in response to a variety of signals from the tip (Combes et al. 2016). In this way, the induction of branching is maintained and the CM can reorganise around newly formed tips, continuing to reshape the growth factor field. This process undoubtedly also plays a role in the positioning and induction of nephrons. Indeed, the process of cell-cell-regulated self-organisation during vertebrate development is so strong that embryonic tissues can be dissociated down to single cells and reform structure after reaggregation via self-organisation (Weiss and Taylor 1960). This is also the case for the kidney (Unbekandt and Davies 2010; Lusis et al. 2010). As a result, we reasoned that if the appropriate cell types were present in cultures of hPSCs differentiated towards UB and MM, the provision of a 3D environment would facilitate spontaneous self-organisation and potentially improve morphogenesis in vitro. In a protocol using fully defined APEL media and an initial 7 days of PS and IM induction, cultures were dissociated to single cells, aggregated and cultured using floating filter cultures as is standard for the *ex vivo* culture of mouse embryonic kidney (Saxén and Lehtonen 1987). This approach resulted in the formation of complex multicellular kidney organoids that, after 11 days in organoid culture, contain not only patterning and segmenting nephrons but also a contiguous collecting duct epithelium. Large numbers of nephrons are present within each organoid with evidence of progression through S-shaped body to capillary loop stages of patterning, including the eventual formation of UMOD$^+$ loops of Henle. Glomeruli are evident and contain early podocytes with tightly interdigitated foot processes. While the majority of these represent early nephrons and hence are avascular, the organoids also contain an extensive network of endothelial cells, themselves adjacent to early PDGFRA$^+$ pericytes and mesangial cells. The endothelium is positive for SOX17, KDR and PECAM and in some instances can be seen within glomeruli, suggesting the initial formation of glomerular capillaries. In addition, the epithelial structures, nephrons and collecting ducts, are surrounded by a stromal population expressing MEIS1$^+$, a reported marker of the renal interstitium (Brunskill and Potter 2012). The presence of both an endothelial population and a renal interstitial population is consistent with these cell types arising from the MM or the IM, although they may also arise from a closely associated paraxial or lateral plate population.

The presence of more than eight identifiable kidney cell types, arranged in an appropriate morphological arrangement, strongly suggests the successful recreation of a developing human kidney from pluripotent stem cells (Fig. 11.3). However, the identity of these cell types has been based on what we know from mouse. Indeed, there are few genes that are specific to only one organ and many genes expressed during metanephros and expressed in mesonephros and gonad, as expected based on their common IM origin. To address whether these organoids represented the kidney, RNAseq expression profiling was performed in total iPSC-derived kidney organoids and compared in an unbiased fashion to the transcriptional profiles of >20 human foetal tissue types using a previously defined algorithm, KeyGenes (Roost et al. 2015). Using this unbiased analysis, hPSC-derived organoids after 11 days of culture most closely matched trimester 1 human kidney. Of note, early after generating the aggregate, these cultures showed a closer alignment to indeterminate gonad, the only other derivative of IM analysed. An analysis of the temporal gene expression changes across organoid culture shows a significant increase in markers of nephron-specific cell types, including podocytes and proximal tubule cells, with time. Electron microscopic analyses showed distinct evidence for proximal and distal tubules and avascular glomeruli. Initially, LTL$^+$ proximal tubular segments were EpCAM$^+$ but CDH1$^-$, as has been reported during development (Combes et al. 2015). However, with time, LTL$^+$ CDH1$^+$ epithelial segments formed. To investigate whether this proximal tubular epithelium was sufficiently mature to selectively respond to nephrotoxic injury, organoids were treated with cisplatin to induce apoptosis. Maturing proximal tubular segments showed a specific and dose-responsive caspase 3 activation, suggesting that they were affected by this nephrotoxicant (Takasato et al. 2015).

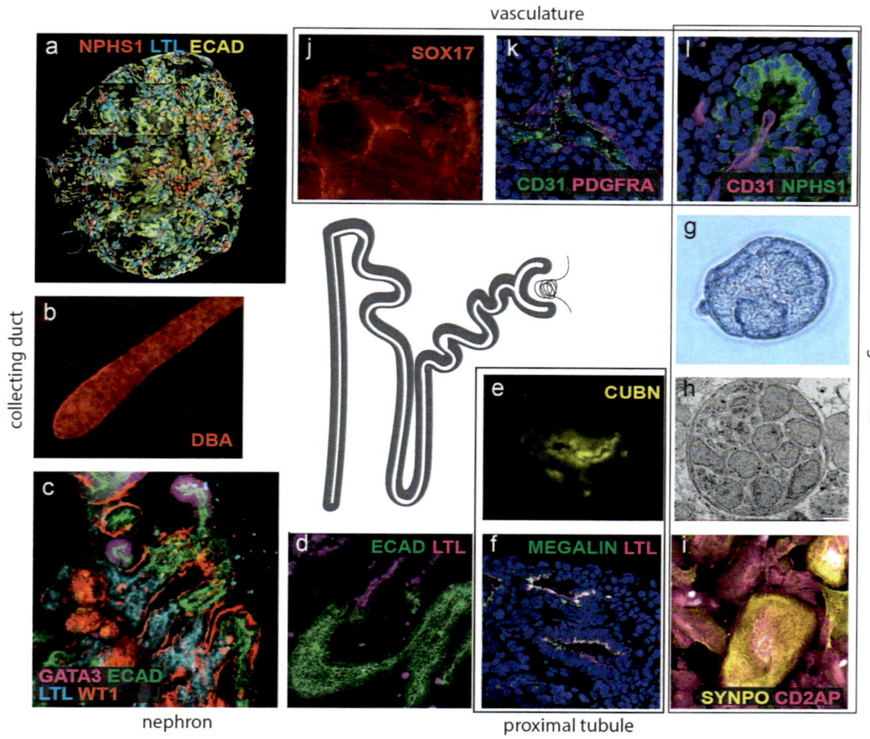

Fig. 11.3 Cells present within kidney organoids. Kidney organoids generated via directed differentiation of human pluripotent stem cells contain multiple cellular components reflecting the cell types present during kidney morphogenesis. Organoids (**a**) contain collecting duct epithelium (**b**) positive for *Dolichos biflorus* agglutinin (DBA) together with patterning and segmenting nephrons (**c**), including distal tubules expressing CDH1 (**d**) and proximal tubules expressing MEGALIN and CUBILIN (CUBN), and positive for *Lotus tetragonolobus* lectin (LTL) (**e**, **f**) and podocytes within maturing glomeruli (**g–i**). Intact glomeruli can be sieved from organoids (**g**) and contain tightly packed podocytes (h) positive for SYNAPTOPODIN (SYNPO) and CD2AP (**i**). Surrounding the nephrons, early endothelial cells ($SOX17^+$ $CD31^+$) and perivascular cells ($PDGFRA^+$) form (**j**, **k**), with some beginning to enter the forming glomeruli (**l**) (Adapted from Takasato et al. 2015 with additional images from Jessica Vanslambrouck and Lorna Hale)

11.3 Alternative Protocols for Directing Differentiation of hPSCs to the Kidney

Alongside the identification of this protocol for the generation of complex kidney organoids, a number of other protocols for the renal differentiation of hPSC in vitro have been reported. In most cases, these directed differentiation protocols also involve stepwise differentiation through embryological stages leading to kidney development. Hence, this again requires the derivation of the posterior PS, followed by IM and then MM and/or UB. Other published protocols demonstrate the differentiation of human iPSC to a single renal cell type or segment (Song et al. 2012; Wang et al. 2013; Sharmin et al. 2016; Ciampi et al. 2016), but in this section we will concentrate on the protocols (i) shown to establish multisegmented kidney organoids and (ii) employing fully defined culture conditions (summarised in Table 11.1).

Initial iPSC renal differentiation protocols were based on monolayer culture and utilised various combinations of CHIR99021, BMP, FGF and activin signals. Taguchi et al. (2014) applied a protocol developed for mouse ESC protocol to the differentiation of human iPSC first generating embryoid bodies and subsequently deriving appropriate MM (WT1$^+$, PAX2$^+$, SALL1$^+$ and SIX2$^+$) which, when cultured with dorsal spinal cord, gave rise to contiguous WT+/nephrin + glomerular structures, CDH6+ proximal tubules and CDH1$^+$ distal tubules. The same group later transplanted nephrin-GFP-expressing glomeruli underneath mouse renal capsules and observed vascularisation of the iPSC-derived glomeruli by murine endothelial cells (Sharmin et al. 2016). During their protocol optimisation, Lam et al.'s (2014b) prolonged CHIR99021 exposure led to lateral plate mesoderm fate (FOXF1$^+$), but introduction of FGF2 signalling diverted CHIR99021-treated cells to PAX2$^+$ IM fate. They also demonstrated that FGF9 and activin treatment of IM gave rise to SIX2$^+$ MM and treatment of this population with further CHIR99021 induced a downregulation of SIX2$^+$ mesenchyme and the genesis of LTL$^+$ proximal tubular structures (Lam et al. 2014b).

Takasato et al. (2014) demonstrated both CHIR99021 and high BMP4/low activin A could effectively induce PS mesendoderm but specified IM fate using FGF9 and heparin which was continued to derive SIX2$^+$ MM. Twelve days following withdrawal of growth factors in CHIR99021-treated cell lines, both MM and UE structures were present and self-organised when aggregated to form 3D structures resembling kidney tubules expressing WT1, PAX2, CDH6, JAG1, aquaporins 1 and 2 and SLC3A1 (Takasato et al. 2014). While this first aggregate protocol utilised foetal calf serum in its culture media, this group would later publish the refined protocol described above to derive organoids displaying NPHS1$^+$/WT$^+$ glomeruli, LTL$^+$/CUBN$^+$ PT, PAX2$^+$/CDH1$^+$/GATA3$^+$ ureteric epithelium, CD31$^+$ endothelium and MEIS1$^+$ stromal components (Takasato et al. 2015).

Self-organisation within the nephrogenic zone of the developing kidney is driven by GDNF stimulation of UE branching morphogenesis by the MM and reciprocal FGF9 and low-level Wnt signalling from the UB required for self-renewal and maintenance of the cap mesenchyme. This self-organisation was also utilised by Morizane et al. (2015) who published a protocol using 4-day CHIR99021+ noggin to establish PS, followed by 3-day activin A to establish posterior IM and 7 day-FGF9 to establish a SIX2$^+$/WT1$^+$/SALL1$^+$ MM. Here nephrogenesis was also induced by 2-day CHIR99021, giving rise to NPHS1$^+$/PODXL$^+$ podocytes, CDH2$^+$/LTL$^+$ proximal tubules, CDH1$^+$/UMOD$^+$ loops of Henle and CDH1$^+$ BRN1$^+$ distal tubule after 4-week culture. Where Takasato et al. (2015) pelleted differentiated cells and cultured the pellets on an air-liquid interface, Morizane et al. (2015) replated cells in an ultra-low attachment, round bottom, 96-well plate. Freedman et al. (2015) utilised a Matrigel sandwich culture technique to derive epiblast

Table 11.1 Fully defined protocols for the in vitro directed differentiation of human iPSC to multisegmented nephron fate

Author, Year	Culture Method	Derivation of Intermediate Mesoderm (IM) via Posterior Primitive Streak (PPS)	Derivation of Metanephric Mesenchyme (MM) and/or Ureteric Bud (UB)	Endpoints and markers by immunofluorescence						Notes
				Glom	PT	DT	CD	Endo	Stroma	
Mac et al. (2013)	Monolayer	*PPS:* 3 days CHIR99021 + Activin A *IM:* 8 days CHIR99021 + BMP7	*MM*: 7 days CHIR99021, BMP7	+	+					Generated an OSR1-GFP reporter line to enable high throughput growth factors screening for IM induction.
Taguchi et al. (2014)	Embryoid Body	*Epiblast:* 2 days Activin A *PPS:* 2 days CHIR99021, BMP4 *Posterior nascent mesoderm:* 4 days CHIR99021, BMP4 *Posterior IM:* 2 days Activin A, BMP4, CHIR99021, RA	*MM:* 3 days CHIR99021, FGF9	+	+	+				MM was derived under fully defined conditions but culture of organoid end points was by culture with dorsal spinal cord.
Lam et al. (2014a, b)	Monolayer	*PPS:* 2 days CHIR99021 *IM:* 4 days FGF2 + RA	*MM:* 3 days FGF9 + activin A		+					IM differentiated into LTL+, N-cadherin+ PT cells with removal of FGF2 and RA and 1 day CHIR99021.
Takasato et al. (2014)	Monolayer	*PS:* 2 days CHIR99021 *IM:* 4 days FGF9 + Heparin	*MM and UB:* 6 days FGF9 + Heparin then 6 days no growth factors.	+	+	+	+			Early podocytes colocalising WT1+ and synaptopodin+ with PCR evidence for upregulation of podocyte genes.
Araoka et al. (2014)	Monolayer	*PPS:* 2 days CHIR99021 + Activin A *IM:* 4 days AM580 or TTNPB	*MM and UB:* 8 days Wnt3a + BMP7	+	+		+?			Also found CHIR99021 alone can give rise to PPS. TTNPB protocol was utilised for the differentiation of end point cultures.
Imberti et al. (2015)	Monolayer	*PPS/IM:* 6 days RA + PI3K inh + RhoA inh + 2 days activin A	*MM:* FGF2 + BMP7 + GDNF	+	+					Integration of human renal progenitors into cisplatin injured mouse nephrons was observed.
Morizane et al. (2015)	Monolayer then 3D	*Late PS:* 4 days CHIR99021 ± noggin *Posterior IM:* 3 days activin A	*MM:* 2 days FGF9 *Self-organising nephrons:* 2 days FGF9 + CHIR99021 then 3 days FGF9 then no growth factors.	+	+	+			a	3D culture performed by replating cells on day 10 to ultra-low-attachment, round bottom 96-well plates and in suspension culture.
Freedman et al. (2015)	Spheroid	*Epiblast spheroid:* 2 days sandwiched between Matrigel in mTeSR1 medium *Mesoderm:* 1.5days CHIR99021	*Self-organising nephrons:* 12 days B27 supplement	+	+	+		+		Generated epiblast spheroids in Matrigel sandwich culture, prior to differentiation. CD31+/vWF+ endothelium also present.
Takasato et al. (2015)	Monolayer then 3D	*PPS:* 4 days CHIR99021 *Ant and Post IM:* 3d FGF9 + Heparin	*Organoid:* 1 hour CHIR99021, then 5 days FGF9 + Heparin, then 7 days no growth factors.	+	+	+	+	+	+	First report of endothelial progenitors (CD31+, SOX17+) with lumen formation, observed invading glomeruli.

IM intermediate mesoderm, *PPS* posterior primitive streak, *RA* retinoic acid, *Glom* glomerular structures, *PT* proximal tubules, *DT* distal tubule, *CD* collecting duct, *Endo* endothelium

[a]Personal communication

spheroids from iPSC and then induced epithelial-to-mesenchymal transition using a short period of high-dose CHIR treatment (1.5 days, 12 μM) followed by culture in B27-supplemented media. This resulted in the development of nephron structures with WT1+/SYNPO+ podocytes, LTL+ proximal tubule, ECAD+ distal tubule and CD31+ vascular progenitors which were subsequently used to model nephrotoxic kidney injury and model disease (Freedman et al. 2015).

Imberti et al. (2015) omitted CHIR99021 from their protocol, instead using small molecules CCG1423 (a Rho pathway inhibitor) and LY294002 (a phosphoinositol 3 kinase inhibitor). Activin A was added for days 2 and 3, guiding differentiation to a T^+/$OSR1^+$ primitive streak, and another 6 days of small-molecule exposure achieved a $WT1^+$/$PAX2^+$ IM. The stage 2 protocol for the differentiation of a $WT1^+$/$PAX2^+$/$SIX2^+$/$SALL1^+$ MM/nephron progenitor population consisted of 6-day culture in FGF2, BMP7 and GDNF. The role of GDNF in this culture is unclear given the absence of collecting duct, and the work in mouse would suggest that the addition of BMP7 may induce differentiation rather than maintaining MM/CM (Brown et al. 2013). Indeed, across the time course investigated in this study, expression of SIX2 was maximal at day 6 but subsequently declined. The outcome of the differentiation was characterised as glomerular ($CD133^+$/$CD24^+$/$claudin1^+$) and proximal tubule ($AQP1^+$/$GGT1^+$) (Imberti et al. 2015). Araoka et al. (2014) also used small molecules AM580 and TTNPB identified during high-throughput screening using an OSR1-GFP reporter hiPSC line, but in contrast Imberti et al. (2015) utilised CHIR99021 instead of activin A to induce IM and demonstrated $OSR1^+$/$WT1^+$/$PAX2^+$/$SALL1^+$ IM monolayer which formed LTL^+/$AQP1^+$ proximal tubular cells, PDX^+/$WT1^+$ podocytes and DBA^+/$CK8^+$/$SALL4^+$ nephric duct alongside adrenal cortex and gonad markers.

In summary, despite the variety of published protocols, the consensus approach from the field currently favours the induction of Wnt and/or activin signalling to derive PS and subsequently drive nephrogenesis followed by FGF signalling to promote maintenance of a MM cell mass. The ability of cells to self-organise in monolayer and 3D culture suggests that once a certain stage of differentiation is reached, cells can guide their own differentiation and maintenance, as well as that of the cultured tissue. Future challenges in protocol development will include the maturation and vascularisation of glomerular structures as well as improving the maturity of tubular expression and collecting duct derivation.

11.4 Applications to the Use of Kidney Organoids

11.4.1 Utility of Kidney Organoids for Analysing Human Kidney Development

While there has been some limited anatomical studies of the developing human kidney (Osathanondh and Potter 1963), and more recently some analysis of gene expression using in situ hybridisation and immunohistochemistry (O'Brien et al. 2016),

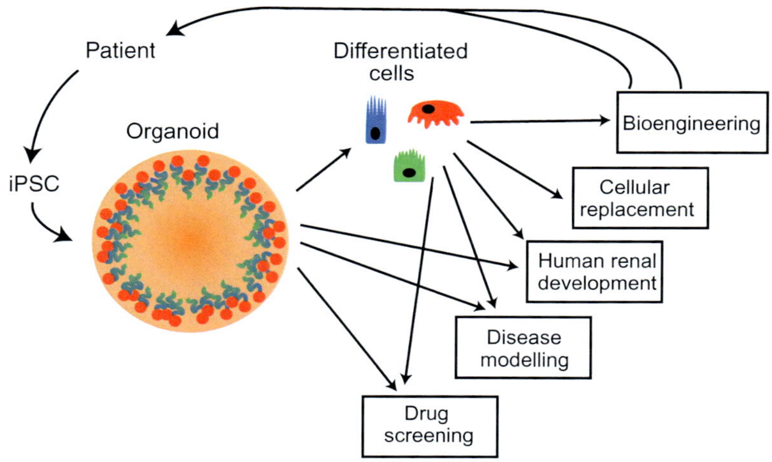

Fig. 11.4 Application for kidney organoids generated from human pluripotent stem cells. Applications include the use of organoids as a whole in nephrotoxicity or drug screening and in disease modelling using gene edited of patient-derived iPSC. They also serve as a model for kidney development in the human. The generation of human kidney cell types using organoid differentiation protocols will also provide individual cell types for cell therapy or bioengineering of replacement renal tissue

access to viable material is limited, particularly from late gestation, and genetic tools available in mouse such as lineage tracing and reporter tagging have not been possible. The capacity to apparently generate human kidney cell types in vitro via the directed differentiation of pluripotent stem cells provides the first opportunity to examine human kidney development (Fig. 11.4). Indeed, the amenability of iPSC to CRISPR/Cas9 gene modification may provide an opportunity for lineage analysis during kidney organoid formation. A number of human iPSC reporter lines have now been established, primarily as part of the process of developing protocols for the generation of human iPSC-derived kidney tissue. These include fluorescently tagged reporters driven from the MIXL1 (Davis et al. 2008; Takasato et al. 2014), OSR1 (Mae et al. 2013) and NPHS1 promoters (Sharmin et al. 2016). An ability to isolate cells derived using such reporters will facilitate the more stringent characterisation of specific cell types to more thoroughly compare mouse and human and to also investigate intermediate stages of cell maturation. There is, however, a 'chicken and egg' aspect to such studies. We are using our existing understanding of mouse development, coupled with validated postnatal markers of mature cell type in both mouse and human, to infer the identity of specific cell types within iPSC-derived kidney tissue cultures. It remains possible, therefore, that what we are characterising is a synthetic cell type generated as a result of the culture conditions that may have no genuine counterpart in normal human development. Better understanding of human kidney development, therefore, will continue to require validation in comparison to human material.

11.4.2 Drug Screening Using Kidney Organoids

Drug toxicity is a major cause of attrition in late-stage clinical trials and of drug withdrawal from the market after registration, leading to injury in patients and massive opportunity costs for drug companies (Schuster et al. 2005). Nephrotoxicity, in particular, is one of the top ten organ-specific toxicities seen in drugs which are withdrawn, showing that the in vitro assays and animal models currently used are not sufficiently predictive (Fung et al. 2001). Most cases of drug-induced nephrotoxicity involve proximal tubule injury, as proximal tubule cells express many transporters and receptors which can take up drugs from the interstitial fluid or filtrate. Animal models often do not predict nephrotoxicity in humans because transporter expression and subcellular localisation differ between species (Koepsell 2013). To eliminate these species-specific differences, two-dimensional cultures of primary human proximal tubule cells or immortalised proximal tubule cell lines have been used to screen for nephrotoxicity. However, these cells have relatively low transporter expression or lose expression after a few passages, which alters their susceptibility to drugs and thus decreases the predictivity of the assay (Qi et al. 2007; Jenkinson et al. 2012; Nieskens et al. 2016). These cells also express injury and dedifferentiation markers such as KIM-1 and vimentin and do not consistently upregulate injury markers in response to known nephrotoxins (Rached et al. 2008; Li et al. 2013; Sohn et al. 2013; DesRochers et al. 2013).

Proximal tubule cells derived from iPSCs have higher transporter expression compared to primary proximal tubule cells, suggesting that they are more functionally mature and will therefore detect nephrotoxins more accurately (Kandasamy et al. 2015). Indeed, using IL-6 and IL-8 production as a readout of injury, a nephrotoxicity screen using iPSC-derived proximal tubule cells distinguished between nephrotoxins and non-nephrotoxins with 87% accuracy (Kandasamy et al. 2015). However, iPSC-derived proximal tubule cells in 2D culture still express injury markers at relatively high levels and express markers associated with other renal cell types, showing that there is still room for improving their differentiation and maturation.

The reciprocal signalling involved in cell differentiation and organisation within organoids may improve the functional maturity of organoid proximal tubules compared to 2D cultures (Fig. 11.4). Unlike proximal tubule cells derived from iPSCs in isolation, organoid proximal tubules do not typically express markers which are normally specific to other renal cell types. As for function, organoids have been produced with proximal tubules capable of endocytosing dextran (Takasato et al. 2015). Proximal tubules in organoids have also been shown by various groups to upregulate KIM-1 and undergo apoptosis in response to cisplatin and gentamicin (Takasato et al. 2015; Morizane et al. 2015). These results suggest that the proximal tubules in organoids are expressing the endocytic receptors and transporters known to mediate drug-induced proximal tubule injury. Further work will be required to confirm that the major drug influx and efflux transporters are expressed at appropriate levels and that drug-induced injury in organoid proximal tubules is indeed mediated by these transporters.

Organoids transplanted under the renal capsule of immunocompromised mice or rats show host-derived vascularisation in their glomeruli (Xinaris et al. 2012; Sharmin et al. 2016). Mice with transplanted organoids may serve as 'humanised' in vivo models that combine the advantages of in vivo tests (e.g. detecting potentially toxic drug metabolites) with the advantages of screening for injury to human instead of animal kidney tissue. If transplanted, organoids can be developed with sufficiently mature vascular function and filtrate production. These may also be able to detect disruptions to vascular function or drug-induced nephrolithiasis, which are also common mechanisms of nephrotoxicity.

11.4.3 Kidney Organoids as Models of Disease

The modelling of heritable kidney disease via the generation of kidney organoids from patient-derived iPSC represents another tractable application of this technology (Fig. 11.4). In the past, the phenotypic validation of novel genetic mutations has relied on the generation of murine disease models (Becker and Hewitson 2013). This previously laborious and time-inefficient model has been optimised to a much faster and more genetically precise process using CRISPR/Cas9 technology (Yang et al. 2013). However, the differences between mouse and human kidney development, both at the anatomical and molecular levels, represent a significant limitation of murine models of human kidney disease. The proposed advantages of patient-derived iPSC-based disease modelling include the opportunity to produce limitless amounts of differentiated renal tissue expressing the gene of interest in the context of the patient's own genome, acknowledging the possibility of polygenic contribution to phenotype and obviating the risk and difficulty of procuring and culturing primary tissue. Gene-editing techniques can be introduced to an iPSC line during reprogramming to generate an otherwise isogenic control line (Howden et al. 2015). Advances in diagnostic genomics is facilitating the identification of mutations in affected patients; however even in those instances where a gene defect is not identified, organoid creation provides an opportunity to characterise the disease processes at the cellular level and develop personalised treatments.

iPSC-derived kidney organoids represent models of the developing organ. As such, a group of kidney diseases most amenable to disease modelling with iPSC techniques are the early-onset, monogenic, heritable kidney diseases, which represent up to 20% of adults and 47% of children with chronic kidney disease (Hildebrandt 2010; Fletcher et al. 2013; Mallett et al. 2014; Devuyst et al. 2014). To date, iPSCs have been derived and validated from patients with autosomal dominant (ADPKD) (Thatava et al. 2011; Wang et al. 2013; Freedman et al. 2013, 2015), autosomal recessive polycystic kidney disease (ARPKD) (Freedman et al. 2013), Alport syndrome (Chen et al. 2015b), Wilms' tumour (Thatava et al. 2011) and systemic lupus erythematosus (Thatava et al. 2011). hESCs have been generated from human embryos with Alport syndrome (Frumkin et al. 2010) and Fabry disease (Tropel et al. 2010) identified during preimplantation genetic screening. Xia et al. (2013) differentiated iPSC from a patient with ADPKD to ureteric bud but did not establish a structural or molecular phenotype. Freedman et al. (2013) identified lower ciliary expression of polycystin-2 in embryoid bodies and differentiated hepatoblasts in patients with *PKD1* mutations, which corrected with transfected overexpression of wild-type *PKD1*. The same group later introduced biallelic *PKD1* or *PKD2* mutations (a genotype associated with embryonic lethality) into a normal hPSC line using CRISPR/Cas9 and differentiated them according to their epiblast spheroid sandwich culture protocol, observing the development of sizeable, translucent cysts arising from tubules expressing proximal tubular markers in 6% of organoids (Freedman et al. 2015). This report represents a reasonable proof of principle, but to date there exists no evidence of a patient-derived iPSC kidney organoid disease model.

While disease modelling using this approach seems promising, there are limitations to this approach. Due to immaturity and lack of a circulation, organoids cannot model inflammatory disease. As they do not have an accurate overall organ structure, they are also limited in their ability to model congenital structural renal diseases such as hypoplasia or duplex kidney. The avascular glomeruli lack a well-developed GBM, currently limiting the study of GBM disorders (e.g. Alport syndrome). Recent work demonstrating the incorporation of murine vasculature into subcapsular transplants of human iPSC-derived glomeruli indicates this impediment is likely to be overcome for disease modelling purposes (Sharmin et al. 2016). Furthermore, current organoid nephrons are immature with their capacity to mature further using existing culture methods unclear (Little and Takasato 2015; Takasato et al. 2015). Thus, organoids have not yet been proven suitable to study diseases without a foetal or infantile phenotype. Conversely, genomic examination of developing renal tissue prior to the development of a phenotype may control for differential gene expression occurring secondary to phenotype rather than the result of the primary genomic imbalance. In contrast, modelling a potential disease causing gene in mouse has the advantage that the organ formed is anatomically correct, can mature into a functional organ and can also reflect the effect of systemic changes on the organ as a result of the mutation. However, the mouse does not model the potentially mutant allele in the context of the patient's genome. Balancing these advantages and limitations, it is likely that both mouse and organoid techniques will be necessary to facilitate informative disease modelling research until such time that differentiation protocols are optimised to generate more mature and complex iPSC-derived tissues.

11.5 Longer-Term Directions and Challenges for Directed Differentiation to the Kidney

The initial isolation of human pluripotent stem cells was seen as opening up the prospect of regenerative medicine. The directed differentiation of human pluripotent stem cells to endpoints such as dopaminergic neuron is now providing opportunities for clinical trials in the treatment of Parkinson's disease (Barker et al. 2012), and the generation of retinal pigmented epithelium from iPSC is being trialled for the treatment of macular degeneration (Ouyang et al. 2016). However, the delivery and integration of iPSC-derived tissues brings with it considerable challenges. It is not clear what residual risk of teratoma formation exists with the delivery of cells differentiated in vitro. The prospect of accumulated genomic instability in pluripotent stem cell lines may also bring with it a risk of cancer in the resulting progeny. From a more practical standpoint, getting the cell type you wish to deliver into the right location and functionally integrating into the desired organ, particularly in the face of existing tissue injury, as is likely in any patient with underlying disease, is a major challenge. Hence, while directed differentiation to cardiomyocyte was first described over a decade ago, the successful integration and electrical coupling of such cells have not been successful (Mummery et al. 2010).

Directed differentiation of human pluripotent stem cells to organoids, while generating a highly complex multicellular structure similar to the developing organ itself, does not at this point in time provide a tissue available for the treatment of renal disease. What it does do is create the progenitor populations for all of the final structures within the kidney. There are, therefore, several pathways from organoid that may ultimately deliver regenerative approaches to the treatment of kidney disease (Fig. 11.4). These include cellular therapy and bioengineering.

11.5.1 Provision of Cell Sources for Cellular Therapy

We would define the cellular therapy of kidney disease as the restoration of kidney function via the delivery and site-appropriate functional integration of specific kidney cell types into a patient. A number of studies have reported that it is feasible to deliver specific cell types into animal models of renal injury and see evidence that these cells reached the kidney and reside in appropriate locations. This has been reported for adult renal progenitor cells (Lazzeri et al. 2007) and was initially proposed for bone marrow-derived stem cells specifically in response to acute renal injury (Poulsom et al. 2001). In the case of the latter studies, it was subsequently shown that such events were extremely rare and involved fusion with existing renal epithelial cells rather than functional integration (Lin et al. 2005). A number of clinical trials have examined the delivery of human mesenchymal stem cells (bone marrow derived) in the treatment of acute kidney injury and the modulation of graft rejection in transplant recipients (Reinders et al. 2013; Fleig and Humphreys 2014). In this instance, in both mouse and man, the MSCs delivered into the patient survive only transiently with the apparent clinical improvement a product of an alteration in the growth factor milieu within the patient rather than any evidence of cellular integration. As such, this is likely to represent a transient modulation rather than a cellular therapy with prolonged functional integration. Now it is possible to direct differentiation of pluripotent stem cells to kidney cell types, it is perhaps now time to systematically reassess whether any given kidney cell type can be successfully reintroduced to a patient and functionally contribute. Two such studies have been reported. The first of these (Imberti et al. 2015) delivered a PAX2/8^{+}OSR1^{+} progenitor population derived using a previously undescribed differentiation protocol (RA/PI3K inhibition, activin A, then BMP7/GDNF/FGF2) into a cisplatin-induced acute injury model performed in NOD-SCID mice via tail vein injection (5 x 10^5 cells/mouse). They report the presence of human cells within the immunocompromised recipient animals, although cell contribution was not quantified, fusion was not investigated and functional integration was not proven. However, evidence was presented to suggest that the delivery of these cells improved BUN, reduced necrosis and showed less protein casts. The second of these studies (Toyohara et al. 2015) employed a distinct protocol for the generation of OSR1^{+}SIX2^{+} renal progenitors which were enriched based upon the insertion of GFP into the OSR1 locus. When isolated, this cell population proved capable of forming nephrons in vitro in response to Wnt signalling. These human iPSC-derived progenitors were delivered via renal subcapsular transplantation into immunodeficient mice subjected to an ischemia/reperfusion injury, BUN and serum creatinine levels were reduced, and histological analyses also suggested less protein casts, reduced tubular dilatation and improved brush borders and less necrosis (Toyohara et al. 2015). However, in this instance, there was no evidence that the cells themselves integrated into the recipient kidney with the effect presumed to arise due to the production of pro-reparative growth factors by the implanted cells (Toyohara et al. 2015). The dissociation and delivery of multiple cell types from an organoid has not been performed. Perhaps the advantage of starting preclinical trials using this as a source of cells would rest with the diversity of cell types present. However, irrespective of the kidney cell type used, there is likely to be much preclinical work to be done to improve the delivery of such cells to optimise integration.

11.5.2 Implantation of Regenerated Tissue

It may also be possible to deliver the entire organoids for in vivo vascularisation. The protocol of Takasato et al. (2015) generates a human kidney organoid comprised of different renal cell types and measuring approximately 5 mm in diameter after 25 days of culture. While these organoids contain SOX17^{+} KDR^{+} CD31^{+} endothelial cells, scaling up the volume of tissue regeneration under such conditions is not likely to be optimal without a blood supply. Indeed, while the formation of glomerular capillaries is present within some forming glomeruli in these cultures, the majority of glomeruli are immature and avascular. Expression profiling also suggests that this tissue represents a trimester 1 human kidney, a time at which nephrogenesis has commenced but there are few maturing nephrons (Little 2015).

It has been demonstrated that the implantation of organoid-like structures under the renal capsule can successfully draw in a vasculature from the recipient animal

(Sharmin et al. 2016). Xinaris et al. (2012) also demonstrated the maturation and vascularisation of organoids aggregated from dissociated foetal mouse kidneys and transplanted into athymic host rats. Such results are in line with what had previously been shown when an embryonic kidney is transplanted into the omentum of a recipient animal (Rogers et al. 1998; Dekel et al. 2002). Here, the formation of glomerular capillaries via the ingrowth of host vasculature resulted in the formation of a urinary filtrate. Supplying the organoid with a functional circulation, and indeed the initiation of filtration of the blood, may well assist in the functional maturation of many of the component epithelial cell types along the forming nephrons, including the proximal tubule cells and the podocytes. Indeed, the transplantation of intestinal organoids for periods of 2–3 months can significantly improve intestinal epithelial maturation and allow the formation of an appropriate smooth musculature within the surrounding mesenchymal layers (Watson et al. 2014). These types of studies with kidney organoids will begin to clarify whether renal tubular maturation requires unidirectional flow of a glomerular filtrate.

Another critical challenge for kidney organoid implantation, following from their vascularisation, will be to ensure a viable exit path for the urinary filtrate that will be produced. In mammalian kidney development, the failure to develop a collecting system results in multicystic dysplasia and involution of the kidney (Hains et al. 2009). Similarly, transplanted, vascularised kidney organoids with no drainage will develop hydronephrosis (Yokote et al. 2015).

While the Takasato et al. (2015) organoid approach appears to contain a ureteric epithelium, the way in which an organoid is aggregated, which involves the complete dissociation of cultures to a single-cell suspension followed by reassociation, will not result in the formation of a single common ureteric tree and certainly not one existing ureter. This was identified early as a major distinction between a kidney organoid and a kidney (Davies 2015). For the formation of a replacement kidney, it is likely that the ureter itself must be patterned first or separately and then surrounded by nephrogenic epithelium. This is counterintuitive based upon our fundamental understanding of kidney morphogenesis. The branching of the ureteric epithelium to form a single unified tree through which the urinary filtrate can flow requires the close apposition of the ureteric tip with the cap mesenchyme. Conversely, the formation and attachment of nephrons occur in a spatially restricted region around the ureteric epithelial tips. While we know that the fusion of the nascent nephrons with the ureteric tips to form a single contiguous lumen occurs at the late renal vesicle stage via a process of invasion on CM-derived cells into the ureteric epithelium (Georgas et al. 2009; Kao et al. 2012), how this process is regulated and whether it can be induced without an initial nephrogenic zone is also not clear. Studies into the process of fusion between new nephrons and existing collecting duct epithelium are being performed in the setting of the zebrafish mesonephros (Diep et al. 2015), but whether this is applicable in the mammalian setting is yet to be determined.

11.5.3 Bioengineering Replacement Kidney Tissue

Finally, the capacity to make many of the founding cell types required for the formation of a functional adult kidney provides a source of cells for use in the bioengineering of a replacement organ. While it would appear inconceivable that we might be able to regenerate an organ of the size and complexity of the human kidney, and certainly impossible to do so with the tight architectural accuracy of the original organ, protocols allowing the generation of kidney organoids from human pluripotent stem cells for the first time provide an expandable and renewable source of material from which to bioengineer some form of replacement organ. There have already been a number of bioengineering attempts to generate kidney tissue prior to this point in time. These have included the recellularisation of decellularised kidney scaffolds (Song et al. 2013; Bonandrini et al. 2014; Caralt et al. 2015), seeding of primary cells along hollow fibres (Jansen et al. 2016), and the generation of a renal assist device designed to work in parallel with dialysis (Humes et al. 2004). To date, the cells used for these attempts have either been primary isolates of human cells acquired from cadaveric tissue (Humes et al. 2004; Song et al. 2013; Jansen et al. 2016) or undifferentiated cell types, including hPSC (Bonandrini et al. 2014). With the accurate patterning of iPSC to specific kidney cell types, either fully matured or in a progenitor state, organoid production would represent an approach that would allow increased scale-up and improved reproducibility.

11.6 Conclusion

In this chapter, we have described the development of approaches for the generation of complex human kidney organoids from human pluripotent stem cell sources. These kidney organoids represent models of the cells and structures forming within the developing organ, including the patterning nephrons, collecting duct epithelium and surrounding interstitium. The capacity to pattern pluripotent cells in a dish to simultaneously induce multiple interacting cell types able to self-organise according to the morphogenetic principles at work within the developing foetus is quite remarkable. The opportunities that arise with such an approach are substantial; however, there remains a long way to go.

Acknowledgements ML is a National Health and Medical Research Council Senior Principal Research Fellow. Her work is supported by the NHMRC, Australian Research Council, Royal Children's Hospital Foundation, Kidney Health Australia and the National Institutes of Health, USA. TF is an NHMRC Postgraduate Scholar and holds an RACP/NHMRC Award for Excellence. JY-CS holds a University of Melbourne Research Scholarship. MCRI is supported by the Victorian Government's Operational Infrastructure Support Program.

References

Araoka T, Mae S, Kurose Y, Uesugi M, Ohta A, Yamanaka S, Osafune K (2014) Efficient and rapid induction of human iPSCs/ESCs into nephrogenic intermediate mesoderm using small molecule-based differentiation methods. PLoS One 9:e84881. doi:10.1371/journal.pone.0084881

Bachiller D, Klingensmith J, Kemp C, Belo JA, Anderson RM, May SR, McMahon JA, McMahon AP, Harland RM, Rossant J, De Robertis EM (2000) The organizer factors Chordin and Noggin are required for mouse forebrain development. Nature 403:658–661. doi:10.1038/35001072

Barak H, Huh S-H, Chen S, Jeanpierre C, Martinovic J, Parisot M, Bole-Feysot C, Nitschké P, Salomon R, Antignac C, Ornitz DM, Kopan R (2012) FGF9 and FGF20 maintain the stemness of nephron progenitors in mice and man. Dev Cell 22:1191–1207. doi:10.1016/j.devcel.2012.04.018

Barker N, Rookmaaker MB, Kujala P, Ng A, Leushacke M, Snippert H, van de Wetering M, Tan S, Van Es JH, Huch M, Poulsom R, Verhaar MC, Peters PJ, Clevers H (2012) $Lgr5^{+ve}$ stem/progenitor cells contribute to nephron formation during kidney development. Cell Rep 2:540–552. doi:10.1016/j.celrep.2012.08.018

Becker GJ, Hewitson TD (2013) Animal models of chronic kidney disease: useful but not perfect. Nephrol Dial Transplant 28:2432–2438. doi:10.1093/ndt/gft071

Ben-Haim N, Lu C, Guzman-Ayala M, Pescatore L, Mesnard D, Bischofberger M, Naef F, Robertson EJ, Constam DB (2006) The nodal precursor acting via activin receptors induces mesoderm by maintaining a source of its convertases and BMP4. Dev Cell 11:313–323. doi:10.1016/j.devcel.2006.07.005

Berger K, Bangen J-M, Hammerich L, Liedtke C, Floege J, Smeets B, Moeller MJ (2014) Origin of regenerating tubular cells after acute kidney injury. Proc Natl Acad Sci U S A 111:1533–1538. doi:10.1073/pnas.1316177111

Biben C, Stanley E, Fabri L, Kotecha S, Rhinn M, Drinkwater C, Lah M, Wang CC, Nash A, Hilton D, Ang SL, Mohun T, Harvey RP (1998) Murine cerberus homologue mCer-1: a candidate anterior patterning molecule. Dev Biol 194:135–151. doi:10.1006/dbio.1997.8812

Bonandrini B, Figliuzzi M, Papadimou E, Morigi M, Perico N, Casiraghi F, Dipl C, Sangalli F, Conti S, Benigni A, Remuzzi A, Remuzzi G (2014) Recellularization of well-preserved acellular kidney scaffold using embryonic stem cells. Tissue Eng Part A 20:1486–1498. doi:10.1089/ten.TEA.2013.0269

Brown AC, Muthukrishnan SD, Guay JA, Adams DC, Schafer DA, Fetting JL, Oxburgh L (2013) Role for compartmentalization in nephron progenitor differentiation. Proc Natl Acad Sci U S A 110:4640–4645. doi:10.1073/pnas.1213971110

Brown AC, Muthukrishnan SD, Oxburgh L (2015) A synthetic niche for nephron progenitor cells. Dev Cell 34:229–241. doi:10.1016/j.devcel.2015.06.021

Brunskill EW, Potter SS (2012) Changes in the gene expression programs of renal mesangial cells during diabetic nephropathy. BMC Nephrol 13:70. doi:10.1186/1471-2369-13-70

Burn SF, Webb A, Berry RL, Davies JA, Ferrer-Vaquer A, Hadjantonakis AK, Hastie ND, Hohenstein P (2011) Calcium/NFAT signalling promotes early nephrogenesis. Dev Biol 352:288–298. doi:10.1016/j.ydbio.2011.01.033

Burridge PW, Keller G, Gold JD, Wu JC (2012) Production of de novo cardiomyocytes: human pluripotent stem cell differentiation and direct reprogramming. Cell Stem Cell 10:16–28. doi:10.1016/j.stem.2011.12.013

Caralt M, Uzarski JS, Iacob S, Obergfell KP, Berg N, Bijonowski BM, Kiefer KM, Ward HH, Wandinger-Ness A, Miller WM, Zhang ZJ, Abecassis MM, Wertheim JA (2015) Optimization and critical evaluation of decellularization strategies to develop renal extracellular matrix scaffolds as biological templates for organ engineering and transplantation. Am J Transplant 15:64–75. doi:10.1111/ajt.12999

Carroll TJ, Park J-S, Hayashi S, Majumdar A, McMahon AP (2005) Wnt9b plays a central role in the regulation of mesenchymal to epithelial transitions underlying organogenesis of the mammalian urogenital system. Dev Cell 9:283–292. doi:10.1016/j.devcel.2005.05.016

Chatterjee R, Ramos E, Hoffman M, VanWinkle J, Martin DR, Davis TK, Hoshi M, Hmiel SP, Beck A, Hruska K, Coplen D, Liapis H, Mitra R, Druley T, Austin P, Jain S (2012) Traditional and targeted exome sequencing reveals common, rare and novel functional deleterious variants in RET-signaling complex in a cohort of living US patients with urinary tract malformations. Hum Genet 131:1725–1738. doi:10.1007/s00439-012-1181-3

Chen S, Brunskill EW, Potter SS, Dexheimer PJ, Salomonis N, Aronow BJ, Hong CI, Zhang T, Kopan R (2015a) Intrinsic age-dependent changes and cell-cell contacts regulate nephron progenitor lifespan. Dev Cell 35:49–62. doi:10.1016/j.devcel.2015.09.009

Chen W, Huang J, Yu X, Lin X, Dai Y (2015b) Generation of induced pluripotent stem cells from renal tubular cells of a patient with Alport syndrome. Int J Nephrol Renov Dis 8:101–109. doi:10.2147/IJNRD.S85733

Ciampi O, Iacone R, Longaretti L, Benedetti V, Graf M, Magnone MC, Patsch C, Xinaris C, Remuzzi G, Benigni A, Tomasoni S (2016) Generation of functional podocytes from human induced pluripotent stem cells. Stem Cell Res 17:130–139. doi:10.1016/j.scr.2016.06.001

Colvin JS, Feldman B, Nadeau JH, Goldfarb M, Ornitz DM (1999) Genomic organization and embryonic expression of the mouse fibroblast growth factor 9 gene. Dev Dyn 216:72–88. doi:10.1002/(SICI)1097-0177(199909)216:1<72::AID-DVDY9>3.0.CO;2-9

Combes AN, Davies JA, Little MH (2015) Chapter fourteen–cell–cell interactions driving kidney morphogenesis. In: Yap AS (ed) current topics in developmental biology. Academic, New York, pp 467–508

Combes AN, Lefevre JG, Wilson S, Hamilton NA, Little MH (2016) Cap mesenchyme cell swarming during kidney development is influenced by attraction, repulsion, and adhesion to the ureteric tip. Dev Biol. doi: 10.1016/j.ydbio.2016.06.028

Costantini F, Kopan R (2010) Patterning a complex organ: branching morphogenesis and nephron segmentation in kidney development. Dev Cell 18:698–712. doi:10.1016/j.devcel.2010.04.008

Davies JA (2015) Self-organized kidney rudiments: prospects for better in vitro nephrotoxicity assays. Biomark Insights 10:117–123. doi:10.4137/BMI.S20056

Davis RP, Ng ES, Costa M, Mossman AK, Sourris K, Elefanty AG, Stanley EG (2008) Targeting a GFP reporter gene to the MIXL1 locus of human embryonic stem cells identifies human primitive streak-like cells and enables isolation of primitive hematopoietic precursors. Blood 111:1876–1884. doi:10.1182/blood-2007-06-093609

Dekel B, Amariglio N, Kaminski N, Schwartz A, Goshen E, Arditti FD, Tsarfaty I, Passwell JH, Reisner Y, Rechavi G (2002) Engraftment and differentiation of human metanephroi into functional mature nephrons after transplantation into mice is accompanied by a profile of gene expression similar to normal human kidney development. J Am Soc Nephrol 13:977–990

DesRochers TM, Suter L, Roth A, Kaplan DL (2013) Bioengineered 3D human kidney tissue, a platform for the determination of nephrotoxicity. PLoS One 8:e59219. doi:10.1371/journal.pone.0059219

Devuyst O, NVAM K, Remuzzi G, Schaefer F, Board of the Working Group for Inherited Kidney Diseases of the European Renal Association and European Dialysis and Transplant Association (2014) Rare inherited kidney diseases: challenges, opportunities, and perspectives. Lancet 383:1844–1859. doi:10.1016/S0140-6736(14)60659-0

Diep CQ, Peng Z, Ukah TK, Kelly PM, Daigle RV, Davidson AJ (2015) Development of the zebrafish mesonephros. Genesis 53:257–269. doi:10.1002/dvg.22846

Fleig SV, Humphreys BD (2014) Rationale of mesenchymal stem cell therapy in kidney injury. Nephron Clin Pract 127:75–80. doi:10.1159/000363680

Fleming BM, Yelin R, James RG, Schultheiss TM (2013) A role for Vg1/Nodal signaling in specification of the intermediate mesoderm. Development 140:1819–1829. doi:10.1242/dev.093740

Fletcher J, McDonald S, Alexander SI, Australian and New Zealand Pediatric Nephrology Association (ANZPNA) (2013) Prevalence of genetic renal disease in children. Pediatr Nephrol 28:251–256. doi:10.1007/s00467-012-2306-6

Freedman BS, Lam AQ, Sundsbak JL, Iatrino R, Su X, Koon SJ, Wu M, Daheron L, Harris PC, Zhou J, Bonventre JV (2013) Reduced ciliary polycystin-2 in induced pluripotent stem cells from polycystic kidney disease patients with PKD1 mutations. J Am Soc Nephrol 24:1571–1586. doi:10.1681/ASN.2012111089

Freedman BS, Brooks CR, Lam AQ, Fu H, Morizane R, Agrawal V, Saad AF, Li MK, Hughes MR, Werff RV, Peters DT, Lu J, Baccei A, Siedlecki AM, Valerius MT, Musunuru K, McNagny KM, Steinman TI, Zhou J, Lerou PH, Bonventre JV (2015) Modelling kidney disease with CRISPR-mutant kidney organoids derived from human pluripotent epiblast spheroids. Nat Commun. doi:10.1038/ncomms9715

Frumkin T, Malcov M, Telias M, Gold V, Schwartz T, Azem F, Amit A, Yaron Y, Ben-Yosef D (2010) Human embryonic stem cells carrying mutations for severe genetic disorders. In Vitro Cell Dev Biol Anim 46:327–336. doi:10.1007/s11626-010-9275-5

Funa NS, Schachter KA, Lerdrup M, Ekberg J, Hess K, Dietrich N, Honoré C, Hansen K, Semb H (2015) B-catenin regulates primitive streak induction through collaborative interactions with SMAD2/SMAD3 and OCT4. Cell Stem Cell 16:639–652. doi:10.1016/j.stem.2015.03.008

Fung M, Thornton A, Mybeck K, Wu JH, Hornbuckle K, Muniz E (2001) Evaluation of the characteristics of safety withdrawal of prescription drugs from worldwide pharmaceutical markets-1960 to 1999*. Drug Inf J 35:293–317

Georgas K, Rumballe B, Valerius MT, Chiu HS, Thiagarajan RD, Lesieur E, Aronow BJ, Brunskill EW, Combes AN, Tang D, Taylor D, Grimmond SM, Potter SS, McMahon AP, Little MH (2009) Analysis of early nephron patterning reveals a role for distal RV proliferation in fusion to the ureteric tip via a cap mesenchyme-derived connecting segment. Dev Biol 332:273–286. doi:10.1016/j.ydbio.2009.05.578

Hains DS, Bates CM, Ingraham S, Schwaderer AL (2009) Management and etiology of the unilateral multicystic dysplastic kidney: a review. Pediatr Nephrol 24:233. doi:10.1007/s00467-008-0828-8

Hart AH, Hartley L, Sourris K, Stadler ES, Li R, Stanley EG, Tam PPL, Elefanty AG, Robb L (2002) Mixl1 is required for axial mesendoderm morphogenesis and patterning in the murine embryo. Development 129:3597–3608

Hartman HA, Lai HL, Patterson LT (2007) Cessation of renal morphogenesis in mice. Dev Biol 310:379–387. doi:10.1016/j.ydbio.2007.08.021

Hildebrandt F (2010) Genetic kidney diseases. Lancet Lond Engl 375:1287–1295. doi:10.1016/S0140-6736(10)60236-X

Hinchliffe SA, Sargent PH, Howard CV, Chan YF, van Velzen D (1991) Human intrauterine renal growth expressed in absolute number of glomeruli assessed by the disector method and Cavalieri principle. Lab Invest J Tech Methods Pathol 64:777–784

Howden SE, Maufort JP, Duffin BM, Elefanty AG, Stanley EG, Thomson JA (2015) Simultaneous reprogramming and gene correction of patient fibroblasts. Stem Cell Rep 5:1109–1118. doi:10.1016/j.stemcr.2015.10.009

Hoy WE, Douglas-Denton RN, Hughson MD, Cass A, Johnson K, Bertram JF (2003) A stereological study of glomerular number and volume: preliminary findings in a multiracial study of kidneys at autopsy. Kidney Int Suppl 83:S31–S37. doi:10.1046/j.1523-1755.63.s83.8.x

Hughson M, Farris AB, Douglas-Denton R, Hoy WE, Bertram JF (2003) Glomerular number and size in autopsy kidneys: the relationship to birth weight. Kidney Int 63:2113–2122. doi:10.1046/j.1523-1755.2003.00018.x

Humes HD, Weitzel WF, Bartlett RH, Swaniker FC, Paganini EP, Luderer JR, Sobota J (2004) Initial clinical results of the bioartificial kidney containing human cells in ICU patients with acute renal failure. Kidney Int 66:1578–1588. doi:10.1111/j.1523-1755.2004.00923.x

Hynes AM, Giles RH, Srivastava S, Eley L, Whitehead J, Danilenko M, Raman S, Slaats GG, Colville JG, Ajzenberg H, Kroes HY, Thelwall PE, Simmons NL, Miles CG, Sayer JA (2014) Murine Joubert syndrome reveals Hedgehog signaling defects as a potential therapeutic target for nephronophthisis. Proc Natl Acad Sci U S A 111:9893–9898. doi:10.1073/pnas.1322373111

Imberti B, Tomasoni S, Ciampi O, Pezzotta A, Derosas M, Xinaris C, Rizzo P, Papadimou E, Novelli R, Benigni A, Remuzzi G, Morigi M (2015) Renal progenitors derived from human iPSCs engraft and restore function in a mouse model of acute kidney injury. Sci Rep 5:8826. doi:10.1038/srep08826

Jackson SA, Schiesser J, Stanley EG, Elefanty AG (2010) Differentiating embryonic stem cells pass through "temporal windows" that mark responsiveness to exogenous and paracrine mesendoderm inducing signals. PLoS One 5:e10706. doi:10.1371/journal.pone.0010706

Jain S (2009) The many faces of RET dysfunction in kidney. Organogenesis 5:177–190

James RG, Schultheiss TM (2005) Bmp signaling promotes intermediate mesoderm gene expression in a dose-dependent, cell-autonomous and translation-dependent manner. Dev Biol 288:113–125. doi:10.1016/j.ydbio.2005.09.025

Jansen J, Fedecostante M, Wilmer MJ, Peters JG, Kreuser UM, van den Broek PH, Mensink RA, Boltje TJ, Stamatialis D, Wetzels JF, van den Heuvel LP, Hoenderop JG, Masereeuw R (2016) Bioengineered kidney tubules efficiently excrete uremic toxins. Sci Rep 6:26715. doi:10.1038/srep26715

Jenkinson SE, Chung GW, van Loon E, Bakar NS, Dalzell AM, Brown CDA (2012) The limitations of renal epithelial cell line HK-2 as a model of drug transporter expression and function in the proximal tubule. Pflugers Arch 464:601–611. doi:10.1007/s00424-012-1163-2

Kandasamy K, Chuah JKC, Su R, Huang P, Eng KG, Xiong S, Li Y, Chia CS, Loo L-H, Zink D (2015) Prediction of drug-induced nephrotoxicity and injury mechanisms with human induced pluripotent stem cell-derived cells and machine learning methods. Sci Rep 5:12337. doi:10.1038/srep12337

Kao RM, Vasilyev A, Miyawaki A, Drummond IA, McMahon AP (2012) Invasion of distal nephron precursors associates with tubular interconnection during nephrogenesis. J Am Soc Nephrol JASN 23:1682–1690. doi:10.1681/ASN.2012030283

Karner CM, Das A, Ma Z, Self M, Chen C, Lum L, Oliver G, Carroll TJ (2011) Canonical Wnt9b signaling balances progenitor cell expansion and differentiation during kidney development. Dev Camb Engl 138:1247–1257. doi:10.1242/dev.057646

Keller G, Zimmer G, Mall G, Ritz E, Amann K (2003) Nephron number in patients with primary hypertension. N Engl J Med 348:101–108. doi:10.1056/NEJMoa020549

Kobayashi A, Valerius MT, Mugford JW, Carroll TJ, Self M, Oliver G, McMahon AP (2008) Six2 defines and regulates a multipotent self-renewing nephron progenitor population throughout mammalian kidney development. Cell Stem Cell 3:169–181. doi:10.1016/j.stem.2008.05.020

Kobayashi A, Mugford JW, Krautzberger AM, Naiman N, Liao J, McMahon AP (2014) Identification of a multipotent self-renewing stromal progenitor population during mammalian kidney organogenesis. Stem Cell Rep 3:650–662. doi:10.1016/j.stemcr.2014.08.008

Koepsell H (2013) The SLC22 family with transporters of organic cations, anions and zwitterions. Mol Asp Med 34:413–435. doi:10.1016/j.mam.2012.10.010

Kopan R, Chen S, Little M (2014) Nephron progenitor cells: shifting the balance of self-renewal and differentiation. Curr Top Dev Biol 107:293–331. doi:10.1016/B978-0-12-416022-4.00011-1

Lacombe C, Da Silva JL, Bruneval P, Casadevall N, Camilleri JP, Bariety J, Tambourin P, Varet B (1991) Erythropoietin: sites of synthesis and regulation of secretion. Am J Kidney Dis 18:14–19

Lam AQ, Freedman BS, Bonventre JV (2014a) Directed differentiation of pluripotent stem cells to kidney cells. Semin Nephrol 34:445–461. doi:10.1016/j.semnephrol.2014.06.011

Lam AQ, Freedman BS, Morizane R, Lerou PH, Valerius MT, Bonventre JV (2014b) Rapid and efficient differentiation of human pluripotent stem cells into intermediate mesoderm that forms tubules expressing kidney proximal tubular markers. J Am Soc Nephrol 25:1211–1225. doi:10.1681/ASN.2013080831

Lazzeri E, Crescioli C, Ronconi E, Mazzinghi B, Sagrinati C, Netti GS, Angelotti ML, Parente E, Ballerini L, Cosmi L, Maggi L, Gesualdo L, Rotondi M, Annunziato F, Maggi E, Lasagni L, Serio M, Romagnani S, Vannelli GB, Romagnani P (2007) Regenerative potential of embryonic renal multipotent progenitors in acute renal failure. J Am Soc Nephrol 18:3128–3138. doi:10.1681/ASN.2007020210

Li Y, Oo ZY, Chang SY, Huang P, Eng KG, Zeng JL, Kaestli AJ, Gopalan B, Kandasamy K, Tasnim F, Zink D (2013) An in vitro method for the prediction of renal proximal tubular toxicity in humans. Toxicol Res 2:352. doi:10.1039/c3tx50042j

Lin F, Moran A, Igarashi P (2005) Intrarenal cells, not bone marrow-derived cells, are the major source for regeneration in postischemic kidney. J Clin Invest 115:1756–1764. doi:10.1172/JCI23015

Lindström NO, Lawrence ML, Burn SF, Johansson JA, Bakker ER, Ridgway RA, Chang C-H, Karolak MJ, Oxburgh L, Headon DJ, Sansom OJ, Smits R, Davies JA, Hohenstein P (2015) Integrated β-catenin, BMP, PTEN, and Notch signalling patterns the nephron. eLife 4:e04000. doi:10.7554/eLife.04000

Little MH (2015) Improving our resolution of kidney morphogenesis across time and space. Curr Opin Genet Dev 32:135–143. doi:10.1016/j.gde.2015.03.001

Little MH, McMahon AP (2012) Mammalian kidney development: principles, progress, and projections. Cold Spring Harb Perspect Biol 4:a008300. doi:10.1101/cshperspect.a008300

Little MH, Takasato M (2015) Generating a self-organizing kidney from pluripotent cells. Curr Opin Organ Transplant 20:178–186. doi:10.1097/MOT.0000000000000174

Lusis M, Li J, Ineson J, Christensen ME, Rice A, Little MH (2010) Isolation of clonogenic, long-term self renewing embryonic renal stem cells. Stem Cell Res 5:23–39. doi:10.1016/j.scr.2010.03.003

Mae S-I, Shono A, Shiota F, Yasuno T, Kajiwara M, Gotoda-Nishimura N, Arai S, Sato-Otubo A, Toyoda T, Takahashi K, Nakayama N, Cowan CA, Aoi T, Ogawa S, McMahon AP, Yamanaka S, Osafune K (2013) Monitoring and robust induction of nephrogenic intermediate mesoderm from human pluripotent stem cells. Nat Commun 4:1367. doi:10.1038/ncomms2378

Mallett A, Patel C, Salisbury A, Wang Z, Healy H, Hoy W (2014) The prevalence and epidemiology of genetic renal disease amongst adults with chronic kidney disease in Australia. Orphanet J Rare Dis 9:98. doi:10.1186/1750-1172-9-98

McNamara BJ, Diouf B, Hughson MD, Douglas-Denton RN, Hoy WE, Bertram JF (2008) Renal pathology, glomerular number and volume in a West African urban community. Nephrol Dial Transplant 23:2576–2585. doi:10.1093/ndt/gfn039

Morizane R, Lam AQ, Freedman BS, Kishi S, Valerius MT, Bonventre JV (2015) Nephron organoids derived from human pluripotent stem cells model kidney development and injury. Nat Biotechnol 33:1193–1200. doi:10.1038/nbt.3392

Mummery CL, Davis RP, Krieger JE (2010) Challenges in using stem cells for cardiac repair. Sci Transl Med 2:27ps17. doi:10.1126/scitranslmed.3000558

Nieskens TTG, Peters JGP, Schreurs MJ, Smits N, Woestenenk R, Jansen K, van der Made TK, Röring M, Hilgendorf C, Wilmer MJ, Masereeuw R (2016) A human renal proximal tubule cell line with stable organic anion transporter 1 and 3 expression predictive for antiviral-induced toxicity. AAPS J 18:465–475. doi:10.1208/s12248-016-9871-8

Nykjaer A, Dragun D, Walther D, Vorum H, Jacobsen C, Herz J, Melsen F, Christensen EI, Willnow TE (1999) An endocytic pathway essential for renal uptake and activation of the steroid 25-(OH) vitamin D3. Cell 96:507–515

O'Brien LL, Guo Q, Lee Y, Tran T, Benazet J-D, Whitney PH, Valouev A, McMahon AP (2016) Differential regulation of mouse and human nephron progenitors by the Six family of transcriptional regulators. Development 143:595–608. doi:10.1242/dev.127175

Obara-Ishihara T, Kuhlman J, Niswander L, Herzlinger D (1999) The surface ectoderm is essential for nephric duct formation in intermediate mesoderm. Development 126:1103–1108

Olbrich H, Fliegauf M, Hoefele J, Kispert A, Otto E, Volz A, Wolf MT, Sasmaz G, Trauer U, Reinhardt R, Sudbrak R, Antignac C, Gretz N, Walz G, Schermer B, Benzing T, Hildebrandt F, Omran H (2003) Mutations in a novel gene, NPHP3, cause adolescent nephronophthisis, tapeto-retinal degeneration and hepatic fibrosis. Nat Genet 34:455–459. doi:10.1038/ng1216

Osathanondh V, Potter EL (1963) Development of human kidney as shown by microdissection. III. Formation and interrelationship of collecting tubules and nephrons. Arch Pathol 76:290–302

Ouyang H, Goldberg JL, Chen S, Li W, Xu G-T, Li W, Zhang K, Nussenblatt RB, Liu Y, Xie T, Chan C-C, Zack DJ (2016) Ocular stem cell research from basic science to clinical application: a report from Zhongshan Ophthalmic Center Ocular Stem Cell Symposium. Int J Mol Sci doi:10.3390/ijms17030415

Perea-Gomez A, Vella FDJ, Shawlot W, Oulad-Abdelghani M, Chazaud C, Meno C, Pfister V, Chen L, Robertson E, Hamada H, Behringer RR, Ang S-L (2002) Nodal antagonists in the anterior visceral endoderm prevent the formation of multiple primitive streaks. Dev Cell 3:745–756

Pereira LA, Wong MS, Mei Lim S, Stanley EG, Elefanty AG (2012) The mix family of homeobox genes–key regulators of mesendoderm formation during vertebrate development. Dev Biol 367:163–177. doi:10.1016/j.ydbio.2012.04.033

Poulsom R, Forbes SJ, Hodivala-Dilke K, Ryan E, Wyles S, Navaratnarasah S, Jeffery R, Hunt T, Alison M, Cook T, Pusey C, Wright NA (2001) Bone marrow contributes to renal parenchymal turnover and regeneration. J Pathol 195:229–235. doi:10.1002/path.976

Qi W, Johnson DW, Vesey DA, Pollock CA, Chen X (2007) Isolation, propagation and characterization of primary tubule cell culture from human kidney (Methods in Renal Research). Nephrology 12:155–159. doi:10.1111/j.1440-1797.2007.00779.x

Rached E, Hoffmann D, Blumbach K, Weber K, Dekant W, Mally A (2008) Evaluation of putative biomarkers of nephrotoxicity after exposure to ochratoxin A in vivo and in vitro. Toxicol Sci 103:371–381. doi:10.1093/toxsci/kfn040

Reinders MEJ, Rabelink TJ, de Fijter JW (2013) The role of mesenchymal stromal cells in chronic transplant rejection after solid organ transplantation. Curr Opin Organ Transplant 18:44–50. doi:10.1097/MOT.0b013e32835c2939

Rinkevich Y, Montoro DT, Contreras-Trujillo H, Harari-Steinberg O, Newman AM, Tsai JM, Lim X, Van-Amerongen R, Bowman A, Januszyk M, Pleniceanu O, Nusse R, Longaker MT, Weissman IL, Dekel B (2014) In vivo clonal analysis reveals lineage-restricted progenitor characteristics in mammalian kidney development, maintenance, and regeneration. Cell Rep 7:1270–1283. doi:10.1016/j.celrep.2014.04.018

Rogers SA, Lowell JA, Hammerman NA, Hammerman MR (1998) Transplantation of developing metanephroi into adult rats. Kidney Int 54:27–37. doi:10.1046/j.1523-1755.1998.00971.x

Roost MS, van Iperen L, Ariyurek Y, Buermans HP, Arindrarto W, Devalla HD, Passier R, Mummery CL, Carlotti F, de Koning EJP, van Zwet EW, Goeman JJ, de Chuva S, Lopes SM (2015) KeyGenes, a tool to probe tissue differentiation using a human fetal transcriptional atlas. Stem Cell Rep 4:1112–1124. doi:10.1016/j.stemcr.2015.05.002

Rumballe BA, Georgas KM, Combes AN, Ju AL, Gilbert T, Little MH (2011) Nephron formation adopts a novel spatial topology at cessation of nephrogenesis. Dev Biol 360:110–122. doi:10.1016/j.ydbio.2011.09.011

Saxén L, Lehtonen E (1987) Embryonic kidney in organ culture. Differentiation 36:2–11

Schuster D, Laggner C, Langer T (2005) Why drugs fail-a study on side effects in new chemical entities. Curr Pharm Des 11:3545–3559

Sharmin S, Taguchi A, Kaku Y, Yoshimura Y, Ohmori T, Sakuma T, Mukoyama M, Yamamoto T, Kurihara H, Nishinakamura R (2016) Human induced pluripotent stem cell–derived podocytes mature into vascularized glomeruli upon experimental transplantation. J Am Soc Nephrol 27:1778–1791. doi:10.1681/ASN.2015010096

Short KM, Combes AN, Lefevre J, Ju AL, Georgas KM, Lamberton T, Cairncross O, Rumballe BA, McMahon AP, Hamilton NA, Smyth IM, Little MH (2014) Global quantification of tissue dynamics in the developing mouse kidney. Dev Cell 29:188–202. doi:10.1016/j.devcel.2014.02.017

Smeets B, Boor P, Dijkman H, Sharma SV, Jirak P, Mooren F, Berger K, Bornemann J, Gelman IH, Floege J, van der Vlag J, Wetzels JFM, Moeller MJ (2013) Proximal tubular cells contain a phenotypically distinct, scattered cell population involved in tubular regeneration. J Pathol 229:645–659. doi:10.1002/path.4125

Sohn S-J, Kim SY, Kim HS, Chun Y-J, Han SY, Kim SH, Moon A (2013) In vitro evaluation of biomarkers for cisplatin-induced nephrotoxicity using HK-2 human kidney epithelial cells. Toxicol Lett 217:235–242. doi:10.1016/j.toxlet.2012.12.015

Song B, Smink AM, Jones CV, Callaghan JM, Firth SD, Bernard CA, Laslett AL, Kerr PG, Ricardo SD (2012) The directed differentiation of human iPS cells into kidney podocytes. PLoS One 7:e46453. doi:10.1371/journal.pone.0046453

Song JJ, Guyette JP, Gilpin SE, Gonzalez G, Vacanti JP, Ott HC (2013) Regeneration and experimental orthotopic transplantation of a bioengineered kidney. Nat Med 19:646–651. doi:10.1038/nm.3154

Sumi T, Tsuneyoshi N, Nakatsuji N, Suemori H (2008) Defining early lineage specification of human embryonic stem cells by the orchestrated balance of canonical Wnt/beta-catenin, Activin/Nodal and BMP signaling. Development 135:2969–2979. doi:10.1242/dev.021121

Taguchi A, Kaku Y, Ohmori T, Sharmin S, Ogawa M, Sasaki H, Nishinakamura R (2014) Redefining the in vivo origin of metanephric nephron progenitors enables generation of complex kidney structures from pluripotent stem cells. Cell Stem Cell 14:53–67. doi:10.1016/j.stem.2013.11.010

Takasato M, Little MH (2015) The origin of the mammalian kidney: implications for recreating the kidney in vitro. Development 142:1937–1947. doi:10.1242/dev.104802

Takasato M, Er PX, Becroft M, Vanslambrouck JM, Stanley EG, Elefanty AG, Little MH (2014) Directing human embryonic stem cell differentiation towards a renal lineage generates a self-organizing kidney. Nat Cell Biol 16:118–126. doi:10.1038/ncb2894

Takasato M, Er PX, Chiu HS, Maier B, Baillie GJ, Ferguson C, Parton RG, Wolvetang EJ, Roost MS, de Sousa C, Lopes SM, Little MH (2015) Kidney organoids from human iPS cells contain multiple lineages and model human nephrogenesis. Nature 526:564–568. doi:10.1038/nature15695

Tanigawa S, Wang H, Yang Y, Sharma N, Tarasova N, Ajima R, Yamaguchi TP, Rodriguez LG, Perantoni AO (2011) Wnt4 induces nephronic tubules in metanephric mesenchyme by a non-canonical mechanism. Dev Biol 352:58–69. doi:10.1016/j.ydbio.2011.01.012

Tanigawa S, Taguchi A, Sharma N, Perantoni AO, Nishinakamura R (2016) Selective in vitro propagation of nephron progenitors derived from embryos and pluripotent stem cells. Cell Rep 15:801–813. doi:10.1016/j.celrep.2016.03.076

Thatava T, Armstrong AS, De Lamo JG, Edukulla R, Khan YK, Sakuma T, Ohmine S, Sundsbak JL, Harris PC, Kudva YC, Ikeda Y (2011) Successful disease-specific induced pluripotent stem cell generation from patients with kidney transplantation. Stem Cell Res Ther 2:48. doi:10.1186/scrt89

Toyohara T, Mae S-I, Sueta S-I, Inoue T, Yamagishi Y, Kawamoto T, Kasahara T, Hoshina A, Toyoda T, Tanaka H, Araoka T, Sato-Otsubo A, Takahashi K, Sato Y, Yamaji N, Ogawa S, Yamanaka S, Osafune K (2015) Cell therapy using human induced pluripotent stem cell-derived renal progenitors ameliorates acute kidney injury in mice. Stem Cells Transl Med 4:980–992. doi:10.5966/sctm.2014-0219

Tropel P, Tournois J, Côme J, Varela C, Moutou C, Fragner P, Cailleret M, Laâbi Y, Peschanski M, Viville S (2010) High-efficiency derivation of human embryonic stem cell lines following pre-implantation genetic diagnosis. In Vitro Cell Dev Biol Anim 46:376–385. doi:10.1007/s11626-010-9300-8

Unbekandt M, Davies JA (2010) Dissociation of embryonic kidneys followed by reaggregation allows the formation of renal tissues. Kidney Int 77:407–416. doi:10.1038/ki.2009.482

Wang B, Ren C, Zhang W, Ma X, Xia B, Sheng Z (2013) Intensified therapy followed by autologous stem-cell transplantation (ASCT) versus conventional therapy as first-line treatment of follicular lymphoma: a meta-analysis. Hematol Oncol 31:29–33. doi:10.1002/hon.2015

Watson CL, Mahe MM, Múnera J, Howell JC, Sundaram N, Poling HM, Schweitzer JI, Vallance JE, Mayhew CN, Sun Y, Grabowski G, Finkbeiner SR, Spence JR, Shroyer NF, Wells JM, Helmrath MA (2014) An in vivo model of human small intestine using pluripotent stem cells. Nat Med 20:1310–1314. doi:10.1038/nm.3737

Weiss P, Taylor AC (1960) Reconstitution of complete organs from single-cell suspensions of chick embryos in advanced stages of differentiation. Proc Natl Acad Sci U S A 46:1177–1185
Wijgerde M, Karp S, McMahon J, McMahon AP (2005) Noggin antagonism of BMP4 signaling controls development of the axial skeleton in the mouse. Dev Biol 286:149–157. doi:10.1016/j.ydbio.2005.07.016
Xia Y, Nivet E, Sancho-Martinez I, Gallegos T, Suzuki K, Okamura D, Wu M-Z, Dubova I, Esteban CR, Montserrat N, Campistol JM, Belmonte JCI (2013) Directed differentiation of human pluripotent cells to ureteric bud kidney progenitor-like cells. Nat Cell Biol 15:1507–1515. doi:10.1038/ncb2872
Xinaris C, Benedetti V, Rizzo P, Abbate M, Corna D, Azzollini N, Conti S, Unbekandt M, Davies JA, Morigi M, Benigni A, Remuzzi G (2012) In vivo maturation of functional renal organoids formed from embryonic cell suspensions. J Am Soc Nephrol 23:1857–1868. doi:10.1681/ASN.2012050505
Xu J, Wong EYM, Cheng C, Li J, Sharkar MTK, Xu CY, Chen B, Sun J, Jing D, Xu P-X (2014) Eya1 interacts with Six2 and Myc to regulate expansion of the nephron progenitor pool during nephrogenesis. Dev Cell 31:434–447. doi:10.1016/j.devcel.2014.10.015
Yang H, Wang H, Shivalila CS, Cheng AW, Shi L, Jaenisch R (2013) One-step generation of mice carrying reporter and conditional alleles by CRISPR/Cas-mediated genome engineering. Cell 154:1370–1379. doi:10.1016/j.cell.2013.08.022
Yokote S, Matsunari H, Iwai S, Yamanaka S, Uchikura A, Fujimoto E, Matsumoto K, Nagashima H, Kobayashi E, Yokoo T (2015) Urine excretion strategy for stem cell-generated embryonic kidneys. Proc Natl Acad Sci U S A 112:12980–12985. doi:10.1073/pnas.1507803112

Chapter 12
Liver Regeneration Using Cultured Liver Bud

Keisuke Sekine, Takanori Takebe, and Hideki Taniguchi

Abstract Human induced pluripotent stem cells (iPSCs) have highly promising applications in regenerative medicine, drug discovery, and disease modeling, as well as in investigations of developmental biology. Although iPSC can theoretically be differentiated into any cell type, further developments are required for application to tissue replacement. In particular, most differentiation protocols produce fully matured populations of single cell types in two-dimensional cultures.

Some tissues, including retinal pigment epithelium, comprise relatively simple structures of single cell types and can be replaced by monocultures of matured cells. However, gastrointestinal and liver tissues comprise multiple cell types in complex architectures. Therefore, the lack of blood perfusion and three-dimensional interactions between multiple cell types are fundamental obstacles to the application of iPSC to regenerative medicine.

K. Sekine
Department of Regenerative Medicine, Yokohama City University Graduate School of Medicine, 3-9 Fukuura, Kanazawa-ku, Yokohama, Kanagawa 236-0004, Japan
e-mail: ksekine@yokohama-cu.ac.jp

T. Takebe
Department of Regenerative Medicine, Yokohama City University Graduate School of Medicine, 3-9 Fukuura, Kanazawa-ku, Yokohama, Kanagawa 236-0004, Japan

Advanced Medical Research Center, Yokohama City University, 3-9 Fukuura, Kanazawa-ku, Yokohama, Kanagawa 236-0004, Japan

PRESTO, Japan Science and Technology Agency, 4-1-8, Honcho, Kawaguchi, Saitama 332-0012, Japan

Department of Pediatrics, Cincinnati Children's Hospital Medical Center, University of Cincinnati, 3333 Burnet Avenue, Cincinnati, OH 45229-3039, USA
e-mail: takebe@yokohama-cu.ac.jp; Takanori.Takebe@cchmc.org

H. Taniguchi (✉)
Department of Regenerative Medicine, Yokohama City University Graduate School of Medicine, 3-9 Fukuura, Kanazawa-ku, Yokohama, Kanagawa 236-0004, Japan

Advanced Medical Research Center, Yokohama City University, 3-9 Fukuura, Kanazawa-ku, Yokohama, Kanagawa 236-0004, Japan
e-mail: rtanigu@yokohama-cu.ac.jp

T. Tsuji (ed.), *Organ Regeneration Based on Developmental Biology*,
DOI 10.1007/978-981-10-3768-9_12

To address these critical technological gaps, we investigated in vivo developmental processes in the context of essential interactions between cell types. In particular, the development of endodermal organs starts with organ specification as monocellular epithelium and subsequent multicellular interactions result in three-dimensional organs. We previously demonstrated the development of three-dimensional functional human liver buds (iPSC-LBs) from iPSC by coculturing iPSC-derived hepatic endodermal cells with endothelial and mesenchymal cells. This "organ bud technology" has high potential for the application of regenerative medicine to complex tissues and provides a unique model system for analyzing complex human tissues that comprise multiple cell types.

In this review, we present the current understanding of iPSC-LB in regenerative medicine and discuss its potential use as a tool for investigating the developmental biology of liver tissues.

Keywords Foregut • Endoderm • Development • iPSC • Organ bud • Liver bud • Multiple cell types • Regenerative medicine

12.1 Basics of Liver Developmental Biology

The developmental biology of the liver has been mainly studied in mice and zebrafish, and previous studies indicate that liver tissues arise from ventral foregut endoderm during embryonic development (Zaret and Grompe 2008; Ichikawa et al. 2014; Miyajima et al. 2014; Gordillo et al. 2015). After gastrulation, gut endodermal cells form multiple tissues of various functions and shapes, including the lung, liver, and pancreas (Fig. 12.1). Gut endoderm that is adjacent to the cardiac mesoderm is the first site of hepatic-specific endoderm, and hepatic diverticulum with endoderm cell lining then forms a pseudo-stratified architecture (Fig. 12.2) (Si-Tayeb et al. 2010a). Subsequently, cells start expressing liver-specific genes, including Prox1, Hhex, and AFP, and then delaminate from surrounding basal membranes with laminin and collagen IV (Sosa-Pineda et al. 2000; Burke and Oliver 2002; Lee et al. 2005; Lokmane et al. 2008; Le lay and Kaestner 2010; Harrelson et al. 2012). Liver tissue development then proceeds with the dissemination of cells into the surrounding stroma and subsequent sprouting of liver buds. During this stage, surrounding stroma other than endoderm is involved in the "budding" of gut endoderm, and the liver forms from single cell-type monolayers of hepatic-specified gut endoderm cells. Concomitantly, multi-lineage interactions are initiated with mesenchymal and endothelial cells surrounding the gut tube (Fig. 12.2) (Matsumoto et al. 2001). These cellular interactions are essential for the proliferation, migration, and differentiation of various liver cell types. After the initial liver bud stage, nascent liver tissues develop into multicellular organs, and blood perfusion starts immediately. Finally, the liver serves as a hematopoietic tissue during the embryonic period. Because blood and liver maturation are coupled, blood cells likely participate in

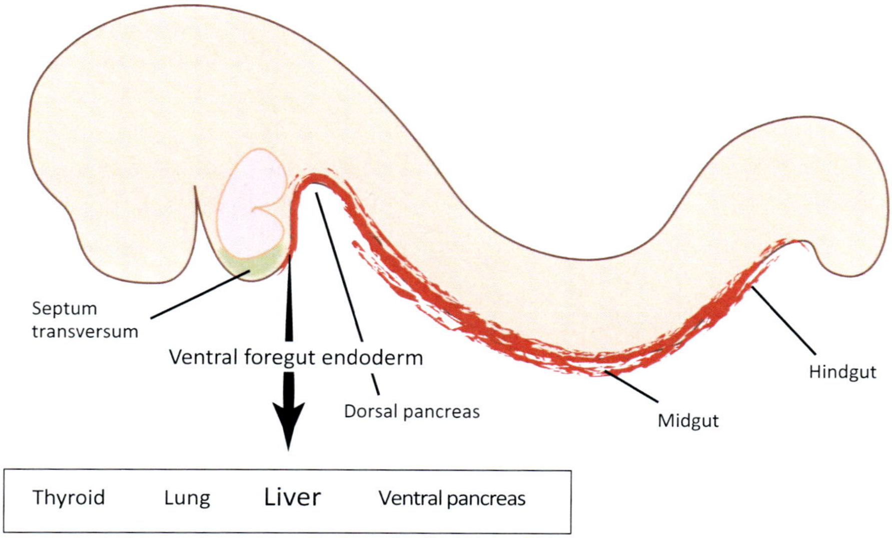

Fig. 12.1 Development of the primitive gut tube and the derivatives of the foregut. Primitive gut gives rise to foregut, midgut, and hindgut endoderm. Foregut endodermal cells form multiple tissues of various functions and shapes, including the thyroid gland, lung, liver, and pancreas

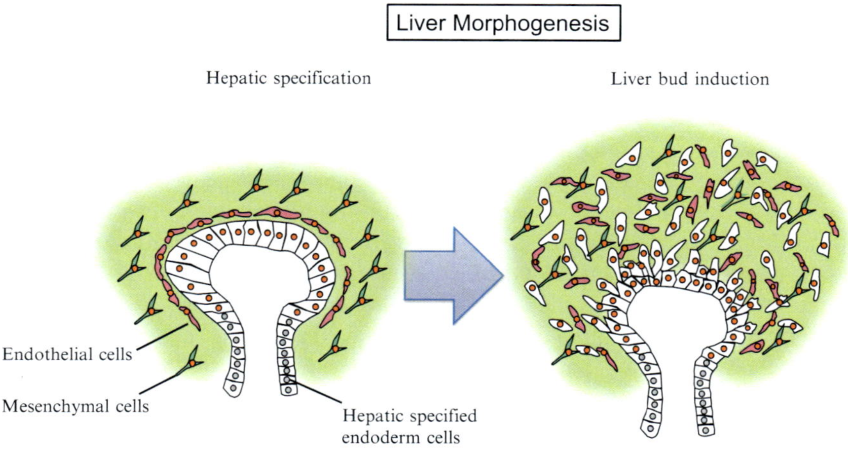

Fig. 12.2 Hepatic specification and liver bud induction in liver morphogenesis. Transverse view of liver budding. Foregut endoderm specified to hepatic endodermal cells lined by immature endothelial cells and surrounded by mesoderm of septum transversum (*left*). Hepatic-specified endoderm cells migrate into septum transversum to form liver bud (*right*). Endothelial cells with mesenchymal cells are required for liver budding

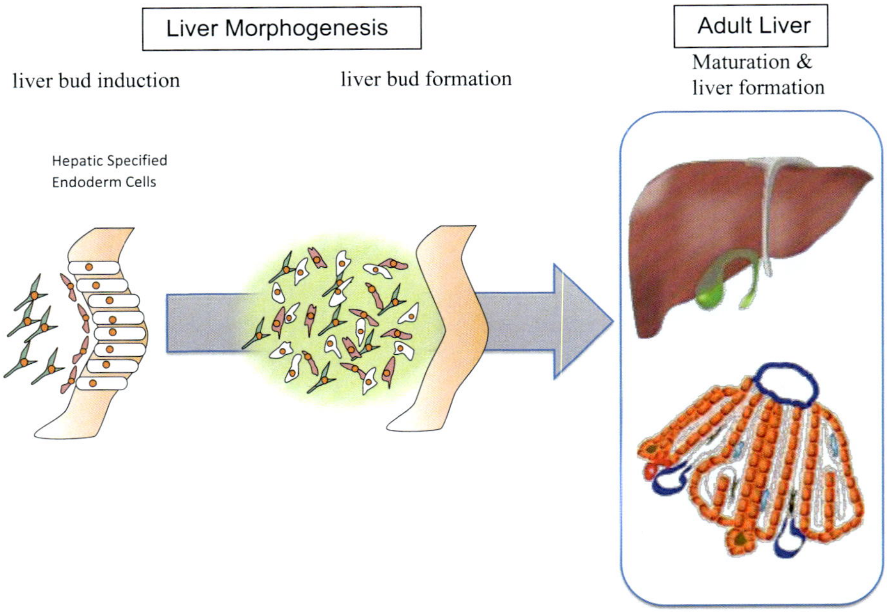

Fig. 12.3 Schematic representation of liver development. Sagittal view of liver region during early development through liver maturation. Foregut endoderm specified to hepatic endodermal cells (*left*). The septum transversum mesenchyme and developing heart send out signals to specify the region of endoderm to become the liver. Hepatic endoderm outgrowth and expand to form the embryonic liver (*middle*). Multicellular interactions including hepatocyte, endothelial cells, mesenchymal cells, and blood cells promote maturation of the liver (*right*)

liver development in concert with endothelial and mesenchymal cells (Fig. 12.3). Neurons, immune cells, and other cell types may also participate in organ development. Accordingly, hepatic differentiation from pluripotent stem cells has been achieved by recapitulating cellular interactions using added cytokines in cell cultures, resulting in certain degrees of two-dimensional hepatocyte differentiation. However, there are several issues with these strategies that need to be addressed first to ensure post-hepatic specification. In particular, current protocols result in liver tissues with poor differentiation efficiency and immature hepatic function, and clinical applications are hampered by poor engraftment rates (Sasai 2013; Collin de l'Hortet et al. 2016; Hannoun et al. 2016; Kopp et al. 2016; Nicolas et al. 2016).

12.2 2D Differentiation of iPSC

The development of iPSC has enabled the use of human pluripotent stem cells without the ethical issues associated with human embryonic stem cells. Consequent advances have led to successful differentiation of human pluripotent stem cells (hPSC) toward various cell types. However, the induction of endoderm differentiation from mouse embryonic stem cells remains inefficient in comparison with that of ectoderm and mesoderm cells, requiring the use of endoderm cells that are induced from embryoid bodies of mouse embryonic stem cells. Embryoid bodies produce various cell types simultaneously, and the induction of endoderm cells remains inefficient.

Recent studies of human pluripotent cells have succeeded in efficiently inducing endoderm differentiation on two-dimensional plastic plates by recapitulating the developmental cues using stepwise additions of cytokines in cell culture media (Si-Tayeb et al. 2010b; Kajiwara et al. 2012; Sudo 2014). Among media supplements, a Rho-associated protein kinase (ROCK) inhibitor prevented cell death during seeding of iPS in single cell suspensions, leading to significant improvements of efficacy, robustness, and purity of differentiation. In addition, activin A was used with and without Wnt3a to induce definitive endoderm cells that express endoderm markers such as CXCR4, FoxA2, and Sox17 and do not express Sox7, which is a marker for extraembryonic endodermal cells. The efficiency of endoderm cell differentiation using this method was over 95% according to flow cytometry analyses of CXCR4 expression. Moreover, during this stage, cells can be differentiated into hepatic cells, pancreatic cells, or intestinal cells, as discussed in the following sections (Tremblay and T 2005; Franklin et al. 2008; Sekine et al. 2012; Loh et al. 2014; Ikonomou and Kotton 2015). From definitive endoderm, hepatic differentiation proceeds more smoothly than pancreatic or intestinal differentiation, reflecting moderate successes of two-dimensional hepatocyte differentiation.

Hepatic specification is generally monitored according to the expression of HNF4A and the transcription factors GATA and FoxA. Other key transcription factors that have been observed during this process include HHEX, PROX1, and TBX3. Moreover, further differentiation leads to the expression of more specific liver signature genes, including alpha-fetoprotein, albumin, several glucose metabolic enzymes such as G6 Pase and PEPCK, and lipid and ammonia metabolic genes. These cells subsequently express various Cyp genes that are specific to mature hepatocytes. Although expression levels of these genes remain lower than those in adult hepatocytes from deceased donors, the quality of these cells varies widely between product lots, likely reflecting differences between donors and cell preparation methods. Nonetheless, these endpoint limitations frustrate the use of iPSC-derived hepatocytes for drug screening and evaluations of hepatic toxicity. Thus, several issues need to be resolved prior to application of the current strategy to posthepatic specification. Among obstacles to clinical application, poor differentiation efficiency, immature hepatic function, and poor engraftment rates of mature hepatocytes are critical issues. Furthermore, freshly isolated adult-donor hepatocytes may not rescue liver failure in the long term. These limitations are also challenges to the application of iPSC to other multicellular organs. Consequently, this research field requires alternative strategies to produce multicellular liver buds that reproduce the three-dimensional cellular interactions observed during development (Ishikawa et al. 2011; Sasai 2013; Ito et al. 2014; Gordillo et al. 2015).

12.3 Liver Buds

Despite current efforts to produce mature cells, generated cells have poor function, and even with functional hepatocytes, the therapeutic efficacy of cell-based interventions remains controversial, as shown in several clinical trials using human adult hepatocytes. Therefore, in vitro generation of complete functional organs offers an alternative to organ transplantation.

In our studies of vascularized organs such as the liver, we assumed that functional cell differentiation from iPS cells may not occur in the absence of stromal cellular interactions. Adult liver tissues comprise about 60% parenchymal hepatic cells and 40% non-parenchymal stromal cells such as liver sinusoidal endothelial cells, mesenchymal stellate cells, and Küpffer cells. Accordingly, spatiotemporal control of dynamic intercellular interactions is essential for efficient and proper cell differentiation from stem cells (Kamiya et al. 1999; Kinoshita et al. 1999; Tanimizu et al. 2007; Suzuki et al. 2008; Okabe et al. 2009). However, we have so far been unable to construct large three-dimensional multicellular tissues from pluripotent stem cells on culture dishes. Because liver development starts with the formation of hepatic endoderm from a thin layer of foregut endoderm, subsequent cellular interactions between endoderm cells, endothelial cells, and mesenchymal cells are essential for the formation of liver buds. Moreover, these interactions are established prior to blood perfusion, which is likely essential to further development of liver buds.

Previously, we attempted to mimic early organogenesis and recapitulate organogenic cellular interactions by generating three-dimensional liver buds from pluripotent stem cells in vitro (Fig. 12.4) (Takebe et al. 2012, 2013, 2014). These liver buds could be transplanted, and subsequent blood perfusion may facilitate maturation to functional liver tissue. Thus, to mimic cellular interactions, we cocultured iPSC-derived hepatic-specified endoderm, mesenchymal stem cells (MSCs), and human umbilical vein endothelial cells (HUVECs) under various conditions. These studies showed that, under defined conditions, large-scale morphogenetic changes were induced by the interactions between these cell types. In particular, condensation of the heterotypic cell mixture led to the formation of a large cellular mass and was initiated by MSCs that depend on substrate stiffness. Hence, mesenchymal cell-driven cellular contraction is essential for the dynamic condensation of heterogeneous cell mixtures.

In subsequent experiments, it appeared that these self-organized cellular masses resembled the proliferation and cellular context of the liver buds observed in mouse embryos. Moreover, transcriptome analyses of these iPSC-LBs revealed similar expression profiles to those in embryonic day (E) 10.5 murine liver buds. The timing and stages during differentiation of iPSC into hepatocytes may be critical to optimizing the functional advantages of liver buds. Potentially, further analyses of iPSC-LB are required to reveal more details of the human hepatic differentiation processes and may indicate dynamic conformational changes of monolayer gut endoderm into three-dimensional LB. In addition, comparative analyses of tran-

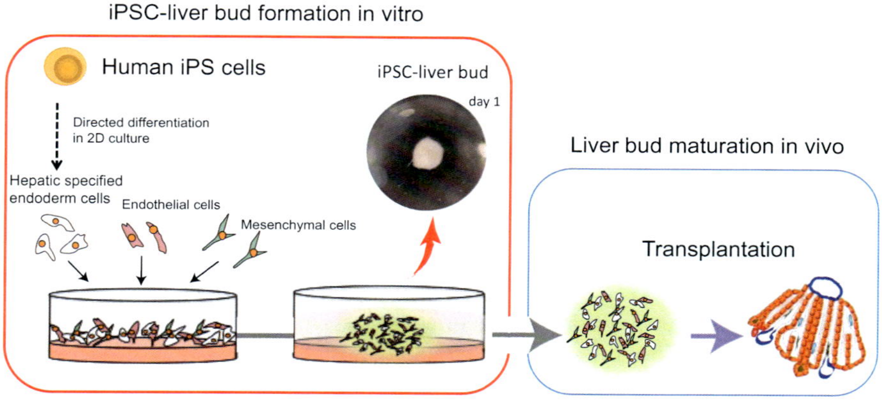

Fig. 12.4 Strategy toward human iPSC-derived liver generation. Mimic early organogenesis and recapitulate organogenic cellular interactions in vitro. Human iPSC-derived hepatic-specified endoderm, mesenchymal cells, and endothelial cells were co-cultivated in vitro. Large-scale morphogenetic changes were induced by the interactions between these cell types. In particular, condensation of the heterotypic cell mixture led to the formation of a large cellular mass and was initiated by mesenchymal cells. The generated liver buds could be transplanted, and subsequent blood perfusion may facilitate maturation to functional liver tissue

scription profiles between iPSC-LB and two-dimensional cultured hepatocytes may identify specific factors that support functional maturation. These analyses will provide critical improvements of liver bud technology using hPSC and will likely reflect the growing importance of multicellular interactions between developing hepatic endodermal cells and surrounding stromal cells.

Recent studies report the use of iPSC-derived hepatocytes as models of liver disease. For example, alpha1-antitrypsin deficiency was the first heritable metabolic disease to be modeled using iPSC-derived hepatocytes, and the resulting phenotypic mimicry of the features of this disease provided the first proof of principle for the use of iPSC-derived hepatocytes as disease models. Moreover, in combination with recently developed gene-editing technologies that can be used to correct causative mutations, iPSC-derived hepatocytes can be used as a drug-screening platform for such diseases. Moreover, three-dimensional multi-lineage organ buds will likely facilitate the establishment of these disease modeling and drug-screening platforms.

12.4 Organ Bud Technology

The development of three-dimensional tissue architectures and artificial organs is a long-standing aim of regenerative medicine. Using scaffolds to build three-dimensional architecture is straightforward, and decellularized tissues have been investigated as elegant scaffolds for seeding cells (Uygun et al. 2010; Bhatia et al. 2014; Sudo 2014). In addition, cell sheet engineering using temperature-responsive culture dishes enables detachment of cell sheets without enzymatic treatments. Cell sheet engineering can hence be used to create three-dimensional structures by layering individual cell sheets, and cell positioning and prepositioning may be achieved using 3D printing. In recent years, these technologies have advanced significantly and have achieved compelling outcomes (Toyoshima et al. 2012; Hirayama et al. 2013; Assawachananont et al. 2014; Oshima et al. 2014; Nagamoto et al. 2016; Takagi et al. 2016).

Current approaches to developing artificial organs are based on the formation of final organ shapes of adult tissues from immature tissues. In contrast, we and others are attempting to develop three-dimensional architectures using intrinsic self-organizing capabilities of cells, as observed during embryonic development. Several studies report organoid technologies that can be used to generate structures such as the optic cup (Eiraku et al. 2011; Nakano et al. 2012), pituitary epithelium (Suga et al. 2011; Ozone et al. 2016), the intestine (Sato et al. 2009, 2011; Spence et al. 2010, 2011; Yui et al. 2012; Watson et al. 2014), the cerebrum (Lancaster et al. 2013; Lancaster and Knoblich 2014; Camp et al. 2015), and the kidney (Chi et al. 2009; Osafune 2010; Takasato et al. 2014, 2015; Toyohara et al. 2015). These reports indicate the critical involvement of intrinsic self-organizing mechanisms in the generation of patterned epithelial architectures. Moreover, these studies suggest that extrinsic signals and forces from external structures, including surrounding mesenchyme, cooperate with the intrinsic self-organizing capabilities of cells during tissue morphogenesis.

Liver bud technology was developed to recapitulate the multicellular interactions that are present at the start of organ morphogenesis and to allow the development of multicell-driven, self-organizing, three-dimensional structures without scaffolds. Using these methods, we developed unprecedented technology for establishing multicellular organ primordia using progenitor cells from human pluripotent stem cells. Multicellular interactions are ubiquitously present in multicellular organisms and are essential for proper homeostasis and are likely descriptive of cancer tissue processes. Hence, formation of MSC-driven self-organized cellular masses may lead to the development of multicellular condensates from other cell sources. Accordingly, we demonstrated the use of "organ bud technology" as a universal culture platform for growing self-organized three-dimensional tissues from a pancreatic beta cell line and from developing kidney, liver, intestine, lung, heart, brain, and even cancer tissues (Takebe et al. 2015).

Previous reports show that endothelial and mesenchymal cells are highly tissue specific. In the liver, stellate cells act as mesenchymal stromal cells, and fibroblastic mesenchymal cells may also be present, although it remains unclear whether these differ from stellate cells. Tissue-specific stromal development must also require immature progenitor-progenitor interactions, warranting studies of the effects of blood cells and/or nerve cells that participate in organ development. Furthermore, in addition to intra-tissue interactions, inter-tissue communication is likely required for proper tissue maturation or organization.

12.5 Toward Clinical Application

Clinical translation of iPSC-LB technology to disease interventions remains complicated by several issues (Daley 2012; Kuroda et al. 2012, 2014). Among these, precise validation of iPSC-LB and their source cells is crucial for safety and functional quality. To this end, substantial improvements of iPSC safety may be achieved using non-genome integrative methods with episomal vectors and confirmation of genomic integration. These manipulations are required to exclude undifferentiated iPSC from differentiated (target) cell sources for LB and to exclude hyperproliferative cells and cells in intermediate stages of differentiation. Undifferentiated iPSCs have been shown to form teratomas from about 1×10^6 cells in immunodeficient animal models. Thus, if transplantation of 1×10^8 cells is required to treat diseases, at least 1% contamination with inappropriate cells could lead to the development of hazardous teratomas. Similarly, if disease interventions require transplantation of 1×10^{10} cells, contamination with inappropriate cells will need to be reduced to less than 0.01%.

Recent advances in next-generation sequencing technology allow interrogation of whole genomes and could be used to exclude mutant cells from transplants. In particular, cell products that have mutation(s) in well-known oncogenic genes must be excluded. However, endogenous cells frequently carry mutations, and these are highly heterogeneous. Thus, mutations in transplant cells may not necessarily result in malignancy. Conversely, if mutant cells are excluded too stringently, patients may not receive desirable therapy, and the cost of iPSC-LB technology will rise significantly. Thus, risk-benefit assessments will be critical during initial human clinical trials of these technologies.

Mass-scale production of LB is also required for therapeutic use, requiring establishment of uniform iPSC-LB and robustness of the manufacturing process. To this end, several previous studies report automated culture systems and mass-scale floating cultures. Clinical application of iPSC-LB technology will also require consideration and optimization of methods for delivery to selected transplantation sites. Currently, cells can be delivered in immunoprotective capsules, although the ensuing methods for engraftment remain poorly validated, and most of these will require further optimization for iPSC-LB transplantations. Finally, preservation of intermediate cell and tissue products will be desirable for precise validation of safety and function.

12.6 Perspectives

The use of hPSC-based cell therapies in regenerative medicine still faces numerous obstacles, including complications of delivery to diverse patients and cost-effectiveness. However, iPSCs remain a promising and unlimited cell source for regenerative medicine and will be further tested in imminent clinical trials.

Accumulated information from these studies will be very important for regenerative medicine interests. Moreover, novel technologies such as iPSC-LB warrant further consideration as models of the multicellular interactions observed in the developmental processes, and even in mature bodies. Organ bud technologies that recapitulate these multicellular interactions will facilitate three- dimensional structure organization without the use of synthetic scaffolds and early blood perfusion after transplantation. Thus, organ bud technology is a next-generation regenerative medicine approach that may offer a future cultured tissue alternative to organ transplantation.

References

Assawachananont J, Mandai M, Okamoto S et al (2014) Transplantation of embryonic and induced pluripotent stem cell-derived 3D retinal sheets into retinal degenerative mice. Stem Cell Reports 2:662–674. doi:10.1016/j.stemcr.2014.03.011

Bhatia SN, Underhill GH, Zaret KS, Fox IJ (2014) STATE OF THE ART REVIEW Cell and tissue engineering for liver disease

Burke Z, Oliver G (2002) Prox1 is an early specific marker for the developing liver and pancreas in the mammalian foregut endoderm. Mech Dev 118:147–155. doi:10.1016/S0925-4773(02)00240-X

Camp JG, Badsha F, Florio M, et al (2015) Human cerebral organoids recapitulate gene expression programs of fetal neocortex development. 1–6. doi: 10.1073/pnas.1520760112

Chi X, Michos O, Shakya R et al (2009) Ret-Dependent Cell Rearrangements in the Wolffian Duct Epithelium initiate ureteric bud morphogenesis. Dev Cell 17:199–209. doi:10.1016/j.devcel.2009.07.013

Collin de l'Hortet A, Takeishi K, Guzman-Lepe J et al (2016) Liver-regenerative transplantation: regrow and reset. Am J Transplant 16:1688–1696. doi:10.1111/ajt.13678

Daley GQ (2012) The promise and perils of stem cell therapeutics. Cell Stem Cell 10:740–749

Eiraku M, Takata N, Ishibashi H et al (2011) Self-organizing optic-cup morphogenesis in three-dimensional culture. Nature 472:51–56. doi:10.1038/nature09941

Franklin V, Khoo PL, Bildsoe H et al (2008) Regionalisation of the endoderm progenitors and morphogenesis of the gut portals of the mouse embryo. Mech Dev 125:587–600. doi:10.1016/j.mod.2008.04.001

Gordillo M, Evans T, Gouon-Evans V (2015) Orchestrating liver development. Development 142:2094–2108. doi:10.1242/dev.114215

Hannoun Z, Steichen C, Dianat N et al (2016) The potential of induced pluripotent stem cell derived hepatocytes. J Hepatol. doi:10.1016/j.jhep.2016.02.025

Harrelson Z, Kaestner KH, Evans SM (2012) Foxa2 mediates critical functions of prechordal plate in patterning and morphogenesis and is cell autonomously required for early ventral endoderm morphogenesis. Biol Open 1:173–181. doi:10.1242/bio.2011040

Hirayama M, Ogawa M, Oshima M et al (2013) Functional lacrimal gland regeneration by transplantation of a bioengineered organ germ. Nat Commun 4:2497. doi:10.1038/ncomms3497

Ichikawa T, Nakazato K, Keller PJ et al (2014) Live imaging and quantitative analysis of gastrulation in mouse embryos using light-sheet microscopy and 3D tracking tools. Nat Protoc 9:575–585. doi:10.1038/nprot.2014.035

Ikonomou L, Kotton DN (2015) Derivation of endodermal progenitors from pluripotent stem cells. J Cell Physiol 230:246–258. doi:10.1002/jcp.24771

Ishikawa M, Sekine K, Okamura A et al (2011) Reconstitution of hepatic tissue architectures from fetal liver cells obtained from a three-dimensional culture with a rotating wall vessel bioreactor. J Biosci Bioeng 111:711–718. doi:10.1016/j.jbiosc.2011.01.019

Ito K, Yanagida A, Okada K et al (2014) Mesenchymal progenitor cells in mouse foetal liver regulate differentiation and proliferation of hepatoblasts. Liver Int 34:1378–1390. doi:10.1111/liv.12387

Kajiwara M, Aoi T, Okita K et al (2012) Donor-dependent variations in hepatic differentiation from human-induced pluripotent stem cells. Proc Natl Acad Sci USA 109:12538–12543. doi:10.1073/pnas.1209979109

Kamiya A, Kinoshita T, Ito Y et al (1999) Fetal liver development requires a paracrine action of oncostatin M through the gp130 signal transducer. EMBO J 18:2127–2136. doi:10.1093/emboj/18.8.2127

Kinoshita T, Sekiguchi T, Xu MJ et al (1999) Hepatic differentiation induced by oncostatin M attenuates fetal liver hematopoiesis. Proc Natl Acad Sci USA 96:7265–7270. doi:10.1073/pnas.96.13.7265

Kopp JL, Grompe M, Sander M (2016) Stem cells versus plasticity in liver and pancreas regeneration. Nat Cell Biol 18:238–245. doi:10.1038/ncb3309

Kuroda T, Yasuda S, Kusakawa S et al (2012) Highly sensitive in vitro methods for detection of residual undifferentiated cells in retinal pigment epithelial cells derived from human iPS cells. PLoS One. doi:10.1371/journal.pone.0037342

Kuroda T, Yasuda S, Sato Y (2014) In vitro detection of residual undifferentiated cells in retinal pigment epithelial cells derived from human induced pluripotent stem cells. Methods Mol Biol 1210:183–192. doi:10.1007/978-1-4939-1435-7_14

Lancaster MA, Knoblich JA (2014) Generation of cerebral organoids from human pluripotent stem cells. Nat Protoc 9:2329–2340. doi: 10.1038/nprot.2014.158

Lancaster MA, Renner M, Martin C et al (2013) Cerebral organoids model human brain development and microcephaly. Nature 501:373–379. doi:10.1038/nature12517

Le lay J, Kaestner KH (2010) The Fox genes in the liver: from organogenesis to functional integration. Physiol Rev 90:1–22. doi:10.1152/physrev.00018.2009

Lee CS, Friedman JR, Fulmer JT, Kaestner KH (2005) The initiation of liver development is dependent on Foxa transcription factors. Nature 435:944–947. doi:10.1038/nature03649

Loh KM, Ang LT, Zhang J et al (2014) Efficient endoderm induction from human pluripotent stem cells by logically directing signals controlling lineage bifurcations. Cell Stem Cell 14:237–252. doi:10.1016/j.stem.2013.12.007

Lokmane L, Haumaitre C, Garcia-Villalba P et al (2008) Crucial role of vHNF1 in vertebrate hepatic specification. Development 135:2777–2786. doi:10.1242/dev.023010

Matsumoto K, Yoshitomi H, Rossant J, Zaret KS (2001) Liver organogenesis promoted by endothelial cells prior to vascular function. Science 294(80):559–563. doi: 10.1126/science.1063889

Miyajima A, Tanaka M, Itoh T (2014) Stem/progenitor cells in liver development, homeostasis, regeneration, and reprogramming. Cell Stem Cell 14:561–574. doi:10.1016/j.stem.2014.04.010

Nagamoto Y, Takayama K, Ohashi K et al (2016) Transplantation of a human iPSC-derived hepatocyte sheet increases survival in mice with acute liver failure. J Hepatol 64:1068–1075. doi:10.1016/j.jhep.2016.01.004

Nakano T, Ando S, Takata N et al (2012) Self-formation of optic cups and storable stratified neural retina from human ESCs. Cell Stem Cell 10:771–785. doi:10.1016/j.stem.2012.05.009

Nicolas C, Wang Y, Luebke-Wheeler J, Nyberg S (2016) Stem cell therapies for treatment of liver disease. Biomedicines 4:2. doi:10.3390/biomedicines4010002

Okabe M, Tsukahara Y, Tanaka M et al (2009) Potential hepatic stem cells reside in EpCAM+ cells of normal and injured mouse liver. Development 136:1951–1960. doi:10.1242/dev.031369

Osafune K (2010) In vitro regeneration of kidney from pluripotent stem cells. Exp. Cell Res. 316:2571–2577

Oshima M, Inoue K, Nakajima K et al (2014) Functional tooth restoration by next-generation bio-hybrid implant as a bio-hybrid artificial organ replacement therapy. Sci Rep 4:1–10. doi:10.1038/srep06044
Ozone C, Suga H, Eiraku M et al (2016) Functional anterior pituitary generated in self-organizing culture of human embryonic stem cells. Nat Commun 7:10351. doi:10.1038/ncomms10351
Sasai Y (2013) Next-generation regenerative medicine: organogenesis from stem cells in 3D culture. Cell Stem Cell 12:520–530. doi:10.1016/j.stem.2013.04.009
Sato T, Vries RG, Snippert HJ et al (2009) Single Lgr5 stem cells build crypt-villus structures in vitro without a mesenchymal niche. Nature 459:262–265. doi:10.1038/nature07935
Sato T, van Es JH, Snippert HJ et al (2011) Paneth cells constitute the niche for Lgr5 stem cells in intestinal crypts. Nature 469:415–418. doi:10.1038/nature09637
Sekine K, Takebe T, Suzuki Y et al (2012) Highly efficient generation of definitive endoderm lineage from human induced pluripotent stem cells. Transplant Proc 44:1127–1129. doi:10.1016/j.transproceed.2012.03.001
Si-Tayeb K, Lemaigre FP, Duncan SA (2010a) Organogenesis and development of the liver. Dev Cell 18:175–189. doi:10.1016/j.devcel.2010.01.011
Si-Tayeb K, Noto FK, Nagaoka M et al (2010b) Highly efficient generation of human hepatocyte-like cells from induced pluripotent stem cells. Hepatology 51:297–305. doi:10.1002/hep.23354
Sosa-Pineda B, Wigle JT, Oliver G (2000) Hepatocyte migration during liver development requires Prox1. Nat Genet 25:254–255. doi:10.1038/76996
Spence JR, Mayhew CN, Rankin SA et al (2010) Directed differentiation of human pluripotent stem cells into intestinal tissue in vitro. Nature 470:105–109. doi:10.1038/nature09691
Spence JR, Mayhew CN, Rankin SA et al (2011) Directed differentiation of human pluripotent stem cells into intestinal tissue in vitro. Nature 470:105–109. doi:10.1038/nature09691
Sudo R (2014) Multiscale tissue engineering for liver reconstruction. Organogenesis 10:1–9. doi:10.4161/org.27968
Suga H, Kadoshima T, Minaguchi M et al (2011) Self-formation of functional adenohypophysis in three-dimensional culture. Nature 480:57–62. doi:10.1038/nature10637
Suzuki K, Tanaka M, Watanabe N et al (2008) p75 Neurotrophin receptor is a marker for precursors of stellate cells and portal fibroblasts in mouse fetal liver. Gastroenterology. doi:10.1053/j.gastro.2008.03.075
Takagi R, Ishimaru J, Sugawara A et al (2016) Bioengineering a 3D integumentary organ system from iPS cells using an in vivo transplantation model. Sci Adv 2:e1500887–e1500887. doi:10.1126/sciadv.1500887
Takasato M, Er PX, Becroft M et al (2014) Directing human embryonic stem cell differentiation towards a renal lineage generates a self-organizing kidney. Nat Cell Biol 16:118–126. doi:10.1038/ncb2894
Takasato M, Er PX, Chiu HS et al (2015) Kidney organoids from human iPS cells contain multiple lineages and model human nephrogenesis. Nature 526:564–568. doi:10.1038/nature15695
Takebe T, Koike N, Sekine K et al (2012) Generation of functional human vascular network. Transplant Proc 44:1130–1133. doi:10.1016/j.transproceed.2012.03.039
Takebe T, Sekine K, Enomura M et al (2013) Vascularized and functional human liver from an iPSC-derived organ bud transplant. Nature 499:481–484. doi:10.1038/nature12271
Takebe T, Zhang R-R, Koike H et al (2014) Generation of a vascularized and functional human liver from an iPSC-derived organ bud transplant. Nat Protoc 9:396–409. doi:10.1038/nprot.2014.020
Takebe T, Enomura M, Yoshizawa E et al (2015) Vascularized and complex organ buds from diverse tissues via mesenchymal cell-driven condensation. Cell Stem Cell 16:556–565. doi:10.1016/j.stem.2015.03.004
Tanimizu N, Miyajima A, Mostov KE (2007) Liver progenitor cells develop CholangiocytetType epithelial polarity in three-dimensional culture. Mol Biol Cell 18:986–994. doi:10.1091/mbc.E06

Toyohara T, Mae S, Sueta S et al (2015) Cell therapy using human induced pluripotent stem cell-derived renal progenitors Ameliorates acute kidney injury in mice. Stem Cells Transl Med 4:980–992. doi:10.5966/sctm.2014-0219

Toyoshima K, Asakawa K, Ishibashi N et al (2012) Fully functional hair follicle regeneration through the rearrangement of stem cells and their niches. Nat Commun 3:784. doi:10.1038/ncomms1784

Tremblay KD, KSZ T (2005) Distinct populations of endoderm cells converge to generate the embryonic liver bud and ventral foregut tissues. 280:87–99. doi: 10.1016/j.ydbio.2005.01.003

Uygun BE, Soto-Gutierrez A, Yagi H et al (2010) Organ reengineering through development of a transplantable recellularized liver graft using decellularized liver matrix. Nat Med 16:814–820. doi:10.1038/nm.2170

Watson CL, Mahe MM, Múnera J et al (2014) An in vivo model of human small intestine using pluripotent stem cells. doi: 10.1038/nm.3737

Yui S, Nakamura T, Sato T et al (2012) Functional engraftment of colon epithelium expanded in vitro from a single adult $Lgr5^+$ stem cell. Nat Med 18:618–623. doi:10.1038/nm.2695

Zaret KS, Grompe M (2008) Generation and regeneration of cells of the liver and pancreas. Science 322:1490–1494. doi:10.1126/science.1161431

Chapter 13
Development and Bioengineering of Lung Regeneration

Andrew Wilson and Laertis Ikonomou

Abstract The limited ability of the lung to respond to devastating degenerative disease entities provides the impetus to develop new cell-based restorative therapies. Approaches to meet this need could include either production of stem/progenitor cells for delivery to the damaged native lung to regenerate damaged tissue or engineering of a de novo transplantable organ. In either case, an understanding of normal lung development provides a road map for directing pluripotent stem cells (PSCs) to differentiate to lung epithelium in vitro. Lung epithelium is derived from the endodermal germ layer, which in turn is formed during gastrulation as cells migrate through the primitive streak. Through precisely coordinated temporospatial exposure to key agonists and antagonists of the WNT, FGF, BMP, and RA pathways, a subset of definitive endoderm cells are induced to express *Nkx2–1*, the earliest known marker of primordial lung progenitor cells, before further differentiation to mature cell types comprising the proximal and distal lung compartments. Novel culture systems, such as decellularized lung scaffolds and in vitro organoids, offer unprecedented opportunities for achieving multilineage differentiation and tissue-like structure formation together with functional evaluation of PSC-derived lung progenitors. Combined with advances in our ability to model lung development in vitro with human PSCs, emerging bioengineering techniques are rapidly transforming the field and are likely both to further our understanding of normal development and to facilitate therapeutic applications of these in the years to come.

Keywords Development • Stem cells • Directed differentiation • NKX2–1 • Decellularization-recellularization • Lung scaffolds • Organoids • 3D bioprinting • Bioartificial lung

---

A. Wilson • L. Ikonomou (✉)
Center for Regenerative Medicine (CReM) of Boston University
and Boston Medical Center, Boston, MA, USA
e-mail: awilson@bu.edu; laertis@bu.edu

T. Tsuji (ed.), *Organ Regeneration Based on Developmental Biology*,
DOI 10.1007/978-981-10-3768-9_13

13.1 Introduction

Degenerative lung diseases, such as COPD and pulmonary fibrosis, are both common and a major cause of morbidity and mortality for afflicted patients. In contrast to some other organs or tissues, the various cell types that comprise the lung are largely quiescent at baseline, with a variety of localized stem/progenitor pools maintaining a low turnover index to collectively support tissue homeostasis. In the setting of either acute or chronic degenerative lung disease, these progenitor populations are either unable to regenerate dying cell types, as occurs in emphysema, or respond to injury with an aberrant repair program, as occurs in pulmonary fibrosis or acute interstitial pneumonia (AIP). In each example, the architecture of the lung is damaged in a manner that prevents it from performing its function of gas exchange and in many cases results in an increased risk of infection. To date, treatment modalities with the capacity to restore lung tissue or functional capacity, short of lung transplant, are not available. The apparent limited ability of the lung to respond to these common and devastating degenerative disease entities provides the impetus to develop new, cell-based, restorative therapies. Approaches to meet this need could include either the production of stem/progenitor cells that could then be delivered to the damaged, native lung to regenerate damaged tissue or the engineering of a de novo transplantable organ. Previous works to accomplish the former approach through the transplantation of bone marrow-derived cells or via tracheal delivery or both have proven to be technically difficult and met with limited success, consistent with cellular transplant results in organ systems outside of the bone marrow and the eye in this regard. While technical advances might ultimately solve the challenge of engrafting exogenously delivered progenitor cells in their local lung niche, the capacity of this approach to restore the architecture of damaged lung tissue could remain a significant challenge. In this chapter, we will focus instead on the goal of building an entire transplantable organ, starting with the derivation in vitro of the cell types that comprise the proximal and distal lung compartments from pluripotent stem cells (PSCs), moving next to the various three-dimensional (3D) systems currently employed for spatial organization and functional characterization of lung lineages, and finally discussing future possibilities for creating a human-scale bioartificial lung based on developmental principles and in vitro organogenesis.

13.2 In Vitro Derivation of Lung Lineages from Pluripotent Stem Cells

The construction of a bioengineered organ requires the ability to produce large quantities of the various differentiated cell types that make up an adult lung. This variety of specialized cells, in turn, needs to come from the same genetic background, ideally that of the eventual organ transplant recipient. PSCs and specifically induced pluripotent stem cells (iPSCs) are the cell type best suited to this task at the current time. While the ability to efficiently derive certain cell types and tissues, such as hepatic or neuronal cells, in vitro from PSCs has been well established, protocols to achieve this goal for lung-derived cells have only recently and incompletely been developed (Green et al. 2011; Huang et al. 2014; Konishi et al. 2016; Longmire et al. 2012). Progress in this approach has largely resulted from the culture of PSCs and their differentiated progeny in conditions designed to mimic those encountered by their in vivo analogues during development. In this section, we will, therefore, review what is known about each stage of lung development in the developing embryo and how this developmental road map is then applied to the in vitro differentiation of PSCs.

13.3 Derivation of the Endodermal Germ Layer

The lung, together with the thyroid, liver, intestines, and pancreas, is derived from the endodermal germ layer. Definitive endoderm is formed during gastrulation, a process that begins at embryonic day (E) 6.5 in the mouse (Tam et al. 2006). During this process, a portion of the epiblast forms a structure termed the "primitive streak." Additional epiblast cells migrate through the primitive streak, exiting the streak in an ordered fashion and in the process giving rise to mesodermal or endodermal cell types based on their specific embryonic location. This process, known as gastrulation, occurs from E7.5 to E8.5 in the mouse and results from the temporospatial expression of key growth factors and inhibitors that establish signaling gradients in the epiblast. Most notably, gradients of the TGF-β, Nodal, and Wnt ligands signal across the anterior/posterior (A/P) axis, while a BMP4 gradient is established along the proximal-distal axis (Fig. 13.1) (Gadue et al. 2005). Cells migrating through the streak experience variable and fate-altering levels of Nodal based on this A/P gradient, with posteriorly emerging cells becoming mesoderm while anteriorly emerging cells form definitive endoderm (DE) (Zorn and Wells 2009). Wnt signaling through β-catenin, meanwhile, helps to maintain Nodal signaling to promote endoderm induction and together with Smad2 promotes the expression of key endodermal genes such as *Foxa2* and *Sox17* (Sinner et al. 2004; Zorn and Wells 2009). Post-gastrulation (approximately E8.5–E9.0), DE then forms the primitive gut tube that is patterned into foregut, midgut, and hindgut regions, each marked by expression of specific transcription factors, along its A/P axis (Cardoso and Lu 2006; Wells and Melton 1999). Reciprocal signaling between the endoderm and mesoderm helps to confer regional fate specification along the A/P axis, with secretion of retinoic acid (RA), Wnts, FGFs, and BMPs by posterior mesoderm inducing intestinal identity, while Wnt and BMP antagonists released anteriorly block these posteriorizing signals in endodermal cells that will ultimately give rise to the lung, liver, and pancreas.

The exposure of PSCs in vitro to key components of these essential signaling pathways can result in the efficient induction of DE, the germ layer that will ultimately give rise to lung epithelium. As during migration of epiblast cells through

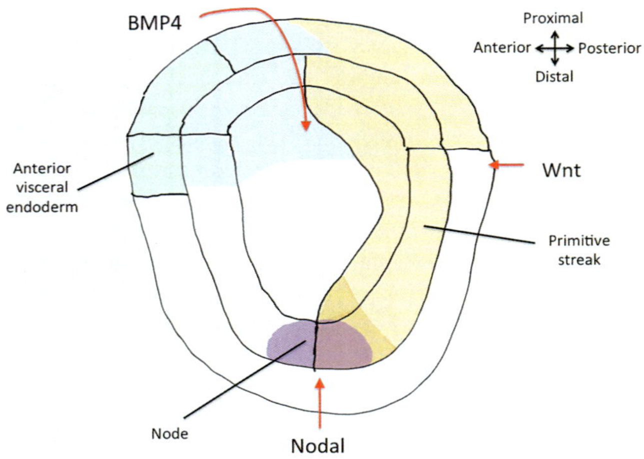

Fig. 13.1 Mouse embryonic gastrulation – the mouse epiblast is depicted at the late primitive streak stage. Epiblast cells migrate through the primitive streak and are exposed to variable concentration of key morphogens. Cells that exit the anterior aspect of the streak form the definitive endoderm (*DE*)

the anterior primitive streak, exposure of PSCs to high levels of activin A, a Nodal analogue that acts through the same receptors, results in induction of DE during in vitro differentiation assays (D'Amour et al. 2005; Gadue et al. 2006; Kubo et al. 2004). Indeed, the recapitulation of key features of this developmental program has been demonstrated by studies featuring mouse ESCs targeted with reporter genes to read-out endodermal gene expression. These studies demonstrate that differentiating ESCs pass through a primitive streak-like intermediate phase (defined by co-expression of *Brachyury* and *Foxa2* together with the cell surface markers C-KIT and CXCR4) before further differentiating into DE (defined by loss of *Brachyury* expression in cells that continue to express *Foxa2* and are C-KIT+/CXCR4+) over approximately 6 days in culture (Christodoulou et al. 2011; Gadue et al. 2006; Gouon-Evans et al. 2006). Protocols established in mouse PSCs have subsequently been adapted for human PSC differentiation with some differences in differentiation kinetics but continued reliance on the expression of cell surface markers including C-KIT, CXCR4, and EPCAM to identify DE (D'Amour et al. 2005; Green et al. 2011). It is from this pool of multipotent cells that lung progenitors are ultimately specified.

13.4 Specification of Lung Progenitors

The endoderm of the primitive foregut develops organ-specific domains representing the primordial thyroid, lung, esophagus, stomach, liver, and pancreas between E8.0 and E9.5 in mouse development (Cardoso and Kotton 2008; Serls et al. 2005). Gradients of WNT2/2B, FGF, and BMP signals from the surrounding mesoderm are established by expression of these factors as well as the BMP antagonist, Noggin, and result in patterning of the anterior foregut along its dorsal/ventral (D/V) axis. During this time period, *Sox2* is expressed dorsally in the future esophagus but downregulated ventrally, where it suppresses *Nkx2–1*. *Nkx2–1* expression, meanwhile, is induced ventrally in the site of the future trachea (Morrisey and Hogan 2010) (Fig. 13.2). While it remains the earliest known marker of primordial lung progenitor cells, *Nkx2–1* is not essential to specify the respiratory lineage, as *Nkx2–1*$^{-/-}$ mice develop lung tissue despite defects in other aspects of lung development

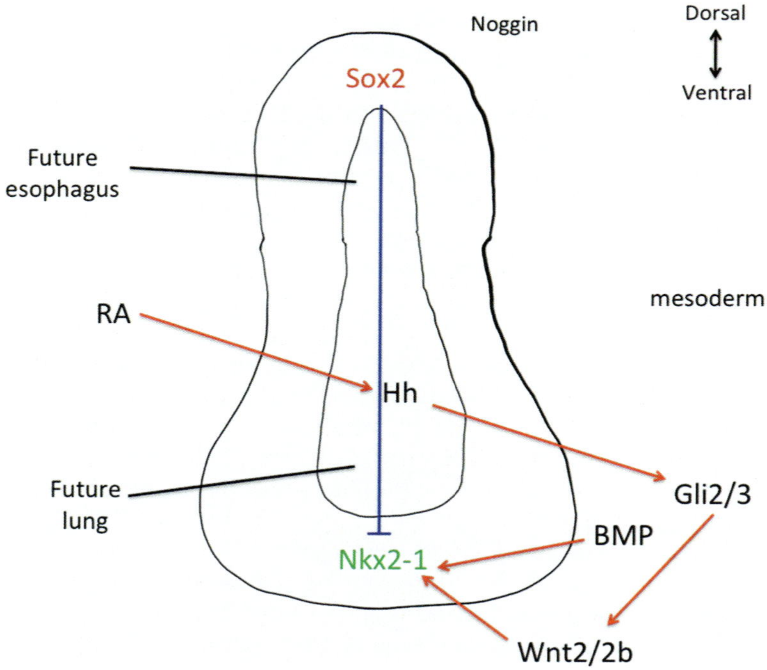

Fig. 13.2 Anterior foregut patterning – this schematic illustrates a section of the developing foregut at approximately E8.5–E9.0 of mouse embryonic differentiation, prior to lung budding and separation of the foregut into the lung and esophagus. Gradients of BMP and Wnt signaling are generated along the dorsal/ventral axis promoting the expression of Sox2 dorsally in the region of the future esophagus and the induction of Nkx2–1+ lung progenitors ventrally in the region that will become the lung

including separation of the trachea and esophagus, branching morphogenesis, and differentiation of lung epithelium (Kimura et al. 1996; Minoo et al. 1999).

The coordinated expression of RA, FGF, Wnt, and BMP signaling pathways is important for the efficient induction of NKX2–1^{+} progenitors. Work featuring a mouse foregut culture system has found that *Fgf10* signals from the mesoderm help to induce *Nkx2*–1 expression and lung bud outgrowth (Chen et al. 2007; Desai et al. 2004). *Wnt2/Wnt2b* expression from the mesoderm surrounding the ventral foregut endoderm is essential for efficient *Nkx2–1* induction as their combined deletion or that of β-catenin in epithelial cells blocks lung formation and allows minimal, transient *Nkx2–1* expression (Goss et al. 2009; Harris-Johnson et al. 2009). The local mesenchymal expression of BMP ligands is likewise essential for robust *Nkx2–1* induction, possibly in part through repression of *Sox2*, with deletion of Bmpr1a and Bmpr1b resulting in tracheal agenesis (Domyan et al. 2011). BMP-induced repression of *Sox2* results in derepression of *Nkx2–1*, potentially facilitating the induction capacity of canonical Wnt signaling. Recently published data in *Xenopus* suggests two temporally distinct roles for RA in this process (Rankin et al. 2016). Through mechanisms that remain to be defined, RA promotes the WNT2/2B–mediated specification of *Nkx2–1*+ lung progenitors via induction of endodermal hedgehog (Hh) signaling at developmental stage NF15–NF25. Inhibition of RA signaling during this stage resulted in loss of *Wnt2b*, *Fgf10*, *Nkx2–1*, and *Sftpc* expression. This process is independent of FGF10, as evidenced by the failure of recombinant FGF10 to rescue the phenotype. Slightly later in *Xenopus* development (NF25–NF38), however, recombinant FGF10 was able to restore expression of *Wnt2b* and *Nkx2–1* together with subsequent expression of *Sftpc*, consistent with stage-specific mechanistic differences in regulation. While additional work by the same group has confirmed that RA acts upstream of Hh/Wnt to specify *Nkx2–1* lung progenitors in mouse explant cultures, the stage-specific relationship between RA and FGF10 has yet to be confirmed in mammals.

At approximately the same time as *Nkx2–1* specification, a gradient of BMP and Wnt signaling along the D/V axis of the developing foregut tube promotes its separation into two separate tracheal and esophageal tubes. BMP, FGF, and Wnt ligands are expressed from the ventral mesoderm, while the BMP antagonist Noggin is expressed dorsally (Rodewald 2008). Together, these signals pattern the foregut resulting in high levels of *Sox2* expression dorsally in the nascent esophagus and *Nkx2–1* ventrally in the nascent trachea. Hh signaling likewise plays a key role in this process, with Shh ligand from the foregut and early respiratory epithelium signaling to the surrounding mesenchyme where they support mesenchymal proliferation and stimulate the expression of Wnt and BMP family members (Pepicelli et al. 1998). This is clearly demonstrated in *Shh*$^{-/-}$ (Litingtung et al. 1998) or *Gli2*$^{-/-}$/*Gli3*$^{+/-}$ (Motoyama et al. 1998) mutant mice that are characterized by profound foregut defects, including esophageal atresia and tracheoesophageal fistula.

To successfully derive *Nkx2–1* expressing respiratory progenitor cells, in vitro directed differentiation protocols seek to recapitulate temporally coordinated expression of these same signaling pathways (Green et al. 2011; Longmire et al. 2012; Mou et al. 2012). Following the generation of C-KIT+/CXCR4+/EPCAM+

DE described above, cells are sequentially exposed to factors promoting first anteriorization and then ventralization. Anteriorization is promoted by combined inhibition of the TGF-β and BMP pathways through the application of pathway inhibitors, such as SB-431542 (pharmacological TGF-β inhibitor) and Noggin (physiological BMP inhibitor) (Green et al. 2011). Together, these inhibitors support the induction of *Sox2* expression, maintenance of *Foxa2*, and suppression of *Cdx2* expression characteristic of anterior foregut endoderm (AFE). Cells that have undergone differentiation to AFE are then exposed to factors known to promote D/V patterning and separation of the developing esophagus and trachea in vivo. These factors include agonists of the Wnt pathway (CHIR99021 or WNT3A) together with BMP4, FGFs (FGF2, FGF7, and/or FGF10), and RA (Chen et al. 2010; Green et al. 2011; Longmire et al. 2012; Mou et al. 2012). Differentiating PSCs exposed in culture to these factors have been shown to have the capacity to subsequently express *Nkx2–1*, the earliest known marker of lung epithelium and identifier of cells with the capacity to further differentiate to the mature parenchymal cells that comprise the adult lung.

13.5 Patterning and Maturation of Developing Lung Epithelium

Between E9.5 and E16.5 in the mouse, also referred to as the "pseudoglandular stage," the primary lung buds emerge and undergo a stereotyped, iterative branching process to generate the primary treelike structure of the lung (Metzger et al. 2008). The primary buds extend into the surrounding mesenchyme, and by E10.5, secondary buds appear at specific locations around the primary bud. This process, known as branching morphogenesis, continues for the remainder of the pseudoglandular stage of lung development, ultimately giving rise to the main stem bronchi and the entire bronchial tree. The branching morphogenesis program is governed by the establishment of a signaling center at the distal lung tip in which FGF10 is expressed from the mesenchyme and, together with BMPs and Wnts, signals to the adjacent epithelium, maintaining proliferation of $SOX9^{+}/ID2^{+}$ epithelial cells. Reciprocal signaling from the epithelium, meanwhile, including expression of *Sprouty*, *Shh*, and BMP4, restricts FGF expression and establishes a feedback loop that controls the size of the progenitor pool as well as the size and shape of the lung bud (Cardoso and Lu 2006). As epithelial cells of the distal bud proliferate, the bud itself elongates resulting in the formation of a stalk that experiences different expression levels of the key factors listed above (see reviews by Cardoso and Lu 2006; Morrisey and Hogan 2010). A repetitive cycle of bud elongation, tip expansion, and bifurcation occurs during this time, during which relatively proximal and distal cells experience varying levels of key morphogens. Components of the Hh pathway play a central role in this process: epithelial *Shh* promotes proliferation of adjacent mesenchymal cells and expression of *Bmp4* while at the same time inhibiting *Fgf10* expression.

The high levels of *Shh* expression at the distal lung bud tip induce expression of hedgehog-interacting protein (*Hhip*) in the adjacent mesenchyme that dampens Gli signaling and thereby allow expression of *Fgf10*. More proximal regions experience lower Shh expression that, by contrast, does not induce *Hhip* resulting in *Fgf10* inhibition. Epithelial cells that become displaced from the distal tip form stalk regions of the airways and adopt a SOX2⁺ proximal epithelial fate. Additional factors not detailed above, including TGF-β signaling, *Wnt7b*/β-catenin signaling, and F*gf9*, likewise play important roles in this process.

A limited understanding of the dynamic epithelial-mesenchymal signaling interactions that occur during the pseudoglandular period together with associated spatial complexity has made the reproduction of this stage of lung development challenging to recapitulate in vitro. A key step in the ability to derive differentiated lung cell types from PSCs is the establishment of the proximal ($Nkx2\text{–}1^+/Sox2^+$) or distal ($Nkx2\text{–}1^+/Sox9^+$) cellular identity that will ultimately produce airway or alveolar epithelial cells (AECs), respectively. While the precise cues required to direct $Nkx2\text{–}1^+$ lung progenitors to adopt a proximal or distal identity have not been established, specific protocols have demonstrated the ability to generate cells expressing markers of differentiated lung cell types. To generate putative proximal lung epithelial cells, several protocols have relied on physical manipulation of the culture environment to assist in cellular maturation. For example, one publication demonstrated the use of subcutaneous injection and subsequent culture of mouse or human progenitor cells to induce a variety of differentiated markers, including P63, CC10, or MUC5AC (Mou et al. 2012). Multiple reports have now demonstrated the use of air-liquid interface culture to further induce expression of differentiated cell markers and/or ciliated cells following directed differentiation of human PSCs to generate putative lung progenitors (Firth et al. 2014; Wong et al. 2012). Most recently, Konishi et al. sorted cells expressing the surface marker carboxypeptidase M (CPM) to identify and purify *Nkx2–1*-expressing cells from ventralized AFE cultures (Konishi et al. 2016). CPM^+ cells were then cultured in 3D conditions in the presence of the WNT agonist CHIR99021 and FGF10 to generate 3D spherical structures before further culture in a commercial bronchial airway cell culture medium together with DAPT, an inhibitor of NOTCH signaling. These culture conditions generated $FOXJ1^+$ motile multiciliated cells as well as cells expressing the neuroendocrine markers CHGA and SYP. Numerous reports have likewise demonstrated the ability to derive lung epithelial cells capable of expressing the type II AEC marker SFTPC from either mouse or human PSCs (Ghaedi et al. 2013; Green et al. 2011; Huang et al. 2014; Longmire et al. 2012). Protocols to induce expression of this key marker feature the induction of $Nkx2\text{–}1^+$ progenitor cells then cultured in the presence of growth factors including Wnt agonists, FGF2 and/or FGF10, KGF, BMP4, EGF, and RA. Increased expressions of SFTPC and SFTPB were achieved in these publications through subsequent culture in conditions including KGF and/or dexamethasone, 8-bromo-cAMP, and isobutylmethylxanthine (together referred to as "DCI") (Huang et al. 2014; Longmire et al. 2012).

13.6 3D Functional Organization of In Vitro Derived Lung Lineages

As discussed in the previous section, there has been considerable progress in the in vitro derivation of putative lung primordial epithelial progenitors in both mouse and human PSC systems (Ikonomou and Kotton 2015). Novel culture systems, such as decellularized lung scaffolds (Song and Ott 2011) and in vitro organoids (Clevers 2016), offer unprecedented opportunities for multilineage differentiation, tissue-like structure formation, and functional evaluation of PSC-derived lung progenitors. Despite their structural differences, both systems are powerful 3D inductive culture environments for the study of morphogenetic/differentiation processes that either are inaccessible – in the case of human lung development – or require complex mouse genetic models. The preponderant role of "dynamic reciprocity" (Bissell et al. 1982) between ECM and cells in these systems reflects the importance of ECM-cell interactions in in vivo lung development, homeostasis, and disease. In this section, we will review recent advances in 3D culture of PSC-derived lung progenitors and discuss how the acquired knowledge will inform future design of bio-artificial lungs, including organ-scale 3D bioprinting.

13.7 Decellularized Lung Scaffolds for In Vitro PSC Lung Differentiation

While grafts based on decellularized 2D tissues, such as the skin, have been in clinical use for many years (van der Veen et al. 2010), it was only recently that solid organ decellularization was introduced as a novel platform for human tissue/organ engineering. The publication of the first proof-of-principle paper on heart decellularization-recellularization (decell-recell) (Ott et al. 2008) was followed by the demonstration of lung scaffold decell-recell (Fig. 13.3a) (Ott et al. 2010; Petersen et al. 2010) and soon after several publications established that this approach can be standardized and generalized for the creation of natural scaffolds from a plethora of organs (Song et al. 2013; Totonelli et al. 2012, 2013; Uygun et al. 2010).

In the following years, various studies have exhaustingly characterized the composition, morphology, and mechanical properties of adult decellularized lung scaffolds from various species. Morphometric and micro-CT analysis has demonstrated that decellularized scaffolds retain intact airway, vascular tree, and alveolar septum architecture without appreciable alveolar surface area loss (Ott et al. 2010; Petersen et al. 2010; Price et al. 2010). The decellularization process does, however, result in the loss of varying degrees of ECM protein. Proteomic analysis by mass spectrometry has provided valuable insights in the matrisome – ECM and ECM-associated proteins – composition (Naba et al. 2012) of decellularized human and porcine lungs (Booth et al. 2012; Gilpin et al. 2014a; Li et al. 2016). White and coworkers

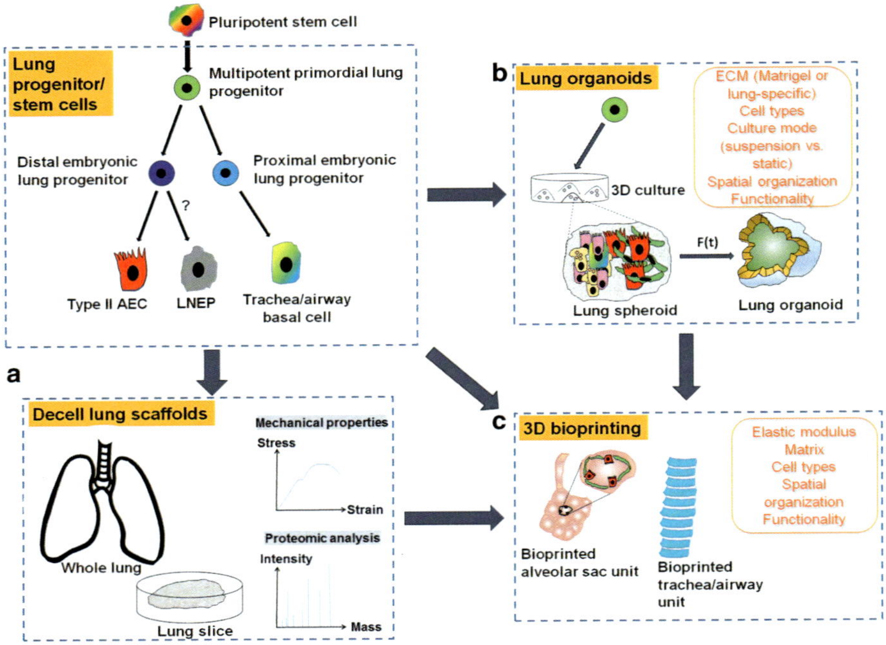

Fig. 13.3 Bioengineering platforms for 3D lung differentiation and functional organization – PSC-derived lung progenitors can be incorporated in various 3D culture systems from decell lung scaffolds to in vitro derived organoids. The resultant cell structures recapitulate certain functional and morphological aspects of the native organ. In (**b**), only a lung multipotent progenitor is shown, but organoids with varying compositions can be derived by any putative lung stem/progenitor cell

demonstrated that there are important differences in ECM composition of decellularized lung tissue from donors with idiopathic pulmonary fibrosis (IPF) compared to controls (Booth et al. 2012). For instance, IPF lungs were enriched in transforming growth factor (TGF)-β-binding protein 1 and depleted in basement membrane proteins, reflecting the pathological role of TGF-β signaling and loss of epithelial cell function in IPF. Although it was initially assumed that detergent-mediated decellularization efficiently removes all cellular material – including non-matrisome proteins – a recent study by Li and colleagues (Li et al. 2016) challenged this notion by demonstrating the presence of 152 matrisome proteins out of 384 total proteins in human decellularized lungs. In addition to the possible immunogenicity introduced by non-matrisome proteins in decell-recell bioartificial lungs, this finding should also be taken into account when interpreting findings from in vitro lung progenitor differentiation on decellularized matrices.

Decellularized lung scaffolds are amenable to a variety of applications from whole-organ tissue engineering to high-throughput screening of culture conditions for PSC lung-directed differentiation. Weiss and coworkers first reported the use of lung tissue slices for in vitro cell culture as an alternative to whole lung decell-recell

(Daly et al. 2012; Wallis et al. 2012). In this technology, decellularized lungs (most commonly of murine origin) are recellularized using a mixture of cells and agarose, sliced to thick pieces (300–2000 μm) and placed into culture media where agarose melts out of the tissue (Daly et al. 2012). Decellularized lung slices retain the original architecture of the airway, alveolar, and endothelial compartments thus offering an expedient and inexpensive system for functional studies of PSC-derived lineages (Gilpin et al. 2014b; Longmire et al. 2012; Shojaie et al. 2015). In the first study reporting the efficient derivation of lung/thyroid lineages from mouse ESCs, purified mouse NKX2–1$^+$ endodermal progenitors retained NKX2–1 expression in mouse lung slice culture giving also rise to NKX2–1$^-$T1α^+ cells, a phenotype reminiscent of type I AECs (Longmire et al. 2012). Niklason and coworkers described similar results with presumed type II AECs derived from human iPSCs (Ghaedi et al. 2013). Recellularization of whole lung scaffolds and rat/human lung slices resulted in proliferating cells that expressed NKX2–1 and markers of distal lung epithelium.

The potential of this culture system to enhance lung epithelial progenitor specification from PSCs was evaluated in two studies (Gilpin et al. 2014b; Shojaie et al. 2015). Ott and coworkers (Gilpin et al. 2014b) reported enhanced lung specification of human iPSCs during directed differentiation on lung slices compared to non-slice controls. Specified progenitors were further differentiated to cells expressing markers of proximal (CC10) and distal lung lineages (T1α), and the recellularized scaffolds were able to support lung function posttransplantation into pneumonectomized rats. Shojaie and colleagues (Shojaie et al. 2015) investigated the inductive properties of lung ECM alone by seeding acellular lung scaffolds with PSC-derived DE cells in growth factor-free media. Intriguingly, about 45% of the cells expressed NKX2–1 after 3 weeks in culture, and these cells seemed to adopt predominantly proximal fates as demonstrated by staining for basal, secretory, ciliated, and mucin-producing cell markers. This report along with an older study that claimed a major contribution of ECM cues to lung differentiation of mouse ESCs (Cortiella et al. 2010) appears to contradict a growing body of literature that has demonstrated derivation of AFE via inhibition of TGF-β/BMP signaling and subsequent use of soluble agonists of BMP and Wnt signaling to be prerequisites of murine and human PSC lung specification (Gotoh et al. 2014; Huang et al. 2014; Longmire et al. 2012; Mou et al. 2012; Rankin et al. 2016). Studies that involve manipulation of TGF-β/BMP/Wnt signals in lung scaffold culture of PSC-derived endodermal cells will be required to settle this controversy. Taken collectively, the abovementioned studies do point to an important role of native lung ECM in lung-directed differentiation of PSCs. Therefore, lung slice culture can be used to systematically test formal hypotheses regarding ECM signaling functions in lung-directed differentiation. For example, the tissue specificity of the observed lung ECM effects can be systematically addressed by seeding of purified PSC-derived multipotent lung progenitors on tissue slices, micro-scaffolds (Shamis et al. 2011), or solubilized matrices from various decellularized organs (e.g., lung, liver, heart, brain) and detailed molecular and phenotypic characterization of the resultant differentiated lineages. Similarly, the hypothesis of ECM "zip codes" for cell attachment and differentiation (Badylak et al. 2011) can be rigorously tested by characterizing the spatial distribution and organization of proximal and distal cell types derived from purified lung progenitors on lung slices and eventually in whole decellularized lungs.

A recent report has partially addressed these questions by using lung embryonic progenitors that were purified using an Nkx2–1^{mCherry} reporter mouse (Bilodeau et al. 2014). The authors recellularized rat lung scaffolds with either freshly isolated or clonally expanded E12.5–E14.5 proximal lung epithelial progenitors. In both cases, the progenitors gave rise to organized and polarized pulmonary epithelia that contain several proximal (SOX2^{+}) cell types, such as secretory, ciliated, and basal cells. There was absent expression of the distal markers SFTPC and PDPN1 even in distal areas of the scaffold, indicating that distalizing ECM signals cannot instruct the differentiation of proximal lung progenitors.

Another type of decellularized scaffold, denuded trachea scaffolds, has been previously used in lung epithelial biology to test the differentiation and proliferation potential of tracheal epithelial cells such as club and basal cells (Hook et al. 1987; Inayama et al. 1988; Randell et al. 1991). In this application, rat tracheas are decellularized by repeated freeze-thawing, recellularized by epithelial tracheal cells, and transplanted subcutaneously into nude mice. Although there are currently no reports on its use in combination with PSC-derived cells, it will undoubtedly become a platform of choice for the functional characterization of PSC-derived proximal lung progenitors and downstream lineages.

As lung-directed differentiation protocols become more refined leading to the efficient derivation of diverse proximal and distal cell types, decellularized lung or tracheal scaffolds will remain an essential tool for the elucidation of the effect of biomechanical signals, including stiffness and ECM, on lung differentiation and maturation.

13.8 3D Lung Organoid Culture

Ex vivo organoids, i.e., multilineage cell assemblies that recapitulate key morphological and functional attributes of an organ, have been used in developmental biology for more than 100 years (Lancaster and Knoblich 2014). In vitro or ex vivo organoid formation is based on and takes advantage of the universal biological principles of self-assembly, self-organization, and self-patterning through mechanisms of cell commitment, proliferation, migration, and differential adhesion (Lancaster et al. 2013; Sasai 2013a, b). For example, when E16.5 lungs were monodispersed and cells were placed on filters in air-liquid interface culture, emergence of two distinct structures that later gave rise to differentiated lung epithelium and connective tissue was observed (Zimmermann 1987). In a similar vein, lung mesenchyme recombinants have been used extensively for the ex vivo study of lung epithelial-mesenchymal interactions and developmental competence of early proximal and distal lung epithelium (Shannon 1994; Shannon et al. 1999; Shannon and Hyatt 2004).

There has been a renewed interest in organoid culture in the context of PSC biology (Clevers 2016; Sasai 2013b). As directed differentiation protocols for an increasing number of cell lineages of all germ layers become more efficient, organoid culture offers an ideal system to study interactions of multiple PSC-derived cell types in vitro.

In the field of lung stem cell biology, organoids have been used primarily to study putative adult stem/progenitor cell populations ex vivo and to establish their differentiation potential and lineage relationships. Hogan and coworkers used a 3D system, wherein single mouse trachea or human airway basal cells were embedded in Matrigel, to assess clonogenicity and multilineage differentiation of these cells (Rock et al. 2009). In both mouse "tracheospheres" and human "bronchospheres," single KRT5$^+$ cells self-renewed and gave rise primarily to ciliated cells. The authors were also able to validate a novel cell surface marker for basal cells, NGFR, in this system by demonstrating that sorted NGFR$^+$ cells were more clonogenic than NGFR$^-$ cells. In a follow-up paper, the tracheosphere assay was used for screening of compounds that promote differentiation of mouse trachea basal cells (Tadokoro et al. 2014). IL-6/STAT3 signaling was found to promote ciliated cells and reduce secretory cell differentiation, and this finding was validated in in vitro human bronchial cell differentiation and in vivo mouse models of airway injury. Work from the same group used "alveolospheres" to demonstrate that type II AECs, or a subset thereof, are stem cells of the alveolar epithelium since they self-renewed and gave rise to type I AECs upon single-cell passaging and sphere formation (Barkauskas et al. 2013). Contrary to tracheospheres, alveolosphere formation required coculture with PDGFRα^+ stromal cells, and these cells were also found to be part of the type II AEC niche in vivo.

Matrigel-induced organoid culture has also been used for the study of putative bronchioalveolar stem cells (BASCs) (Kim et al. 2005; Lee et al. 2014) or novel ($\beta4^+$) (Chapman et al. 2011; McQualter et al. 2010) stem cell populations or subpopulations of facultative airway progenitors (Chen et al. 2012). In the case of BASCs, primary mouse lung endothelial cells were able to support BASC proliferation and differentiation in three different organoid types of distinct morphology and composition (called bronchiolar, alveolar, and bronchoalveolar) (Lee et al. 2014). The effect of endothelial cells on alveolar differentiation of BASCs was found to be organ specific (coculture with liver endothelial cells only supported bronchiolar colonies) and thrombospondin 1 mediated.

Similarly, $\beta4^+$ lung cells, currently known as lineage-negative cells (LNEPs) (Vaughan et al. 2015), were extensively characterized in organoid culture (Chapman et al. 2011). Originally, these cells were found to mainly reside within the alveolar compartment, expressing very low levels of differentiation markers. The authors reported the development of a novel in vivo organoid assay, in which labeled $\beta4^+$ cells were mixed with embryonic lung epithelial cells and transplanted under the kidney capsule of immunodeficient mice. The resulting organoids contained distal (SFTPC$^+$) and proximal (CC10$^+$) areas derived entirely from $\beta4^+$ cells, confirming their multipotency.

It can be concluded from these studies that the use of organoids in lung stem cell biology has been a valuable tool for the discovery and mapping of stem/progenitor cells in the adult lung. The knowledge gained can be easily applied to the development of organoid systems for PSC-derived lung progenitors.

Since the field of lung-directed differentiation is still in its infancy, it is not surprising that the number of published studies on organoid use remains small. Bishop and coworkers (van Vranken et al. 2005) first reported coculture of murine ESCs with dissected E11.5 lung mesenchyme. The presence of NKX2–1$^+$ and SFTPC$^+$ cells in the derived organoids was interpreted as ESC lung differentiation. Nevertheless, no DE derivation took place before organoid formation, and alternative explanations, such as neuroectodermal NKX2–1 differentiation and stochastic SFTPC expression, may explain the observed results. More recently, organoids were employed for the culture of lung lineages derived from human PSCs (Dye et al. 2015; Gotoh et al. 2014). As mentioned above, CPM was identified as a putative marker of human NKX2–1$^+$ distal lung progenitors during directed differentiation of hPSCs by microarray analysis (Gotoh et al. 2014). The authors engineered SFPTC-GFP hPSC reporter lines and then used 3D Matrigel coculture of sorted CPM$^+$ AFE progenitors with fetal human lung fibroblasts in DCI + KGF media to investigate the differentiation potential of CPM$^+$ cells. The derived luminal structures contained mainly AQP5$^+$ (type I AEC-like cells) and SFPTC-GFP$^+$ cells with the latter being ultrastructurally similar to fetal human type II AECs. In a more recent paper, Spence and coworkers first derived lung-fated AFE spheroids by TGF-β/BMP signaling inhibition and addition of Hh (SAG), Wnt (CHIR99021), and FGF (FGF4) agonists (Dye et al. 2015). Embedding the spheroids in Matrigel and culturing for a prolonged period of time in FGF10 resulted in the formation of so-called human lung organoids (HLOs). The latter comprised alveolar- and airway-like structures as demonstrated by detailed immunohistochemical and transcriptomic analysis of markers for type II AECs (SFPTC, SFTPB), type I AECs (PDPN, HOPX), ciliated cells (FOXJ1), and basal cells (P63). Although the global HLO and human fetal lung transcriptomes were similar, the absence of any sorting steps in the protocol does not exclude the possibility that some of these lineages are not derived from lung progenitors.

As with decellularized lung scaffolds, organoid culture is bound to become a widely used platform for multiple applications such as refinement and validation of lung-directed differentiation protocols, in vitro modeling of human lung development and disease, and drug screening (Fig. 13.3b).

13.9 Toward a Human-Scale Bioartificial, Transplantable Lung

The last years have witnessed significant progress in the fields of lung stem cell and developmental biology and lung tissue engineering. Yet, significant hurdles must be overcome before these advances are translated into safe and efficacious cell/organ replacement therapies for respiratory diseases. Emerging research areas that are likely to contribute toward this goal are summarized below.

13.10 Decellularized Scaffold Characterization and Repopulation

Decellularized lung scaffold will be most probably the platform of choice for whole lung organ engineering in the near future. A variety of techniques for measurement of micro and macromechanical properties of tissues are available (Suki 2014) and if employed regularly will help to determine whether the mechanical properties of fully recellularized lungs correspond to the ones of the native organ. Standardization of decellularization protocols and proteomic analysis will also be essential for ensuring experimental reproducibility between different lung tissue engineering labs.

The choice of cell types to be used for recellularization of lung scaffolds will be equally important. Regarding the epithelial compartment, PSC-derived multipotent proximal, such as basal cells, and distal, such as type II AEC or LNEPs, stem/progenitor cells will, most probably, be the best options. In this case, novel protocols for targeted delivery of these cells to the corresponding areas of the decell constructs will be needed.

Restoration of endothelial barrier function will be indispensable for lung scaffold functionality, and recent reports of efficient re-endothelialization of decell human and murine lungs (Ren et al. 2015; Stabler et al. 2016) (about 75% in the case of human lungs) indicate that this problem may be solved soon. Additional cell lineages with specific functions, such as lung fibroblasts and pleural cells, will also need to be derived from PSCs and incorporated to the recellularized scaffolds.

13.11 3D Bioprinting

The decell-recell model for bioartificial lung generation, while promising, presents certain limitations including availability of and access to native scaffolds, inefficient and uncontrolled engraftment of epithelial cells, and loss of ECM proteins during decellularization. An alternative approach that may overcome these limitations is 3D bioprinting for "de novo" generation of functional tissue units. Although the size of 3D bioprinted constructs had been a major limitation in the past, a recent publication (Kang et al. 2016) suggests that 3D bioprinting of human-scale, structurally stable constructs may be within reach.

Despite its complex cellular composition, the respiratory system is characterized by two main functional compartments, namely, the trachea/airways whose main role is to conduct air and the alveolar space where gas exchange takes place. The simplest functional units that can model the respective in vivo compartments are a single trachea or airway, a tube-like structure lined by a pseudostratified epithelium and a single alveolar sac, a grape-like structure composed of clusters of alveoli, and hollow cavity-like structures lined by type I and II AECs. In both cases, 3D bioprinting can be used to initially reconstruct these units in laboratory scale by determining the optimal starting ECM composition, elastic modulus, and growth factor environment of the bioprinted construct (Fig. 13.3c). As with decell scaffolds, multipotent proximal or distal progenitors can be used for seeding of the epithelial surface, and additional cell types such as PSC-derived fibroblasts or chondrocytes can be incorporated at defined positions to provide structural stability.

Eventually, 3D bioprinting of a human-scale, bioartificial lung may become possible by scaling up and assembling of functional proximal and distal lung units with optimal biomechanical and cellular properties.

In summary, advances in our ability to model lung development in vitro with human PSCs combined with emerging bioengineering techniques are rapidly transforming the field and are likely both to further our understanding of normal development and to facilitate therapeutic applications of these in the years to come.

Acknowledgments Andrew A. Wilson is supported by R01DK101501 and Laertis Ikonomou by grants R01 HL111574 and R01 HL124280.

References

Badylak SF, Taylor D, Uygun K (2011) Whole-organ tissue engineering: decellularization and recellularization of three-dimensional matrix scaffolds. In: Yarmush ML, Duncan JS, Gray ML (eds) Annual review of biomedical engineering, vol 13. Annu Rev, Palo Alto, pp 27–53

Barkauskas CE, Cronce MJ, Rackley CR, Bowie EJ, Keene DR, Stripp BR, Randell SH, Noble PW, Hogan BLM (2013) Type 2 alveolar cells are stem cells in adult lung. J Clin Invest 123(7):3025–3036. doi:10.1172/jci68782

Bilodeau M, Shojaie S, Ackerley C, Post M, Rossant J (2014) Identification of a proximal progenitor population from murine fetal lungs with clonogenic and multilineage differentiation potential. Stem Cell Rep 3(4):634–649. doi:10.1016/j.stemcr.2014.07.010

Bissell MJ, Hall HG, Parry G (1982) How does the extracellular-matrix direct gene-expression. J Theor Biol 99(1):31–68

Booth AJ, Hadley R, Cornett AM, Dreffs AA, Matthes SA, Tsui JL, Weiss K, Horowitz JC, Fiore VF, Barker TH, Moore BB, Martinez FJ, Niklason LE, White ES (2012) Acellular normal and fibrotic human lung matrices as a culture system for in vitro investigation. Am J Respir Crit Care Med 186(9):866–876. doi:10.1164/rccm.201204-0754OC

Cardoso WV, Kotton DN (2008) Specification and patterning of the respiratory system stem book. The Stem Cell Research Community. Harvard Stem Cell Institute, Cambridge, MA

Cardoso WV, Lu JN (2006) Regulation of early lung morphogenesis: questions, facts and controversies. Development 133(9):1611–1624

Chapman HA, Li XP, Alexander JP, Brumwel A, Lorizio W, Tan K, Sonnenberg A, Wei Y, Vu TH (2011) Integrin alpha 6 beta 4 identifies an adult distal lung epithelial population with regenerative potential in mice. J Clin Invest 121(7):2855–2862. doi:10.1172/jci57673

Chen F, Desai TJ, Qian J, Niederreither K, Lu JN, Cardoso WV (2007) Inhibition of Tgf beta signaling by endogenous retinoic acid is essential for primary lung bud induction. Development 134(16):2969–2979. doi:10.1242/dev.006221
Chen F, Cao YX, Qian J, Shao FZ, Niederreither K, Cardoso WV (2010) A retinoic acid-dependent network in the foregut controls formation of the mouse lung primordium. J Clin Invest 120(6):2040–2048. doi:10.1172/jci40253
Chen HY, Matsumoto K, Brockway BL, Rackley CR, Liang JR, Lee JH, Jiang DH, Noble PW, Randell SH, Kim CF, Stripp BR (2012) Airway epithelial progenitors are region specific and show differential responses to bleomycin-induced lung injury. Stem Cells 30(9):1948–1960. doi:10.1002/stem.1150
Christodoulou C, Longmire TA, Shen SS, Bourdon A, Sommer CA, Gadue P, Spira A, Gouon-Evans V, Murphy GJ, Mostoslavsky G, Kotton DN (2011) Mouse ES and iPS cells can form similar definitive endoderm despite differences in imprinted genes. J Clin Invest 121(6):2313–2325
Clevers H (2016) Modeling development and disease with organoids. Cell 165(7):1586–1597. doi:10.1016/j.cell.2016.05.082
Cortiella J, Niles J, Cantu A, Brettler A, Pham A, Vargas G, Winston S, Wang J, Walls S, Nichols JE (2010) Influence of acellular natural lung matrix on murine embryonic stem cell differentiation and tissue formation. Tissue Eng Part A 16(8):2565–2580. doi:10.1089/ten.tea.2009.0730
Daly AB, Wallis JM, Borg ZD, Bonvillain RW, Deng B, Ballif BA, Jaworski DM, Allen GB, Weiss DJ (2012) Initial binding and recellularization of decellularized mouse lung scaffolds with bone marrow-derived mesenchymal stromal cells. Tissue Eng Part A 18(1–2):1–16. doi:10.1089/ten.tea.2011.0301
D'Amour KA, Agulnick AD, Eliazer S, Kelly OG, Kroon E, Baetge EE (2005) Efficient differentiation of human embryonic stem cells to definitive endoderm. Nat Biotechnol 23(12):1534–1541
Desai TJ, Malpel S, Flentke GR, Smith SM, Cardoso WV (2004) Retinoic acid selectively regulates Fgf10 expression and maintains cell identity in the prospective lung field of the developing foregut. Dev Biol 273(2):402–415. doi:10.1016/j.ydbio.2004.04.039
Domyan ET, Ferretti E, Throckmorton K, Mishina Y, Nicolis SK, Sun X (2011) Signaling through BMP receptors promotes respiratory identity in the foregut via repression of Sox2. Development 138(5):971–981. doi:10.1242/dev.053694
Dye BR, Hill DR, Ferguson MAH, Tsai YH, Nagy MS, Dyal R, Wells JM, Mayhew CN, Nattiv R, Klein OD, White ES, Deutsch GH, Spence JR (2015) In vitro generation of human pluripotent stem cell derived lung organoids. eLife 4:25. doi:10.7554/eLife.05098
Firth AL, Dargitz CT, Qualls SJ, Menon T, Wright R, Singer O, Gage FH, Khanna A, Verma IM (2014) Generation of multiciliated cells in functional airway epithelia from human induced pluripotent stem cells. Proc Natl Acad Sci USA 111(17):E1723–E1730. doi:10.1073/pnas.1403470111
Gadue P, Huber TL, Nostro MC, Kattman S, Keller GM (2005) Germ layer induction from embryonic stem cells. Exp Hematol 33(9):955–964. doi:10.1016/j.exphem.2005.06.009
Gadue P, Huber TL, Paddison PJ, Keller GM (2006) Wnt and TGF-beta signaling are required for the induction of an in vitro model of primitive streak formation using embryonic stem cells. Proc Natl Acad Sci USA 103(45):16806–16811. doi:10.1073/pnas.0603916103
Ghaedi M, Calle EA, Mendez JJ, Gard AL, Balestrini J, Booth A, Bove PF, Gui LQ, White ES, Niklason LE (2013) Human iPS cell-derived alveolar epithelium repopulates lung extracellular matrix. J Clin Invest 123(11):4950–4962. doi:10.1172/jci68793
Gilpin SE, Guyette JP, Gonzalez G, Ren X, Asara JM, Mathisen DJ, Vacanti JP, Ott HC (2014a) Perfusion decellularization of human and porcine lungs: bringing the matrix to clinical scale. J Heart Lung Transplant 33(3):298–308. doi:10.1016/j.healun.2013.10.030
Gilpin SE, Ren X, Okamoto T, Guyette JP, Mou HM, Rajagopal J, Mathisen DJ, Vacanti JP, Ott HC (2014b) Enhanced lung epithelial specification of human induced pluripotent stem cells on

decellularized lung matrix. Ann Thorac Surg 98(5):1721–1729. doi:10.1016/j.athoracsur.2014.05.080

Goss AM, Tian Y, Tsukiyama T, Cohen ED, Zhou D, Lu MM, Yamaguchi TP, Morrisey EE (2009) Wnt2/2b and beta-Catenin signaling are necessary and sufficient to specify lung progenitors in the foregut. Dev Cell 17(2):290–298. doi:10.1016/j.devcel.2009.06.005

Gotoh S, Ito I, Nagasaki T, Yamamoto Y, Konishi S, Korogi Y, Matsumoto H, Muro S, Hirai T, Funato M, Mae S, Toyoda T, Sato-Otsubo A, Ogawa S, Osafune K, Mishima M (2014) Generation of alveolar epithelial spheroids via isolated progenitor cells from human pluripotent stem cells. Stem Cell Rep 3(3):394–403. doi:10.1016/j.stemcr.2014.07.005

Gouon-Evans V, Boussemart L, Gadue P, Nierhoff D, Koehler CI, Kubo A, Shafritz DA, Keller G (2006) BMP-4 is required for hepatic specification of mouse embryonic stem cell-derived definitive endoderm. Nat Biotechnol 24(11):1402–1411. doi:10.1038/nbt1258

Green MD, Chen A, Nostro MC, d'Souza SL, Schaniel C, Lemischka IR, Gouon-Evans V, Keller G, Snoeck HW (2011) Generation of anterior foregut endoderm from human embryonic and induced pluripotent stem cells. Nat Biotechnol 29(3):267–U153. doi:10.1038/nbt.1788

Harris-Johnson KS, Domyan ET, Vezina CM, Sun X (2009) beta-Catenin promotes respiratory progenitor identity in mouse foregut. Proc Natl Acad Sci USA 106(38):16287–16292. doi:10.1073/pnas.0902274106

Hook GER, Brody AR, Cameron GS, Jetten AM, Gilmore LB, Nettesheim P (1987) Repopulation of denuded tracheas by Clara cells isolated from the lungs of rabbits. Exp Lung Res 12(4):311–329. doi:10.3109/01902148709062843

Huang SXL, Islam MN, O'Neill J, Hu Z, Yang YG, Chen YW, Mumau M, Green MD, Vunjak-Novakovic G, Bhattacharya J, Snoeck HW (2014) Efficient generation of lung and airway epithelial cells from human pluripotent stem cells. Nat Biotechnol 32(1):84. doi:10.1038/nbt.2754

Ikonomou L, Kotton DN (2015) Derivation of endodermal progenitors from pluripotent stem cells. J Cell Physiol 230(2):246–258. doi:10.1002/jcp.24771

Inayama Y, Hook GER, Brody AR, Cameron GS, Jetten AM, Gilmore LB, Gray T, Nettesheim P (1988) The differentiation potential of tracheal basal cells. Lab Investig 58(6):706–717

Kang HW, Lee SJ, Ko IK, Kengla C, Yoo JJ, Atala A (2016) A 3D bioprinting system to produce human-scale tissue constructs with structural integrity. Nat Biotechnol 34(3):312. doi:10.1038/nbt.3413

Kim CFB, Jackson EL, Woolfenden AE, Lawrence S, Babar I, Vogel S, Crowley D, Bronson RT, Jacks T (2005) Identification of bronchioalveolar stem cells in normal lung and lung cancer. Cell 121(6):823–835. doi:10.1016/j.cell.2005.03.032

Kimura S, Hara Y, Pineau T, FernandezSalguero P, Fox CH, Ward JM, Gonzalez FJ (1996) The T/ebp null mouse thyroid-specific enhancer-binding protein is essential for the organogenesis of the thyroid, lung, ventral forebrain, and pituitary. Genes Dev 10(1):60–69

Konishi S, Gotoh S, Tateishi K, Yamamoto Y, Korogi Y, Nagasaki T, Matsumoto H, Muro S, Hirai T, Ito I, Tsukita S, Mishima M (2016) Directed induction of functional multi-ciliated cells in proximal airway epithelial spheroids from human pluripotent stem cells. Stem Cell Rep 6(1):18–25. doi:10.1016/j.stemcr.2015.11.010

Kubo A, Shinozaki K, Shannon JM, Kouskoff V, Kennedy M, Woo S, Fehling HJ, Keller G (2004) Development of definitive endoderm from embryonic stem cells in culture. Development 131(7):1651–1662. doi:10.1242/dev.01044

Lancaster MA, Knoblich JA (2014) Organogenesis in a dish: modeling development and disease using organoid technologies. Science 345(6194):10. doi:10.1126/science.1247125

Lancaster MA, Renner M, Martin CA, Wenzel D, Bicknell LS, Hurles ME, Homfray T, Penninger JM, Jackson AP, Knoblich JA (2013) Cerebral organoids model human brain development and microcephaly. Nature 501(7467):373. doi:10.1038/nature12517

Lee JH, Bhang DH, Beede A, Huang TL, Stripp BR, Bloch KD, Wagers AJ, Tseng YH, Ryeom S, Kim CF (2014) Lung stem cell differentiation in mice directed by endothelial cells via a BMP4-NFATc1-Thrombospondin-1 axis. Cell 156(3):440–455. doi:10.1016/j.cell.2013.12.039

Li QY, Uygun BE, Geerts S, Ozer S, Scalf M, Gilpin SE, Ott HC, Yarmush ML, Smith LM, Welham NV, Frey BL (2016) Proteomic analysis of naturally-sourced biological scaffolds. Biomaterials 75:37–46. doi:10.1016/j.biomaterials.2015.10.011

Litingtung Y, Lei L, Westphal H, Chiang C (1998) Sonic hedgehog is essential to foregut development. Nat Genet 20(1):58–61

Longmire TA, Ikonomou L, Hawkins F, Christodoulou C, Cao YX, Jean JC, Kwok LW, Mou HM, Rajagopal J, Shen SS, Dowton AA, Serra M, Weiss DJ, Green MD, Snoeck HW, Ramirez MI, Kotton DN (2012) Efficient derivation of purified lung and thyroid progenitors from embryonic stem cells. Cell Stem Cell 10(4):398–411. doi:10.1016/j.stem.2012.01.019

McQualter JL, Yuen K, Williams B, Bertoncello I (2010) Evidence of an epithelial stem/progenitor cell hierarchy in the adult mouse lung. Proc Natl Acad Sci USA 107(4):1414–1419. doi:10.1073/pnas.0909207107

Metzger RJ, Klein OD, Martin GR, Krasnow MA (2008) The branching programme of mouse lung development. Nature 453(7196):745–7U1. doi:10.1038/nature07005

Minoo P, Su GS, Drum H, Bringas P, Kimura S (1999) Defects in tracheoesophageal and lung morphogenesis in Nkx2.1(−/−) mouse embryos. Dev Biol 209(1):60–71

Morrisey EE, Hogan BLM (2010) Preparing for the first breath: genetic and cellular mechanisms in lung development. Dev Cell 18(1):8–23. doi:10.1016/j.devcel.2009.12.010

Motoyama J, Liu J, Mo R, Ding Q, Post M, Hui CC (1998) Essential function of Gli2 and Gli3 in the formation of lung, trachea and oesophagus. Nat Genet 20(1):54–57

Mou HM, Zhao R, Sherwood R, Ahfeldt T, Lapey A, Wain J, Sicilian L, Izvolsky K, Musunuru K, Cowan C, Rajagopal J (2012) Generation of multipotent lung and airway progenitors from mouse ESCs and patient-specific cystic fibrosis iPSCs. Cell Stem Cell 10(4):385–397. doi:10.1016/j.stem.2012.01.018

Naba A, Clauser KR, Hoersch S, Liu H, Carr SA, Hynes RO (2012) The matrisome: in silico definition and in vivo characterization by proteomics of normal and tumor extracellular matrices. Mol Cell Proteomics 11(4):18. doi:10.1074/mcp.M111.014647

Ott HC, Matthiesen TS, Goh SK, Black LD, Kren SM, Netoff TI, Taylor DA (2008) Perfusion-decellularized matrix: using nature's platform to engineer a bioartificial heart. Nat Med 14(2):213–221. doi:10.1038/nm1684

Ott HC, Clippinger B, Conrad C, Schuetz C, Pomerantseva I, Ikonomou L, Kotton D, Vacanti JP (2010) Regeneration and orthotopic transplantation of a bioartificial lung. Nat Med 16(8):927–U131. doi:10.1038/nm.2193

Pepicelli CV, Lewis PM, McMahon AP (1998) Sonic hedgehog regulates branching morphogenesis in the mammalian lung. Curr Biol 8(19):1083–1086

Petersen TH, Calle EA, Zhao LP, Lee EJ, Gui LQ, Raredon MB, Gavrilov K, Yi T, Zhuang ZW, Breuer C, Herzog E, Niklason LE (2010) Tissue-engineered lungs for in vivo implantation. Science 329(5991):538–541. doi:10.1126/science.1189345

Price AP, England KA, Matson AM, Blazar BR, Panoskaltsis-Mortari A (2010) Development of a decellularized lung bioreactor system for bioengineering the lung: the matrix reloaded. Tissue Eng Part A 16(8):2581–2591. doi:10.1089/ten.tea.2009.0659

Randell SH, Comment CE, Ramaekers FCS, Nettesheim P (1991) Properties of rat tracheal epithelial-cells separated based on expression of cell-surface alpha-galactosyl end groups. Am J Respir Cell Mol Biol 4(6):544–554

Rankin SA, Han L, McCracken KW, Kenny AP, Anglin CT, Grigg EA, Crawford CM, Wells JM, Shannon JM, Zorn AM (2016) A retinoic acid-hedgehog cascade coordinates mesoderm-inducing signals and endoderm competence during lung specification. Cell Rep 16(1):66–78. doi:10.1016/j.celrep.2016.05.060

Ren X, Moser PT, Gilpin SE, Okamoto T, Wu T, Tapias LF, Mercier FE, Xiong LJ, Ghawi R, Scadden DT, Mathisen DJ, Ott HC (2015) Engineering pulmonary vasculature in decellularized rat and human lungs. Nat Biotechnol 33(10):1097. doi:10.1038/nbt.3354

Rock JR, Onaitis MW, Rawlins EL, Lu Y, Clark CP, Xue Y, Randell SH, Hogan BLM (2009) Basal cells as stem cells of the mouse trachea and human airway epithelium. Proc Natl Acad Sci USA 106(31):12771–12775. doi:10.1073/pnas.0906850106
Rodewald HR (2008) Thymus organogenesis. Annu Rev Immunol 26:355–388. doi: 10.1146/annurev.immunol.26.021607.090408
Sasai Y (2013a) Cytosystems dynamics in self-organization of tissue architecture. Nature 493(7432):318–326. doi:10.1038/nature11859
Sasai Y (2013b) Next-generation regenerative medicine: organogenesis from stem cells in 3D culture. Cell Stem Cell 12(5):520–530
Serls AE, Doherty S, Parvatiyar P, Wells JM, Deutsch GH (2005) Different thresholds of fibroblast growth factors pattern the ventral foregut into liver and lung. Development 132(1):35–47
Shamis Y, Hasson E, Soroker A, Bassat E, Shimoni Y, Ziv T, Sionov RV, Mitrani E (2011) Organ-specific scaffolds for in vitro expansion, differentiation, and organization of primary lung cells. Tissue Eng Part C-Methods 17(8):861–870. doi:10.1089/ten.tec.2010.0717
Shannon JM (1994) Induction of alveolar type-ii cell-differentiation in fetal tracheal epithelium by grafted distal lung mesenchyme. Dev Biol 166(2):600–614
Shannon JM, Hyatt BA (2004) Epithelial-mesenchymal interactions in the developing lung. Annu Rev Physiol 66:625–645. doi:10.1146/annurev.physiol.66.032102.135749
Shannon JM, Gebb SA, Nielsen LD (1999) Induction of alveolar type II cell differentiation in embryonic tracheal epithelium in mesenchyme-free culture. Development 126(8):1675–1688
Shojaie S, Ermini L, Ackerley C, Wang J, Chin S, Yeganeh B, Bilodeau M, Sambi M, Rogers I, Rossant J, Bear CE, Post M (2015) Acellular lung scaffolds direct differentiation of endoderm to functional airway epithelial cells: requirement of matrix-bound HS proteoglycans. Stem Cell Rep 4(3):419–430. doi:10.1016/j.stemcr.2015.01.004
Sinner D, Rankin S, Lee M, Zorn AM (2004) Sox17 and beta-catenin cooperate to regulate the transcription of endodermal genes. Development 131(13):3069–3080. doi:10.1242/dev.01176
Song JJ, Ott HC (2011) Organ engineering based on decellularized matrix scaffolds. Trends Mol Med 17(8):424–432. doi:10.1016/j.molmed.2011.03.005
Song JJ, Guyette JP, Gilpin SE, Gonzalez G, Vacanti JP, Ott HC (2013) Regeneration and experimental orthotopic transplantation of a bioengineered kidney. Nat Med 19(5):646–651. doi:10.1038/nm.3154
Stabler CT, Caires LC, Mondrinos MJ, Marcinkiewicz C, Lazarovici P, Wolfson MR, Lelkes PI (2016) Enhanced re-endothelialization of decellularized rat lungs. Tissue Eng Part C-Methods 22(5):439–450. doi:10.1089/ten.tec.2016.0012
Suki B (2014) Assessing the functional mechanical properties of bioengineered organs with emphasis on the lung. J Cell Physiol 229(9):1134–1140. doi:10.1002/jcp.24600
Tadokoro T, Wang Y, Barak LS, Bai YS, Randell SH, Hogan BLM (2014) IL-6/STAT3 promotes regeneration of airway ciliated cells from basal stem cells. Proc Natl Acad Sci USA 111(35):E3641–E3649. doi:10.1073/pnas.1409781111
Tam PPL, Loebel DAF, Tanaka SS (2006) Building the mouse gastrula: signals, asymmetry and lineages. Curr Opin Genet Dev 16(4):419–425. doi:10.1016/j.gde.2006.06.008
Totonelli G, Maghsoudlou P, Garriboli M, Riegler J, Orlando G, Burns AJ, Sebire NJ, Smith VV, Fishman JM, Ghionzoli M, Turmaine M, Birchall MA, Atala A, Soker S, Lythgoe MF, Seifalian A, Pierro A, Eaton S, De Coppi P (2012) A rat decellularized small bowel scaffold that preserves villus-crypt architecture for intestinal regeneration. Biomaterials 33(12):3401–3410. doi:10.1016/j.biomaterials.2012.01.012
Totonelli G, Maghsoudlou P, Georgiades F, Garriboli M, Koshy K, Turmaine M, Ashworth M, Sebire NJ, Pierro A, Eaton S, De Coppi P (2013) Detergent enzymatic treatment for the development of a natural acellular matrix for oesophageal regeneration. Pediatr Surg Int 29(1):87–95. doi:10.1007/s00383-012-3194-3
Uygun BE, Soto-Gutierrez A, Yagi H, Izamis ML, Guzzardi MA, Shulman C, Milwid J, Kobayashi N, Tilles A, Berthiaume F, Hertl M, Nahmias Y, Yarmush ML, Uygun K (2010) Organ reengi-

neering through development of a transplantable recellularized liver graft using decellularized liver matrix. Nat Med 16(7):814–U120. doi:10.1038/nm.2170

van der Veen VC, van der Wal MBA, van Leeuwen MCE, Ulrich MMW, Middelkoop E (2010) Biological background of dermal substitutes. Burns 36(3):305–321. doi:10.1016/j.burns.2009.07.012

van Vranken BE, Romanska HM, Polak JM, Rippon HJ, Shannon JM, Bishop AE (2005) Coculture of embryonic stem cells with pulmonary mesenchyme: a microenvironment that promotes differentiation of pulmonary epithelium. Tissue Eng 11(7–8):1177–1187

Vaughan AE, Brumwell AN, Xi Y, Gotts JE, Brownfield DG, Treutlein B, Tan K, Tan V, Liu FC, Looney MR, Matthay MA, Rock JR, Chapman HA (2015) Lineage-negative progenitors mobilize to regenerate lung epithelium after major injury. Nature 517(7536):621–U211. doi:10.1038/nature14112

Wallis JM, Borg ZD, Daly AB, Deng B, Ballif BA, Allen GB, Jaworski DM, Weiss DJ (2012) Comparative assessment of detergent-based protocols for mouse lung de-cellularization and re-cellularization. Tissue Eng Part C-Methods 18(6):420–432. doi:10.1089/ten.tec.2011.0567

Wells JM, Melton DA (1999) Vertebrate endoderm development. Annu Rev Cell Dev Biol 15:393–410

Wong AP, Bear CE, Chin S, Pasceri P, Thompson TO, Huan LJ, Ratjen F, Ellis J, Rossant J (2012) Directed differentiation of human pluripotent stem cells into mature airway epithelia expressing functional CFTR protein. Nat Biotechnol 30(9):876–U108. doi:10.1038/nbt.2328

Zimmermann B (1987) Lung organoid culture. Differentiation 36(1):86–109. doi:10.1111/j.1432-0436.1987.tb00183.x

Zorn AM, Wells JM (2009) Vertebrate endoderm development and organ formation. Annu Rev Cell Dev Biol 25:221–251. doi:10.1146/annurev.cellbio.042308.113344

MIX
Papier | Fördert gute Waldnutzung
FSC® C083411

Kolibri 360 GmbH
Ferdinand-Jühlke-Straße 7
99095 Erfurt, Deutschland
produktsicherheit@kolibri360.de